F. Gherardelli (Ed.)

Complex Analysis

Lectures given at a Summer School of the
Centro Internazionale Matematico Estivo (C.I.M.E.),
held in Bressanone (Bolzano), Italy,
June 3-12, 1973

Springer

FONDAZIONE
CIME
ROBERTO CONTI

C.I.M.E. Foundation
c/o Dipartimento di Matematica "U. Dini"
Viale margagni n. 67/a
50134 Firenze
Italy
cime@math.unifi.it

ISBN 978-3-642-10963-8 e-ISBN: 978-3-642-10964-5
DOI:10.1007/978-3-642-10964-5
Springer Heidelberg Dordrecht London New York

Reprint of the 1st ed. C.I.M.E., Ed. Cremonese, Roma 1974
With kind permission of C.I.M.E.

Printed on acid-free paper

Springer.com

CENTRO INTERNAZIONALE MATEMATICO ESTIVO
(C.I.M.E.)

I Ciclo - Bressanone - dal 3 al 12 Giugno 1973

« COMPLEX ANALYSIS »

Coordinatore: Prof. F. Gherardelli

CENTRO INTERNAZIONALE MATEMATICO ESTIVO
(C. I. M. E.)

NINE LECTURES ON COMPLES ANALYSIS

ALDO ANDREOTTI .

Corso tenuto a Bressanone dal 3 al 12 giugno 1973

A. Andreotti

Preface.

In the spring of 1972 I had the opportunity to lecture at Lund University and more extensively at Amsterdam University and at the C.I.M.E. session in the summer of 1973 on some topics of complex analysis of my choice. The subject has been chosen within the limited range of my personal knowledge and is intended for a non excessively specialized audience. We have tried therefore not to obscure the ideas, in the attempt to obtain the most general statements, with an excess of technical details; for this reason, for instance, our main attention is devoted to complex manifolds, and we have recalled basic facts and definitions when needed. The purpose was not to overcome the listeners with admiration for the preacher but to share with hime the pleasure of inspecting some beautiful facets of this field. Indeed I was very grateful to receive many valuable suggestions; in particular I am indebted to L. Garding, L. Hormander, F. Oort, A.J.H.M. van de Ven and especially to P. de Paepe who undertook the heroic task of writing the notes.

The material deals with the theory of Levi convexity and its applications, with the duality theorem of Serre and Malgrange, and with the Hans Lewy problem. The limited time at our disposal may account for some conciseness that, however we hope, will turn to the advantage of the reader.

P. de Paepe has corrected several mistakes of mathematics and presentation; probably only few remained undected.

San Pellegrino al Cassero, September 1973,

Aldo Andreotti.

A. Andreotti

CONTENTS

A. Andreotti

Chapter I. Elementary theory of holomorphic convexity.

1.1 Preliminaries.

a) Let Ω be an open set in $\mathbb{C}^n$, the Cartesian product of n copies of the complex field $\mathbb{C}$, with coordinate functions $z_1,\ldots,z_n$.

A function $f : \Omega \to \mathbb{C}$ is called holomorphic if for every point $z_o \in \Omega$ there exists a neighborhood $U(z_o)$ of z_o in Ω on which f admits an absolutely convergent power series expansion

$$f = \Sigma a_\alpha (z-z_o)^\alpha \quad \text{for every} \quad z \in U(z_o).$$

Here $\alpha = (\alpha_1,\ldots,\alpha_n) \in \mathbb{N}^n$ ($\mathbb{N}$ = natural numbers including 0), $\quad a_\alpha = a_{\alpha_1,\ldots,\alpha_n}$, $z^\alpha = z_1^{\alpha_1} \ldots z_n^{\alpha_n}$.

A map $f = (f_1,\ldots,f_m) : \Omega \to \mathbb{C}^m$ is said to be holomorphic if each component f_i, $1 \leq i \leq m$, is holomorphic.
The composition of two holomorphic maps is (where it is defined) a holomorphic map.

b) We will write $z = x + iy$ with x, y $\mathbb{R}^n$, $i = \sqrt{-1}$. Then $x = \frac{1}{2}(z+\bar{z})$, $y = \frac{1}{2i}(z-\bar{z})$ where the bar denotes complex conjugation, and we will write $dx = \frac{1}{2}(dz+d\bar{z})$, $dy = \frac{1}{2i}(dz-d\bar{z})$. For any function $f : \Omega \to \mathbb{C}$ of class C^1 we have

$$df = \sum_{j=1}^{n} \frac{\partial f}{\partial z_j} dz_j + \sum_{j=1}^{n} \frac{\partial f}{\partial \bar{z}_j} d\bar{z}_j$$

where

$$\frac{\partial f}{\partial z_j} = \frac{1}{2}\left(\frac{\partial f}{\partial x_j} + \frac{1}{i}\frac{\partial f}{\partial y_j}\right)$$

$$\frac{\partial f}{\partial \bar{z}_j} = \frac{1}{2}\left(\frac{\partial f}{\partial x_j} - \frac{1}{i}\frac{\partial f}{\partial y_j}\right)$$

we define

$$\partial f = \sum_{j=1}^{n} \frac{\partial f}{\partial z_j} dz_j$$

$$\bar{\partial} f = \sum_{j=1}^{n} \frac{\partial f}{\partial \bar{z}_j} d\bar{z}_j .$$

The following theorem establishes a criterion for a function of class C^1 to be a holomorphic function.

Theorem. A function $f : \Omega \to \mathbb{C}$ of class C^1 is holomorphic iff at every point of Ω f satisfies the Cauchy-Riemann equations:

$$\bar{\partial} f = 0$$

(i.e. $\frac{\partial f}{\partial \bar{z}_1} = 0, \ldots, \frac{\partial f}{\partial \bar{z}_n} = 0$, i.e.

$$\frac{\partial f}{\partial x_j} = \frac{1}{i} \frac{\partial f}{\partial y_j} \quad \text{for } 1 \leq j \leq n) .$$

c) Let f be holomorphic in a neighborhood of the closed polycilinder

$$P = \{ z \in \mathbb{C}^n \mid |z_j| \leq 1 \quad \text{for} \quad 1 \leq j \leq n$$

then for every $z \in \mathring{P}$, the interior of P, we have the Cauchy integral formula

$$f(z) = \frac{1}{(2\pi i)^n} \int_{|\xi_1|=1} \cdots \int_{|\xi_n|=1} \frac{f(\xi_1, \ldots, \xi_n)}{(\xi_1 - z_1) \ldots (\xi_n - z_n)} d\xi_1 \ldots d\xi_n .$$

From this formula it follows easily (by expansion of the kernel of the integral in power series) that a continuous function $f : \Omega \to \mathbb{C}$ which is separately holomorphic in each variable is a holomorphic function (Osgood's lemma). This is even true if the condition that f is continuous is removed (Hartogs' theorem) but this is much more difficult to prove.

d) We recall the following result:
<u>the set of points where a holomorphic function has a zero of infinite order is open and closed</u> (<u>principle of analytic continuation</u>).

In particular if f is defined in Ω, if Ω is connected and if f vanishes at some point of Ω of infinite order then f is identically zero on Ω.

e) We denote by $\mathcal{H}(\Omega)$ the set of all holomorphic functions in Ω. It is a vector space over $\mathbb{C}$.

We can provide $\mathcal{H}(\Omega)$ with a locally convex topology defined by the family of seminorms

$$||f||_K = \sup_K |f|$$

where K is a compact subset of Ω. A fundamental system of neighborhoods of the origin is then given by the sets

$$V(K, \varepsilon) = \{ f \in \mathcal{H}(\Omega) \mid ||f||_K < \varepsilon \}$$

for K compact in Ω and $\varepsilon > 0$. This topology is the topology of uniform convergence on compact subsets of Ω.

If $K_1 \subset K_2 \subset K_3 \subset \dots$ is a sequence of compact sets such that $K_m \subset \overset{\circ}{K}_{m+1}$ for $m = 1, 2, \dots$, and $\Omega = \bigcup_{m=1}^{\infty} K_m$, one easily verifies that the countable set of seminorms $|| \; ||_{K_m}$ defines the same topology. Thus $\mathcal{H}(\Omega)$ is a metrizable space, one can take for instance as a distance the function

$$d(f,g) = \sum_{m=1}^{\infty} \frac{1}{2^m} \frac{||f-g||_{K_m}}{1 + ||f-g||_{K_m}}, \quad f, g \in \mathcal{H}(\Omega).$$

Since continuous functions satisfying the Cauchy integral formula are necessarily holomorphic, it follows that $\mathcal{H}(\Omega)$ is a complete metric space (i.e. a Frechet space) and therefore a Baire space.

We also note that bounded sets $B \subset \mathcal{H}(\Omega)$ are relatively compact. This is a consequence of

Vitali's theorem: If $\{f_\nu\}$ is a sequence of holomorphic functions on Ω such that for every compact set $K \subset \Omega$ there exists a constant $C(K)$ for which we have

$$\|f_\nu\|_K \leq C(K) \qquad \nu = 1,2,\dots ,$$

then we can extract from $\{f_\nu\}$ a subsequence $\{f_{\nu}\}$ which converges uniformly on any compact subset of Ω.

In particular the unit ball in the norm $\| \ \|_{K_{m+1}}$ is relacompact with respect to the norm $\| \ \|_{K_m}$ i.e. $\mathcal{H}(\Omega)$ is a space of Frechet-Schwartz (cf. [14]). For more details we refer to [28] , [31] , [43].

1.2 Hartogs domains.

Consider the subset in $\mathbb{R}^3$ (coordinates x, y, t, $z = x + iy$)

$$T = \{|z| < b , \ 0 \leq t < c\} \cup \{a < |z| < b, \ 0 \leq t < d\}$$

where $0 < a < b$ and $0 < c < d$.

Because of its shape we call T a "top hat". A top hat (or Hartogs domain) in $\mathbb{C}^n = \mathbb{C} \times \mathbb{C}^{n-1}$, $n \geq 2$, is the set of all points $z = (z_1,\dots,z_n)$ in $\mathbb{C}^n$ for which $(z_1, (\sum_{j=2}^n |z_j|^2)^{\frac{1}{2}})$ is contained in a top hat in $\mathbb{R}^3$.

Theorem (1.2.1). (Hartogs). Let f be holomorphic in the top hat

$$T = \{|z_1| < b, \ \sum_{j=2}^n |z_j|^2 < c^2\} \cup \{a < |z_1| < b, \ \sum_{j=2}^n |z_j|^2 < d^2\}$$

in $\mathbb{C}^n$, $n \geq 2$.

Then f extends holomorphically to the filled up top hat

$$\hat{T} = \{|z_1| < b, \ \sum_{j=2}^n |z_j|^2 < d^2\}.$$

<u>Proof</u>. Let the functions g_p , $p \in \mathbb{N}$, be defined by the Cauchy integral

$$g_p(z_1,\dots,z_n) = \frac{1}{2\pi i} \int_{|\xi| = b-\frac{1}{p}} \frac{f(\xi, z_2,\dots,z_n)}{\xi - z_1} d\xi$$

If p is large g_p is well-defined, continuous and holomorphic in each variable for $|z_1|$ b-1/p and $\sum_{j=2}^{n} |z_j|^2 < d^2$. Therefore g_p is holomorphic. Moreover g_p is independent of p. Set $g = g_p$, then g is defined in $\hat{T}$, is holomorphic there and $g|T = f$ because $g-f$ is holomorphic in T , is zero on $\{|z_1| < b, \sum_2^n |z_j|^2 < c^2\}$ and T is connected. Q.E.D.

Let $h : \hat{T} - \mathbb{C}^n$ be a biholomorphic map onto an open set $h(\hat{T})$ of $\mathbb{C}^n$, (i.e. h is invertible and h and h^{-1} are both holomorphic). Then any holomorphic function on $h(T)$ extends holomorphically to $h(\hat{T})$.

This is a consequence of Hartogs' theorem and the fact that the composition of two holomorphic maps is a holomorphic map. It is sometimes called the "<u>disc theorem</u>".

We quote some simple consequences of Hartogs' theorem.

Let $n \geq 2$ and f holomorphic on the punctured ball $\{0 < \sum_1^n |z_j|^2 < r^2\}$ then f extends holomorphically to the ball $\{\sum_1^n |z_j|^2 < r^2\}$. In fact we can put a top hat T in the punctured ball so that $\hat{T}$ covers the origin. In particular it follows that <u>a holomorphic function f in $n \geq 2$ variables cannot have an "isolated singularity" nor an isolated zero</u> (since this would be an isolated singularity for $1/f$).

1.3 <u>Open sets of holomorphy</u>

a) <u>Open sets with a smooth boundary</u>. An open set Ω in $\mathbb{C}^n$ has a <u>smooth boundary</u> if for every point $z_o \in \partial\Omega = \bar{\Omega} - \Omega$ we can find a neighborhood $U(z_o)$ and a C^∞ function $\phi : U(z_o) = \mathbb{R}$ such that

$$(d\phi)_{z_o} \neq 0 , \Omega \cap U(z_o) = \{ z \in U(z_o) \mid \phi(z) < \phi(z_o) \}.$$

This amounts to say that by a local diffeomorphism near z_o $\Omega \cap U(z_o)$ can be transformed into an open subset of a half space. Indeed we can select a set of real C^∞ coordinates in which $\phi - \phi(z_o) = x_1$ is the first coordinate. Thus $\Omega \cap U(z_o)$ is an open subset of the halfspace $\{ x_1 < 0 \}$.

Let Ω be an open subset of $\mathbb{C}^n$ $(= \mathbb{R}^{2n})$ with a smooth boundary and let $z_o \in \partial\Omega$ be a boundary point of Ω. Given $f \in \mathcal{H}(\Omega)$ we will say that f is <u>holomorphically extendable over</u> z_o if we can find a neighborhood $V(z_o)$ of z_o and a holomorphic function $\hat{f} \in \mathcal{H}(\Omega \cup V(z_o))$ such that

$$\hat{f}|_\Omega = f .$$

<u>Definition</u>. Let Ω be an open subset of $\mathbb{C}^n$ with a smooth boundary. We say that Ω is an <u>open set of holomorphy</u> if for every boundary point $z_o \in \partial\Omega = \bar{\Omega} - \Omega$ we can find a holomorphic function $f \in \mathcal{H}(\Omega)$ which cannot be extended holomorphically over z_o.

<u>Examples</u>.

1. Every open subset $\Omega \subset \mathbb{C}$ with smooth boundary is an open set of holomorphy. Indeed for every $z_o \in \partial\Omega$ $f = (z-z_o)^{-1}$ is not extendable over z_o.

2. The ball $\Omega = \{ \sum_{j=1}^{n} |z_j|^2 < b^2 \}$ in $\mathbb{C}^n$ is an open set of holomorphy. Indeed $f = (z_1 - b)^{-1}$ is not extendable over $(b,0,\ldots,0)$. Since the unitary group $U(n)$ acts transitively on $\partial\Omega$ by holomorphic transformations, the assertion follows.

3. The circular shell $\Omega = \{ a^2 < \sum_{j=1}^{n} |z_j|^2 < b^2 \}$ for $0 < a < b$, if $n \geq 2$ is not an open set of holomorphy. Indeed for every point z_o of the inner boundary $\sum_{j=1}^{n} |z_j|^2 = a^2$ we can place a top hat T in Ω such that $z_o \in \hat{T}$.

b) <u>Open sets with arbitrary boundary</u>.

By a domain we mean an open connected set.

Let $\Delta \subset \Delta$ be two domains in $\mathbb{C}^n$ and let $S \subset \mathcal{H}(\Delta)$ be a set of holomorphic functions in Δ. We say that $\hat{\Delta}$ is an S - <u>completion</u> of Δ if

$$\operatorname{Im}\{\mathcal{H}(\hat{\Delta}) \to \mathcal{H}(\Delta)\} \supset S$$

i.e. if every $f \in S$ extends holomorphically to $\hat{\Delta}$.
Note that by the principle of analytic continuation the extension of f to $\hat{f} \in \mathcal{H}(\hat{\Delta})$ is unique.

For instance for a top hat T, $\hat{T}$ is an $\mathcal{H}(T)$-completion of T $(n)\geq 2)$.

<u>Definition</u>.

Let Ω be an open set in $\mathbb{C}^n$. We say that Ω is an <u>open set of holomorphy</u> if:

for every domain $\Delta \subset \Omega$ every $\mathcal{H}(\Omega)|_{\Delta}$ - completion $\hat{\Delta}$ of Δ is contained in Ω.

<u>Remark</u>: Open sets of holomorphy with a smooth boundary are necessarily open sets of holomorphy in the sense of this general definition. We will see later that the converse is also true. We will refer for the moment to the definition given before for open sets of holomorphy with a smooth boundary as the "provisorial definition of open sets of holomorphy".

c) <u>Holomorphic convexity, characterization of open sets of holomorphy</u>.

An open set Ω in $\mathbb{C}^n$ is called <u>holomorphically convex</u> if for every compact subset $K \subset \Omega$ the holomorphically convex envelope $\hat{K}$ of K in Ω, defined by

$$\hat{K} = \{z \in \Omega \mid |f(z)| \leq \|f\|_K \text{ for every } f \in \mathcal{H}(\Omega)\},$$

is also compact.

<u>Theorem</u> (1.3.1). <u>An open set $\Omega \subset \mathbb{C}^n$ is holomorphically convex if for every divergent</u>[1] <u>sequence $\{x_\nu\} \subset \Omega$ there exists an $f \in \mathcal{H}(\Omega)$ such that</u>

$$\sup_\nu |f(x_\nu)| = +\infty \text{ ("condition D").}$$

<u>Proof</u>. Condition D implies that Ω is holomorphically convex. Indeed, if K is compact and $\hat{K}$ is not we can find a divergent sequence $\{x_\nu\}$ in Ω, $\{x_\nu\} \subset K$.
But then for every $f \in \mathcal{H}(\Omega)$, $|f(x_\nu)| \leq \|f\|_K < \infty$ which contradicts condition D.

Conversely let Ω be holomorphically convex. We want to show that condition D holds. Of this fact we will give two proofs.

1st proof: By absurdity; suppose that there exists a divergent sequence $\{x_\nu\} \subset \Omega$ such that for every $f \in \mathcal{H}(\Omega)$ $\sup_\nu |f(x_\nu)| < \infty$. By passing to a subsequence we may assume $\{x_\nu\}$ contained in a connected component of Ω. Without loss of generality we may thus assume Ω connected. Set

$$A = \{f \in \mathcal{H}(\Omega) \mid \sup_\nu |f(x_\nu)| \leq 1\}.$$

Then

$$\mathcal{H}(\Omega) = \bigcup_{m=1}^{\infty} mA.$$

Now A is a closed subset of $\mathcal{H}(\Omega)$. Thus by the Baire category theorem A must contain an interior point. But A is convex and symmetric $(A = -A)$ thus A must contain a neighborhood of the origin say

$$V(K, \varepsilon) = \{f \in \mathcal{H}(\Omega) \mid \|f\|_K < \varepsilon\} \subset A$$

(for a compact $K \subset \Omega$ and some $\varepsilon > 0$). We may assume as well that K has non-empty interior.

(1) By this we mean a sequence $\{x_\nu\}$ with no accumulation point in Ω.

Now for every $f \neq 0$, $f \in \mathcal{H}(\Omega)$, $\frac{\varepsilon}{\|f\|_K} \cdot f \in A$, therefore for every $f \in \mathcal{H}(\Omega)$

$$\sup_{\nu} |f(x_\nu)| \leq \frac{1}{\varepsilon} \|f\|_K .$$

In particular, replacing f by f^m, we get

$$\sup_{\nu} |f^m(x_\nu)| \leq \frac{1}{\varepsilon} \|f^m\|_K$$

i.e.

$$\sup_{\nu} |f(x_\nu)| \leq (\tfrac{1}{\varepsilon})^{\frac{1}{m}} \|f\|_K .$$

This shows that

$$\sup |f(x_\nu)| \leq \|f\|_K$$

for every $f \in \mathcal{H}(\Omega)$. Hence $\{x_\nu\} \subset \hat{K}$. This contradicts the holomorphic convexity of Ω.

2nd <u>proof</u>: Select a sequence $\{K_m\}$ of compact subsets of such that

$$K_m \subset \overset{\circ}{K}_{m+1}, \quad \bigcup_{m=1}^{\infty} K_m = \Omega, \quad \hat{K}_m = K_m .$$

Let $\{x_\nu\}$ be a divergent sequence in Ω. Replacing $\{K_m\}$ and $\{x_\nu\}$ by subsequences we may assume that

$$x_m \notin K_m , \quad x_m \in K_{m+1} \quad \text{for } m = 1,2,\ldots .$$

Since $x_m \notin K_m = \hat{K}_m$ we can find $g_m \in \mathcal{H}(\Omega)$ such that

$$\|g_m\|_{K_m} < 1 \quad |g_m(x_m)| .$$

Choose positive integers λ_m successively so that

$$\| g_m^{\lambda_m} \|_{K_m} < 2^{-m}$$

$$| g_m^{\lambda_m} (x_m) | > m + \sum_{l=m+1}^{\infty} \frac{1}{2^l} + | \sum_{i=1}^{m-1} g_i^{\lambda_i} (x_m) | .$$

Now $f = \sum g_m^{\lambda_m}$ converges uniformly on every K_m and thus on every compact subset of Ω as any such set is contained in some K_m.
Thus f is a holomorphic function in Ω. But from the last inequality we derive that $|f(x_m)| > m$.
Therefore $\sup_\nu |f(x_\nu)| = +\infty$.

<u>Theorem</u> (1.3.2). (Cartan - Thullen). <u>An open set $\Omega \subset \mathbb{C}^n$ is an open set of holomorphy iff Ω is holomorphically convex.</u>

<u>Proof</u>. If Ω is holomorphically convex, then condition D holds, thus clearly Ω is an open set of holomorphy.

Conversely suppose that Ω is an open set of holomorphy. We want to prove that Ω is holomorphically convex. If this is not the case, then there exists a compact subset $K \subset \Omega$ such that $\hat{K}$ is not compact.
Because for each coordinate function we have $\|z_j\|_K = \|z_j\|_{\hat{K}}$, $\hat{K}$ is bounded. Let $\{x_\nu\} \subset \hat{K}$ be a divergent sequence in Ω such that $x_\nu \to z_o \in \partial\Omega$. Let $\|z\| = \sup |z_j|$ denote the polycilindrical norm in $\mathbb{C}^n$, and let

ρ = polycilindrical distance of K and $\partial\Omega$.

Certainly $\rho > 0$. If $\rho = \infty$ then $\Omega = \mathbb{C}^n$ which is clearly holomorphically convex. We may assume $\rho < \infty$. Let K' be the set of points in Ω whose polycilindrical distance from K is $\leq \frac{1}{2}\rho$. Then K' is a compact subset of Ω. For every n - tuple α of non-negative integers and every $f \in \mathcal{H}(\Omega)$

$\frac{1}{\alpha!} D^{\alpha} f \in \mathcal{H}(\Omega)$ (1),

therefore for any $x \in \hat{K}$

(1) $\quad |\frac{1}{\alpha!} D^{\alpha} f(x)| \leq \|\frac{1}{\alpha!} D^{\alpha} f\|_K$.

For $z \in K$ we have by the Cauchy formula

$$|D^{\alpha} f(z)| = |\frac{\alpha!}{(2\pi i)^n} \int_{|\xi_1 - z_1| = \frac{\rho}{2}} \cdots \int_{|\xi_n - z_n| = \frac{\rho}{2}} \frac{f(\xi)}{(\xi_1 - z_1)^{\alpha_1+1} \cdots (\xi_n - z_n)^{\alpha_n+1}} d\xi_1 \ldots d\xi_n ,$$

and therefore

(2) $\quad |D^{\alpha} f(z)| \leq \alpha! \quad \|f\|_{K'} \frac{1}{(\frac{\rho}{2})^{|\alpha|}}$

From (1) and (2) it follows that the Taylor series of f at a point $x \in \hat{K}$

$$\frac{1}{\alpha!} D^{\alpha} f(x) (z-x)^{\alpha}$$

is majorized by the series

$$\|f\|_{K'} \quad |\frac{z-x}{\rho/2}|^{\alpha}$$

and therefore is absolutely convergent in $Q(x) = \{\|z-x\| < \rho/4\}$.

Now for ν sufficiently large $Q(x_{\nu})$ contains the point z_o . Let Δ be the connected component of $Q(x_{\nu}) \cap \Omega$ containing x_{ν}.

---

(1) As usual for $\alpha = (\alpha_1, \ldots, \alpha_n)$ $\mathbb{N}^n$ we set $D^{\alpha} = \frac{\partial^{\alpha_1 + \ldots + \alpha_n}}{\partial z_1^{\alpha_1} \ldots z_n^{\alpha_n}}$, $\alpha! = \alpha_1! \ldots \alpha_n!$, $|\alpha| = \alpha_1 + \ldots + \alpha_n$.

Then $Q(x_\nu)$ is an $\mathcal{H}(\Omega)_{|\Delta}$-completion of Δ. But $Q(x_\nu) \not\subset \Omega$ as it contains the point z_o. Therefore Ω cannot be an open set of holomorphy.

Remark. Let Ω be an open set in $\mathbb{C}^n$ with a smooth boundary. If Ω is an open set of holomorphy then by the Cartan-Thullen theorem Ω is holomorphically convex and therefore satisfies condition D. Hence Ω is an open set of holomorphy in the provisorial sense. For open sets with a smooth boundary the provisorial and the general definition of open set of holomorphy are equivalent.

1.4 Levi[1] - convexity.

a) Let Ω be an open subset of $\mathbb{C}^n$ and let $\phi : \Omega \to \mathbb{R}$ be a C^∞ function. At a point $a \in \Omega$ we consider the Taylor expansion of ϕ; with obvious notations for the partial derivatives, we have,

$$\begin{aligned}\phi(z) = \phi(a) &+ \Sigma\, \partial_\alpha\phi(a)(z_\alpha - a_\alpha) + \Sigma\, \partial_{\bar\alpha}\phi(a)(\bar z_\alpha - \bar a_\alpha) + \\ &+ \tfrac{1}{2}\Sigma\, \partial_{\alpha\beta}\phi(a)(z_\alpha - a_\alpha)(z_\beta - a_\beta) \\ &+ \tfrac{1}{2}\Sigma\, \partial_{\bar\alpha\bar\beta}\phi(a)(\bar z_\alpha - \bar a_\alpha)(\bar z_\beta - \bar a_\beta) \\ &+ \Sigma\, \partial_{\alpha\bar\beta}\phi(a)(z_\alpha - a_\alpha)(\bar z_\beta - \bar a_\beta) + O(\|z-a\|^3) .\end{aligned}$$

Because $\overline{\dfrac{\partial f}{\partial \bar z_j}} = \dfrac{\partial \bar f}{\partial z_j}$ and because ϕ is real-valued, we must have $\overline{\partial_\alpha \phi(a)} = \partial_{\bar\alpha}\phi(a)$; $\overline{\partial_{\alpha\beta}\phi(a)} = \partial_{\bar\alpha\bar\beta}\phi(a)$; $\partial_{\alpha\bar\beta}\phi(a) = \overline{\partial_{\bar\alpha\beta}\phi(a)}$.

In particular the quadratic form

$$L(\phi)_a(v) = \Sigma\,\partial_{\alpha\bar\beta}\phi(a)\; v_\alpha \bar v_\beta$$

---

(1) Eugenio Elia Levi, 1883 - 1917.

is hermitian; it is called the <u>Levi-form</u> of ϕ at a.

A biholomorphic change of coordinates near a acts on $L(\phi)_a$ with a linear change of variables

$$v \to J(a)v$$

where $J(a)$ is the Jacobian matrix of the change of variables at a.

It follows that <u>the number of positive and the number of negative eigenvalues of the Levi-form at a does not depend on the choice of local coordinates.</u>

<u>Remark.</u> If $(d\phi)_a \neq 0$ we can perform a change of coordinates in which a is at the origin and in which the new z_1 - coordinate is

$$\sum \partial_\alpha \phi(a)(z_\alpha - a_\alpha) + \tfrac{1}{2}\sum \partial_{\alpha\beta}\phi(a)(z_\alpha - a_\alpha)(z_\beta - a_\beta).$$

Then ϕ takes the following Taylor expansion:

$$\phi(z) = \phi(0) + 2\,\mathrm{Re}\, z_1 + L(\phi)_0(z) + O(\|z\|^3) .$$

b) Let us assume that $(d\phi)_a \neq 0$ and, for simplicity of notations that a is at the origin. Set

$$U = \{z \in \Omega \mid \phi(z) < \phi(0) \quad .$$

Then $\partial U = \bar{U} - U$ is smooth near $a = 0$ and the real tangent plane to U at the origin is given by

$$\sum \frac{\partial \phi}{\partial x_\alpha}(0)x_\alpha + \sum \frac{\partial \phi}{\partial y_\alpha}(0)\, y_\alpha = 0 .$$

This plane contains the (n-1) - dimensional complex plane with equation

$$\sum \partial_\alpha \phi(0)\, z_\alpha = 0 .$$

This is called the <u>analytic tangent plane</u> to ∂U at a and will be denoted by $T_a(\partial U)$.

Consider the Levi-form of ϕ at a restricted to $T_a(\partial U)$,

$$L(\phi)_{a|T_a(\partial U)} = \begin{cases} \sum \partial_{\alpha\bar{\beta}}\phi(a)v_\alpha \bar{v}_\beta \\ \sum \partial_\alpha \phi(a)v_\alpha = 0 \end{cases}$$

We obtain in this way a hermitian form in n-1 variables and again we realize that the number of positive and negative eigenvalues is independent of the choice of local holomorphic coordinates.

Suppose now that U is defined in a neighborhood of a by another C^∞ function ψ with $(d\psi)_a \neq 0$:

$$U = \{z \in \Omega \mid \psi(z) < \psi(a)\}.$$

By subtracting constants from ϕ and ψ we may assume that $\phi(a) = \psi(a) = 0$. Then either ϕ or ψ can be taken among a set of C^∞ real local coordinates (cf. 1.3. a)). Applying the Taylor formula with the rest in integral form we realize that in a neighborhood of a $\phi = h\psi$ with h a C^∞ function and invertible (i.e. $h(a) \neq 0$). Since $\phi > 0$ where $\psi > 0$ we must have $h(a) > 0$.

Now

$$\partial.\bar{\partial}\phi = \partial(h\,\bar{\partial}\psi + \bar{\partial}h.\psi)$$

$$= h\partial\bar{\partial}\psi + \partial h.\bar{\partial}\psi + \bar{\partial}h.\partial\psi + \partial\bar{\partial}h.\psi$$

and therefore

$$L(\phi)_{a|T_a(\partial U)} = h(a)\, L(\psi)_{a|T_a(\partial U)}.$$

This shows that the signature (i.e. the number of positive and negative eigenvalues) of the Levi-form restricted to the analytic tangent plane to ∂U at a is independent also of the choice of the defining function ϕ for U near a.

<u>Proposition</u> (1.4.1). <u>Let</u> U <u>be an open subset of</u> $\mathbb{C}^n$ <u>with a smooth boundary</u>. <u>At any point</u> $a \in \partial U$ <u>the Levi-form of any defining function for</u> ∂U <u>restricted to the analytic tangent plane to</u> ∂U <u>at</u> a <u>has a signature which is independent of local holomorphic coordinates and of the choice of the defining function</u>.

Let $p(a)$ $(q(a))$ be the number of strictly positive (strictly negative) eigenvalues of $L(\phi)_{a}|_{T_a(\partial U)}$. These are biholomorphic invariants of the triple $(U, \partial U, a)$. Note that we must have

$$p(a) + q(a) \leq n - 1 .$$

As an exercise we can show that there is an analytic disc of dimension p

$$\tau : D^p \to \mathbb{C}^n$$

(i.e. the biholomorphic image of the unit ball $D^p = \{t \in \mathbb{C}^p \mid \sum_1^p |t_i|^2 < 1$ in $\mathbb{C}^p)$ such that

$$\tau(0) = a$$

$$\tau(D^p) - \{a\} \subset \Omega - \bar{U} .$$

Analogously there is an analytic disc $\sigma : D^q \to \mathbb{C}^n$ of dimension q such that

$$\sigma(0) = a$$

$$\sigma(D^q) - \{a\} \subset U .$$

Indeed we can choose coordinates at the origin such that

$$\phi(z) = \text{Re}\, z_1 + L(\phi)_0(z) + O(\|z\|^3)$$

with

$$L(\phi)_0\big|_{T_0(\partial U)} = \sum_2^{p+1} \lambda_j \, |z_j|^2 - \sum_{p+2}^{p+q+1} \lambda_j \, |z_j|^2$$

where the λ_j's are > 0.
Therefore near 0, for $\varepsilon > 0$ sufficiently small, if

$$0 < |z_2|^2 + \dots + |z_{p+1}|^2 < \varepsilon \quad \text{and} \quad z_1 = z_{p+2} = \dots = z_n = 0$$

then $\phi(z) > 0$.
This proves the first statement. The second one is proved with a similar argument.

c) Theorem (1.4.2). (E.E.Levi [36]). Let Ω be an open set of holomorphy with a smooth boundary. Then the Levi-form at each boundary point restricted to the analytic tangent plane is positive semidefinite.

Proof. Assume, if possible, that $L(\phi)_0\big|_{T_0(\partial\Omega)}$ has a negative eigenvalue at the point $0 \in \partial\Omega$, ϕ being a defining function for $\partial\Omega$ with $\phi(0) = 0$. By suitable choice of the holomorphic coordinates we may write near 0

$$\phi(z) = 2 \operatorname{Re} z_1 \Big(1 + \sum_{j=1}^{n} a_j \bar{z}_j\Big) - z_2 \bar{z}_2 + \sum_{j=3}^{n} \lambda_j z_j \bar{z}_j + O(\|z\|^3).$$

First restrict ϕ to $\mathbb{R}^3 = \{\operatorname{Im} z_1 = 0, \ z_3 = \dots = z_n = 0\}$.
There exists $\varepsilon > 0$ such that for $\|z\| < 2\varepsilon$ on the region $\mathbb{R}^3 \cap \{x_1 \leq 0\}$, $(z_1 = x_1 + iy_1)$, we have

$$|O(\|z\|^3)| < \tfrac{1}{2} \, |2 \operatorname{Re} x_1(1 + a_1 x_1 + a_2 \bar{z}_2) - z_2 \bar{z}_2| .$$

Therefore : for ε sufficiently small, $\phi < 0$ on the discs

$$D_r = \{z_1 = r, \ |z_2| < \varepsilon, \ z_3 = \dots = z_n = 0\} \ ; \ -\varepsilon < r < 0 .$$

i.e. $D_r \subset \Omega$.

Also if ε is sufficiently small,

$$\{z_1 = 0,\ \frac{\varepsilon}{2} < |z_2| < \varepsilon,\ z_3 = \dots = z_n = 0\} \subset \Omega .$$

Hence there exists δ, $0 < \delta < \varepsilon$, such that

$$A = \{\frac{\varepsilon}{2} < |z_2| < \varepsilon,\ |z_1|^2 + |z_3|^2 + \dots + |z_n|^2 < \delta\} \subset \Omega ,$$

and there exists η, $0 < \eta < \delta$, such that

$$B = \{|z_1 + \frac{\delta}{2}| < \eta,\ |z_2| < \varepsilon,\ |z_3|^2 + \dots + |z_n|^2 < \eta\} \subset \Omega .$$

Let $A \cup B = \Delta$ and let

$$\hat{\Delta} = \{|z_2| < \varepsilon,\ |z_1|^2 + |z_3|^2 + \dots + z_n{}^2 < \delta\} .$$

By the disc-theorem $\hat{\Delta}$ is an $\mathcal{H}(\Omega)|_\Delta$ -completion of Δ. But $\hat{\Delta}$ contains the origin $0 \notin \Omega$, thus Ω is not an open set of holomorphy.

It is natural to ask if the above necessary condition for an open set Ω in $\mathbb{C}^n$ with a smooth boundary to be an open set of holomorphy is also sufficient (Levi-problem). The answer is affirmative for open sets in $\mathbb{C}^n$ but not for open sets on complex manifolds. We will return later to this question.

__Exercises.__

1. Prove that every convex domain in $\mathbb{C}^n$ is a domain of holomorphy.

2. Suppose that Ω has a smooth boundary and that at a point $a \in \partial\Omega$ the Levi-form restricted to the analytic tangent plane at a to $\partial\Omega$ is strictly positive. Prove that we can choose local holomorphic coordinates at a such that Ω is locally elementary convex at a.

Hint: we can replace the defining function ϕ by an increasing convex function ψ of ϕ so that $L(\psi)_a$ is strictly positive (for instance take $\psi = e^{c\phi}$ with $c >> 0$, see [28], p. 263) and then use the remark in a).

3. Under the same assumption of the previous exercise, prove that there is a fundamental system of neighborhoods $B(a)$ of a which are domains of holomorphy such that $B(a) \cap \Omega$ is an open set of holomorphy.

Hint: in the above specified local coordinates take for $B(a)$ any small coordinate ball with center in a, then apply the first exercise.

The material of this chapter is covered in all standard books on complex analysis as [28], [31].

Chapter II. Pseudoconcave manifolds.

2.1 Preliminaries.

a) Presheaves. A presheaf on a topological space X is a contravariant functor from the category of open subsets U of X to the category of abelian groups i.e.

for every U an abelian group $S(U)$ is given and
for every inclusion of open sets $V \subset U$ a homomorphism

$$r^U_V : S(U) \to S(V)$$

is given such that for every chain of inclusions $W \subset V \subset U$ of open subsets of X we have

$$r^V_W \circ r^U_V = r^U_W .$$

A presheaf $\mathcal{S} = S(U) \; ; \; r^U_V$ is called a sheaf if for every open set $\Omega \subset X$ and every open covering $\mathcal{U} = \{U_i\}_{i\in I}$ of the following sequence is exact

$$0 \to S(\Omega) \overset{\varepsilon}{\to} \prod_{i\in I} S(U_i) \overset{\delta}{\to} \prod_{(i,j)\in I^2} S(U_i \cap U_j)$$

where ε is defined by

$$\varepsilon(f)_{U_i} = r^{\Omega}_{U_i}(f) \quad , \quad f \in S(\Omega)$$

and where δ is defined by

$$\delta(f)_{U_i \cap U_j} = r^{U_j}_{U_i \cap U_j} f_{U_j} - r^{U_j}_{U_i \cap U_j} f_{U_i} \quad , \; f = \{f_{U_i}\} \in \prod_{i\in I} S(U_i).$$

Example:

$\mathcal{S} = \{\text{Homcont}(U, \mathbb{C}),\; r^U_V\}$, where Homcont $(U, \mathbb{C})$ denotes the space of continuous functions on U with values in $\mathbb{C}$ and where r^U_V are the natural restriction maps, is a presheaf and also a sheaf.

In a similar way one defines sheaves of rings and also sheaves of modules over a sheaf of rings.

b) a **stack** $\mathfrak{F}$ over X of abelian groups is the data of a topological space $\mathfrak{F}$, a continuous surjective map $\pi : \mathfrak{F} \to X$ such that

α) π is a local homeomorphism i.e. every point f $\mathfrak{F}$ has an open neighborhood $s = s(f)$ such that $\pi|_s$ is a homeomorphism of s onto an open subset of X ;

β) for each point $x \in X$, $\mathfrak{F}_x = \pi^{-1}(x)$ has the structure of an abelian group in such a way that the map

$$\mathfrak{F} \times_X \mathfrak{F}^{(1)} \to \mathfrak{F}$$

given by

$$(\alpha, \beta) \to \alpha - \beta \quad \text{is continuous.}$$

Given a stack $(\mathfrak{F}, \pi, X)$ of abelian groups, for every open set $U \subset X$ we can consider the abelian group

$$\Gamma(U, \mathfrak{F}) = \{ s : U \to \mathfrak{F} \mid s \text{ continuous, } \pi \circ s = \text{identity on } U \}$$

of all "**sections**" s of $\mathfrak{F}$ over U. If $V \subset U$, the natural restriction map $r^U_V : \Gamma(U, \mathfrak{F}) \to \Gamma(V, \mathfrak{F})$ is defined and one obtains in this way a presheaf which is also a sheaf.

Conversely, given a presheaf $\mathcal{S} = \{ S(U) ; r^U_V \}$ one can associate to it a stack $(\mathfrak{F}, \pi, X)$ as follows.
We set for every $x \in X$:

$$\mathfrak{F}_x = \varinjlim_{U \ni x} S(U) \text{ , i.e. an element of}$$

$\mathfrak{F}_x$ is a class of equivalence of couples (U, f) with

---

(1) the "fibered product" $\mathfrak{F} \times_X \mathfrak{F}$ is defined as the part of $\mathfrak{F} \times \mathfrak{F}$ lying above the diagonal Δ of $X \times X$ by projection $\pi \times \pi : \mathfrak{F} \times \mathfrak{F} \to X \times X$;
$\mathfrak{F} \times_X \mathfrak{F} = (\pi \times \pi)^{-1} (\Delta)$.

$x \in U$, $f \in S(U)$ under the relation

$$(U_1, f_1) \sim (U_2, f_2)$$

if there exists $U_3 \ni x$, $U_3 \subset U_1 \cap U_2$ such that

$$r^{U_1}_{U_3} f_1 = r^{U_2}_{U_3} f_2 .$$

The equivalence class in $\mathcal{F}_x$ of (U, f) is denoted by f_x and it is called the germ of f at x.
We then define $\mathcal{F} = \bigcup_{x \in X} \mathcal{F}_x$ and π by $\pi(\mathcal{F}_x) = x$.
If we take on $\mathcal{F}$ as a basis for open sets the sets of the form $\bigcup_{x \in U} f_x$ for all $f \in S(U)$, we obtain, as one verifies, a stack of abelian groups $(\mathcal{F}, \pi, X)$.

Starting in this construction with a sheaf, constructing the corresponding stack and then the corresponding sheaf of sections we get back the original sheaf. We thus have a one-to-one correspondence between sheaves of abelian groups and stacks of abelian groups. Although this could generate some confusion it is customary to represent a sheaf by the associated stack (see for instance [25] and [30] or [18]).

c) <u>Meromorphic functions</u>. Let now X be a complex manifold and let $\mathcal{O}$ be the sheaf of germs of holomorphic functions on X. For every open set $U \subset X$ it is defined by the space $H(U)$ and the natural restriction maps. The space $H(U)$ is a ring. Let $D(U)$ be the subset of $H(U)$ of divisors of zero, i.e. $D(U)$ is the set of those holomorphic functions on U vanishing on some connected component of U. Let $Q(U)$ be the quotient ring of $H(U)$ with respect to $D(U)$ i.e. $Q(U)$ is the set of quotients $\frac{f}{g}$ with $f \in H(U)$, $g \in H(U) - D(U)$ with obvious identifications:

$$\frac{f}{g} = \frac{f'}{g'} \quad \text{iff} \quad fg' = f'g .$$

If $V \subset U$ is an inclusion of open sets, the restriction map ${r^U}_V: \mathcal{H}(U) \to \mathcal{H}(V)$ sends $\mathcal{H}(U) - \mathcal{D}(U)$ into $\mathcal{H}(V) - \mathcal{D}(V)$ and thus induces a homomorphism of rings

$${r^U}_V: \mathcal{Q}(U) \to \mathcal{Q}(V)$$

We obtain in this way a presheaf. The corresponding sheaf $\mathcal{M}$ is called the <u>sheaf of germs of meromorphic functions</u> on X. the ring $\mathcal{K}(X) = \Gamma(X, \mathcal{M})$ is called the <u>ring of meromorphic functions</u> on on X. Note that $\mathcal{Q}(X) \in \mathcal{K}(X)$ but $\mathcal{Q}(X)$ may be actually smaller than $\mathcal{K}(X)$.

<u>Example</u>: Take $X = \mathbb{P}_1(\mathbb{C})$, the Riemann sphere. Then $\mathcal{H}(X) = \mathbb{C}$ thus $\mathcal{Q}(X) = \mathbb{C}$ while $\mathcal{K}(X)$ is isomorphic to the field of all rational functions in one variable t, $\mathcal{K}(X) \simeq \mathbb{C}(t)$.

If X is connected then $\mathcal{K}(X)$ and $\mathcal{Q}(X)$ are <u>fields</u>.

In the sequel <u>we will always assume that</u> X <u>is a connected manifold</u>.

2.2. <u>Mermorphic functions and holomorphic line bundles</u>.

a) <u>Holomorphic line bundles</u>. Let X be a complex manifold; by a holomorphic line bundle on X we mean a triple (F, π, X) where F is a complex manifold, $\pi : F \to X$ a holomorphic surjective map such that

i) π is of maximal rank

ii) for every $x \in X$ $\pi^{-1}(x)$ is isomorphic to the complex field $\mathbb{C}$ in such a way that

α) the map

$$F \times_X F \to F$$

given by $(u, v) \to v+v$ is holomorphic

β) the map

$$\mathbb{C} \times F \to F$$

given by $(\lambda, v) \to \lambda v$ is holomorphic.

Given two holomorphic line bundles (F, π, X), (E, ω, X) over X a <u>morphism</u> (or bundle map) is a holomorphic map $f : F \to E$ such that

i) $\pi = \omega \circ f$

ii) for every $x \in X$ the induced map $f_x : \pi^{-1}(x) \to \omega^{-1}(x)$ is $\mathbb{C}$-linear.

A holomorphic line bundle (F, π, X) is said to be trivial if it is isomorphic to the bundle $(X \times \mathbb{C}, pr_X, X)$.

Every holomorphic line bundle is locally trivial (as it follows from the implicit function theorem). Therefore there exist an open covering $\mathcal{U} = U_i$ of X and biholomorphic maps

$$\phi_i : \pi^{-1}(U_i) \to U_i \times \mathbb{C}$$

such that $\pi \circ \phi_i^{-1}(x, y) = x$, $(x, y) \in U_i \times \mathbb{C}$. On $U_i \cap U_j$ we have two trivializations of F and thus

$$\phi_i \circ \phi_j^{-1}(x, v) = (x, g_{ij}(x) v)$$

with

(1) $g_{ij} : U_i \cap U_j \to \mathbb{C}^*$ is holomorphic and never zero

(2) On $U_i \cap U_j \cap U_k$ we must have $g_{ij} g_{jk} = g_{ik}$ (consistency conditions). The collection of functions $\{g_{ij}\}$ are called the <u>transition functions</u> of the bundle F (relative to the local trivializations ϕ_i).

Conversely, given on an open covering $\mathcal{U} = \{U_i\}$ a system of transition functions (1) satisfying the consistency conditions (2) one can construct a holomorphic line bundle with local trivializations on the sets U_i having the given systems as a system of transition functions.

Two holomorphic line bundles given on the same covering $\mathcal{U} = U_i$ by transition functions $\{g_{ij}\}$, $\{f_{ij}\}$ are isomorphic if there exist holomorphic maps

$$\lambda_i : U_i \to \mathbb{C}^* \quad \text{such that}$$

$$g_{ij} = \lambda_i \, f_{ij} \, \lambda_j^{-1} \quad \text{on } U_i \cap U_j .$$

Given a holomorphic line bundle (F, π, X) we can consider the space of holomorphic sections :

$$\Gamma(X, F) = \{s : X \to F \mid s \text{ holomorphic}, \pi \circ s = id_X\} .$$

In terms of local trivializations of F on the covering $\mathcal{U} = U_i$ a holomorphic section is given by a collection

$$s_i : U_i \to \mathbb{C}$$

of holomorphic functions such that

$$s_i = g_{ij} s_j \quad \text{on} \quad U_i \quad U_j \qquad (\text{cf. } 30).$$

b) Given two holomorphic sections $s_1 = \{s_{1i}\}$ and $s_0 = \{s_{0i}\}$ of the bundle F, if s_0 is not identically zero on any open set we can construct a meromorphic function

$$\frac{s_1}{s_0} \quad \text{on } X, \text{ given locally by } \frac{s_{1i}}{s_{0i}} .$$

In this way we can obtain all meromorphic functions of $\mathcal{K}(x)$: indeed we have

<u>Proposition</u> (2.2.1). <u>Every meromorphic function</u> m <u>on a complex manifold</u> X <u>is the quotient of two holomorphic sections of an appropriate holomorphic line bundle on</u> X.

<u>Proof.</u> For every point $x \in X$ we can find a neighborhood V such that

$$m|_V = \frac{p}{q} \quad \text{with } p, q \in \mathcal{H}(V), \quad q \in \mathcal{H}(V) - \mathcal{D}(V).$$

Since the ring $\mathcal{O}_x$ is a unique factorization domain (cf. [28]), if we take V sufficiently small we may assume that the germs p_x and q_x of p and q at x are coprime. But if p_x and q_x are coprime and if V is sufficiently small, then also the germs p_y and q_y of p and q at any point $y \in V$ are coprime (cf. [29], [48]).

Let $\{V(x_i)\}_{i\in I}$ be a covering of X with such neighborhoods. Then on $V(x_i)$

$$m|V(x_i) = \frac{p_i}{q_i}, \quad p_i \in \mathcal{H}(V_i), \quad q_i \in \mathcal{H}(V_i) - \mathcal{D}(V_i)$$

and on $V(x_i) \cap V(x_j)$

$$\frac{p_i}{q_i} = \frac{p_j}{q_j}, \text{ i.e. } p_i q_j = p_j q_i .$$

By the Euclid lemma p_i must divide P_j and p_j must divide p_i, i.e.

$$P_i = g_{ij} P_j \text{ with } g_{ij} \text{ a unit in } \mathcal{H}(V_i) \cap \mathcal{H}(V_j);$$

this means that g_{ij} on $V(x_i) \cap V(x_j)$ is holomorphic and never zero. It follows then that we also have

$$q_i = g_{ij} q_j .$$

Moreover on $V(x_i) \cap V(x_j) \cap V(x_k)$ we must have

$$g_{ij} g_{jk} g_{ki} = 1$$

This shows that the collection $\{g_{ij}\}$ is a set of transition functions of a holomorphic line bundle F over X, the collection $\{p_i\}$ gives a holomorphic section s_1 of F and the collection $\{q_i\}$ gives a holomorphic section s_0 of F with s_0 not identically zero on any open set. So

$$m = \frac{s_1}{s_0} \text{ as required.}$$

c) Let (F, π, X) be a holomorphic line bundle on X given on a covering $\mathcal{U} = \{U_i\}_{i\in I}$ of X by transition functions $\{g_{ij}\}$. One can consider the "l-th tensor power of F", (F^l, π_l, X) which is given by the transition functions $\{g_{ij}^l\}$.

We can then consider the graded ring

$$\mathcal{A}(X, F) = \bigcup_{l=0}^{\infty} \Gamma(X, F^l)$$

of the holomorphic sections of the different tensor powers of F (F^0 = trivial bundle). Note that if $s \in \Gamma(X, F^l)$, $t \in \Gamma(X, F^m)$ then $st \in \Gamma(X, F^{l+m})$. If X is connected, as we always assume, then $\mathcal{A}(X, F)$ is an integral domain and one can consider the field of quotients

$$\mathcal{Q}(X, F) = \left\{ \frac{s_1}{s_0} \;\middle|\; s_1, s_0 \in \Gamma(X, F^l) \text{ for some } l, \; s_0 \neq 0 \right\}.$$

We have

$$\mathcal{Q}(X, F) \subset \mathcal{K}(X),$$

in particular $\mathcal{O}(X) = \mathcal{O}(X, \text{trivial bundle})$.

<u>Theorem</u> (2.2.2). <u>For every holomorphic line bundle F the field $\mathcal{Q}(X, F)$ is an algebraically closed subfield of $\mathcal{K}(X)$</u> (X <u>connected</u>).

<u>Proof</u>. Let $h \in \mathcal{K}(X)$ be algebraic over $\mathcal{Q}(X, F)$ i.e. h satisfies an equation

$$h^{\nu} + k_1 h^{\nu-1} + \ldots + k_{\nu} \equiv 0$$

where $k_i \in \mathcal{Q}(X, F)$. Let $k_i = \frac{s_i}{t_i}$ with $s_i, t_i \in \Gamma(X, F^{l_i})$; multiplying the above equation by $\prod_{i=1}^{\nu} t_i$ we obtain an equation

$$\sigma_0 h^{\nu} + \sigma_1 h^{\nu-1} + \ldots + \sigma_{\nu} \equiv 0$$

where $\sigma_i \in \Gamma(X, F^l)$ for a suitable l ($l = \sum_{i=1}^{\nu} l_i$) and where $\sigma_0 \not\equiv 0$. After multiplication by $\sigma_0^{\nu-1}$ the above equqtion can be written as follows:

$$(\sigma_0 h)^{\nu} + \sigma_1 (\sigma_0 h)^{\nu-1} + \ldots + \sigma_0^{\nu-1} \sigma_{\nu} \equiv 0.$$

At each point $x \in X$ $\sigma_0 h$ satisfies an equation with holomorphic coefficients and with the coefficient of the highest power equal to one. This shows that $\sigma_0 h$ is meromorphic at x and integral over $\mathcal{O}_x$; since $\mathcal{O}_x$ is integrally closed $\sigma_0 h = \tau$ must be holomorphic at x.

Hence $\sigma_0 \in \Gamma(X, F^l)$ and also $\sigma_0 h = \tau \in \Gamma(X, F^l)$, thus

$$h = \frac{\tau}{\sigma_0} = \frac{\sigma_0 h}{\sigma_0} \in \mathcal{Q}(X, F).$$

thus $h = \frac{\tau}{\sigma_0} = \frac{\sigma_0 h}{\sigma_0} \in \mathcal{Q}(X, F)$.

2.3 Pseudoconcave manifolds. A connected complex manifold X is called pseudoconcave if we can find a non-empty open subset Y X with the following properties

i) Y is relatively compact in X, $Y \subset\subset X$.

ii) $\partial Y = \bar{Y} - Y$ is smooth and the Levi form of ∂Y restricted to the analytic tangent plane has at least one negative eigenvalue at each point of ∂Y[(1)].

In particular for any point $z_0 \in \partial Y$ there is an analytic disc of dimension ≥ 1 which is tangent at z_0 to ∂Y and is contained in Y except for the point z_0.

Examples.

1) Every compact connected manifold is pseudoconcave (take $y = X$ then $\partial Y = \emptyset$ thus condition ii) is void).

2) Let Z be a compact connected manifold of $\dim_{\mathbb{C}} Z \geq 2$. Let $\{a_1, \ldots, a_m\}$ be a finite subset of Z. Then $X = Z - \{a_1, \ldots, a_m\}$ is pseudoconcave (take for Y the complement of a set of disjoint coordinate balls centered at the points a_i).

3) Not every pseudoconcave manifold is compactibiable (i.e. isomorphic to an open subset of a compact manifold). For instance if we take $\mathbb{P}_2(\mathbb{C}) - \{0\} \supset \mathbb{C}^2 - \{0\}$ and if z_1, z_2 are the holomorphic coordinates on $\mathbb{C}^2$, we can consider the exterior form

$$\phi_\varepsilon = dz_1 \wedge dz_2 + \varepsilon \partial \bar{\partial} \log (|z_1|^2 + |z_2|^2) \quad \text{with } \varepsilon \neq 0 ,$$

and define a function f to be holomorphic if it satisfies the

---

(1) As usual the defining function for ∂Y is chosen so that it is < 0 on Y and > 0 outside of Y.

differential equation

$$df \wedge \phi_{\varepsilon} = 0 .$$

In this way we define a complex structure on $\mathbb{C}^2 - \{0\}$ (which agrees with the natural one if $\varepsilon = 0$). One can show that this complex structure can be extended to $\mathbb{P}_2(\mathbb{C}) - \{0\}$, that if $\varepsilon \neq 0$ is small it provides $\mathbb{P}_2(\mathbb{C}) - \{0\}$ with a pseudoconcave structure and that it is not compactifiable (cf. section 4.4 and [10]).

Remark: Every holomorphic function on a pseudoconcave manifold is constant.

In fact let f be holomorphic and non-constant on X and let $z_o \in \bar{Y}$ such that $|f(z_o)| = \sup_{\bar{Y}} |f|$. By the maximum modulus principle, $z_o \in \partial Y$. If D is a 1-dimensional disc tangent to ∂Y at z_o and except z_o contained in Y then $|f|_D$ has a maximum on an interior point of D. Thus f is constant on D and thus there is an interior point z_1 of $\bar{Y}$ such that $|f(z_1)| = |f(z_o)| = \sup_{\bar{Y}} |f|$. This is a contradiction.

In particular a pseudoconcave manifold X (not reduced to a single point) cannot be isomorphic to any locally closed submanifold of numerical space $\mathbb{C}^N$ (otherwise there will be a p olynomial on $\mathbb{C}^N$ inducing on X a non-constant holomorphic function).

More generally one can prove the following

Theorem (2.3.1). For any holomorphic line bundle F on a pseudoconcave manifold X we have

$$\dim_{\mathbb{C}} \Gamma(X, F) < \infty .$$

Lemma (2.3.2). Let F be a holomorphic line bundle over a pseudoconcave manifold X. There exists a finite number of points $a_1, \ldots, a_k$ in X and an integer $h = h(F)$ such that if $s \in \Gamma(X, F)$ vanishes at each point a_i of order $\geq h$ then $s = 0$.

<u>Proof</u>. Let Y be as in the definition of pseudoconcave manifolds. For every point $x \in \bar{Y}$ we can choose a coordinate polycilinder P_x, coordinates p_i, $i = 1,\dots, n$, with center x and of radius r_x such that

i) $F|_{\bar{P}}$ is trivial

ii) $S(P_x^x) = \{y \in U \mid |p_i(y) - p_i(x)| = r_x,\ 1 \leq i \leq n\} \subset Y$

where U is the coordinate patch on which p_i are coordinates. This is possible in view of the pseudoconcavity of Y.

Let P_x' be the concentric polycilinder to P_x with radius $r_x e^{-1}$.

We can select a finite number of points $a_1,\dots, a_k$ such that

iii) $\cup P'_{a_i} \supset \bar{Y}$.

Let F be given by transition functions

$$f_{ij} : \bar{P}_{a_i} \cap \bar{P}_{a_j} \to \mathbb{C}^*$$

and set

$$\|F\| = \sup_{i,j} \sup_{\bar{P}_{a_i} \cap \bar{P}_{a_j}} |f_{ij}| = e^{\mu}.$$

Note that since $f_{ij} = f_{ji}^{-1}$ we must have $\mu \geq 0$.

Now choose an integer h with $h > \mu$, for instance $h = [\mu] + 1$ where $[\mu]$ denotes the integral part of μ.

Let $s \in \Gamma(X, F)$, vanishing at the points a_i of order $\geq h$. The section s is given by holomorphic functions $s_i : \bar{P}_{a_i} \to \mathbb{C}$. We set

$$M = \sup_i \sup_{\bar{P}_{a_i}} |s_i|.$$

There exists a point $z_o \in S(\bar{P}_{a_{i_o}})$, for some a_{i_o}, such that

$$|s_{i_o}(z_o)| = M$$

(indeed $S(\bar{P}_{a_i})$ is the Silov boundary of $\bar{P}_{a_i}$).

Since $z_o \in Y$ there exists a $P'_{a_{j_o}}$ containing z_o. Certainly $j_o \neq i_o$ and we will have

$$s_{i_o}(z_o) = f_{i_o j_o}(z_o)\, s_{j_o}(z_o) .$$

Therefore

$$M = |s_{i_o}(z_o)| = |f_{i_o j_o}(z_o)|\, |s_{j_o}(z_o)|$$

$$\leq \|F\|\, |s_{j_o}(z_o)| .$$

By Schwarz's lemma

$$|s_{j_o}(z_o)| \leq M \frac{\|z_o\|^h}{r^h_{a_{j_o}}}$$

where

$$\|z_o\| = \sup_i |p_i(z_o) - p_i(a_{j_o})| \leq r_{a_{j_o}} e^{-1} ,$$

the functions p_i being the coordinates on $P_{a_{j_o}}$. Hence

$$M \leq \|F\|\, M e^{-h} = e^{\mu - h} M .$$

Since $\mu - h < 0$ we must have $M = 0$ which implies $s \equiv 0$.

<u>Proof of theorem (2.3.1).</u>

The natural map $\Gamma(X, F) \to \prod_{i=1}^{k} \frac{\mathcal{O}_{a_i}}{\mathcal{M}^h_{a_i}}$,

which associates to each section $s \in \Gamma(X, F)$ the Taylor expansion of s up to order $h - 1$ at each point a_i, is an injective map by the previous lemma. The righthand space is a finite-dimensional vector space over $\mathbb{C}$ (dimension $\leq k\binom{n+h}{h}$).

<u>Remark</u>. Let X be pseudoconcave and $Y \subset X$ as in the definition. Then $\Gamma(X, F) \to \Gamma(Y, F)$ is injective. Using

Hartogs' theorem and the pseudoconcavity of Y we can construct an open neighborhood $\tilde{Y}$ of $\bar{Y}$ such that the restriction map $r^{\tilde{Y}}_Y : \Gamma(\tilde{Y}, F) \to \Gamma(Y, F)$ is an isomorphism. Now $r^{\tilde{Y}}_Y$ is a compact map for the Fréchet topology of $\Gamma(\tilde{Y}, F)$ and $\Gamma(Y, F)$. Thus the Fréchet space $\Gamma(Y, F)$ is locally compact and therefore finite-dimensional. This would give a more direct proof of theorem (2.3.1). However the previous proof has the merit to give an estimate for the dimension of $\Gamma(X, F)$ which will be useful in the sequel.

2.4. Analytic and algebraic dependence of meromorphic functions.

Let X be a connected complex manifold. Let $f_1, \ldots, f_k \in \mathcal{K}(X)$. We say that these meromorphic functions are analytically dependent if

$$df_1 \wedge \ldots \wedge df_k \equiv 0 \quad \text{wherever this is defined.}$$

In other words $f_1, \ldots, f_k$ are analytically dependent if at any point where each one of these functions is holomorphic the Jacobian $\frac{\partial(f_1, \ldots, f_k)}{\partial(z_1, \ldots, z_n)}$ with respect to a system $z_1, \ldots, z_n$ of local holomorphic coordinates, has rank $< k$.

The meromorphic functions $f_1, \ldots, f_k$ are said to be algebraically dependent if there exists a non-identically zero polynomial $p(x_1, \ldots, x_k)$ in k variables and with complex coefficients such that

$$p(f_1, \ldots, f_k) \equiv 0 \quad \text{wherever it is defined.}$$

Algebraic dependence implies analytic dependence. In fact if $k > n = \dim_{\mathbb{C}} X$ there is nothing to prove. Assume $k \leq n$. Without loss of generality we may also assume that $f_1, \ldots, f_{k-1}$ are algebraically independent. Let $p(x_1, \ldots, x_k)$ be a polynomial $\not\equiv 0$ of minimal degree in x_k such that

$$p(f_1, \ldots, f_k) = 0.$$

Differentiating this identity we get

$$\sum \frac{\partial p}{\partial x_i}(f)\, df_i = 0.$$

But $\frac{\partial p}{\partial x_k}(f) \not\equiv 0$, thus we get a non-trivial linear relation between the differentials df_i in an open dense subset of X. This implies that $df_1 \wedge \ldots \wedge df_k \equiv 0$ wherever defined on X.

The converse, of this statement (except for $k = 1$) is not true in general. For instance the functions

$f_s(x) = e^{z^s}$, $s = 1,2,3,\ldots$ in $\mathcal{K}(\mathbb{C})$ are all algebraically independent while any two of them are analytically dependent.

The converse is however true for pseudoconcave manifolds; we have in fact the following

__Theorem__ (2.4.1). __Let__ S __be a pseudoconcave manifold.__ __If__ $f_1,\ldots, f_k, f \in \mathcal{K}(X)$ __are analytically dependent then they are also algebraically dependent__ (i.e. __on pseudoconcave manifolds analytic dependence__ = __algebraic dependence__).

__Proof.__ It is not restrictive to assume that $f_1,\ldots, f_k$ are analytically independent. Otherwise replace $f_1,\ldots, f_k$ by a maximal subset of $f_1,\ldots, f_k, f$ of analytically independent functions and take for f one of the remaining functions. There exists a holomorphic line bundle F on X and holomorphic sections $s_i \in \Gamma(X, F)$, $0 \leq i \leq k$, with $s_0 \not\equiv 0$ such that

$$f_i = \frac{s_i}{s_0}.$$

Indeed for each f_i there exists a holomorphic line bundle F_i and holomorphic sections $t_0^{(i)}$, $t_1^{(i)}$ with $t_0^{(i)} \not\equiv 0$ such that

$$f_i = \frac{t_1^{(i)}}{t_0^{(i)}}$$

Taking $F = F_1 \dots F_k$ then $s_0 = \prod_{i=1}^{k} t_0^{(i)} \in \Gamma(X, F)$ and $s_0 \not\equiv 0$.

Moreover $s_i = t_0^{(1)} \dots t_1^{(i)} \dots t_0^{(k)} \in \Gamma(X, F)$

and we have $f_i = \frac{s_i}{s_0}$.

We can choose a covering of $\bar{Y}$ by coordinate concentric polycilinders $P_{a_i} \supset P'_{a_i}$, $1 \leq i \leq N$, as in the lemma (2.3.2) such that

i) $F|\bar{P}_{a_i}$ is trivial

ii) $S(\bar{P}_{a_i}) \subset Y$

iii) $\cup P'_{a_i} \supset \bar{Y}$

iv) at each point a_i the functions $f_1, \dots, f_k$ are holomorphic and $f_1 - f_1(a_i) = \xi_1^{(i)}, \dots, f_k - f_k(a_i) = \xi_k^{(i)}$ can be taken among a set of local holomorphic coordinates.

This can be done by small translations in the coordinate patches of the polycilinders $P_{a_i} \supset P'_{a_i}$ as conditions i), ii), iii) are not effected by these translations and as the set of points where condition iv) cannot be satisfied has an open dense complement.

Also there exists a holomorphic line bundle G on X and holomorphic sections $\sigma_0, \sigma_1 \in \Gamma(X, G)$ with $\sigma_0 \not\equiv 0$ such that

$$f = \frac{\sigma_1}{\sigma_0}.$$

we may assume that

v) $G|\bar{P}_{a_i}$ is trivial

vi) f is holomorphic at each point a_i.

As in the lemma (2.3.2) we define

$$\|F^k\| = e^k \quad (\text{where } F^k = F \dots F \ (k \text{ times}))$$

$$\|G\| = e^{\omega}.$$

Consider a generic polynomial in $k+1$ variables of degree r in each one of the variables $x_1, \dots, x_k$ and of degree s in x_{k+1},

$$P(x_1, \dots, x_k, x_{k+1}) = \sum c_{\alpha_1, \dots, \alpha_k, \alpha_{k+1}} x_1^{\alpha_1} \dots x_k^{\alpha_k} x_{k+1}^{\alpha_{k+1}}$$

where $0 \leq \alpha_i \leq r$ for $1 \leq i \leq k$ and $0 \leq \alpha_{k+1} \leq s$. Let

$$\Pi(x_0, \dots, x_k, y_0, y_1) = x_0^{kr} \ y_0^s \ P(\frac{x_1}{x_0}, \dots, \frac{x_k}{x_0}, \frac{y_1}{y_0})$$

be the corresponding homogeneous polynomial.
These polynomials form a vector space $W(r, s)$ over $\mathbb{C}$ of dimension $(r+1)^k(s+1)$.
Now note that

$$\Pi(s_0, s_1, \dots, s_k, \sigma_0, \sigma_1) \in \Gamma(X, F^{kr} \cdot G^s).$$

The theorem will be proved if we show that $\operatorname{Ker} \varepsilon \neq \{0\}$, where ε is the above natural linear map $W(r,s) \to \Gamma(X, F^{kr} . G^s)$.
Let h be the smallest integer $> kr\mu + s\omega$. The map which associates to $\Pi(s_0, \dots, s_k, \sigma_0, \sigma_1)$ the Taylor expansion up to order $h-1$ of the function $P(f_1, \dots, f_k, f)$ at each point a_i gives a linear map :

$$\operatorname{Im} \varepsilon \to \prod_{i=1}^{N} \frac{\mathbb{C}\{\xi_1^{(i)}, \dots, \xi_k^{(i)}\}}{\mathfrak{m}_{a_i}^h},$$

where $\mathfrak{m}_{a_i}$ is the maximal ideal of the local ring $\mathbb{C}\{\xi_1^{(i)}, \dots, \xi_k^{(i)}\}$ of convergent power series in the variables $\xi^{(i)}$. By lemma (2.3.2) this map is injective.

Now the target space has a dimension

$$\delta = N \begin{bmatrix} [kr\mu + s\omega] + 1+k \\ k \end{bmatrix}$$

$$= N \left(\frac{[kr\mu + s\omega] + 1+k}{k}\right) \left(\frac{[kr\mu + s\omega] + 1 + k - 1}{k-1}\right) \cdots \left(\frac{[kr\mu + s\omega] + 1 + 1}{1}\right)$$

$$\leq N([kr\mu + s\omega] + 2)^k$$

$$\leq Nk^k\mu^k r^k + \text{lower order terms in } r.$$

If we select s such that

$$s+1 > Nk^k\mu^k ,$$

then, if r is sufficiently large, we get

$$\dim_{\mathbb{C}} W(r, s) > \dim_{\mathbb{C}} \operatorname{Im} \mathcal{E} \quad \text{and therefore} \quad \operatorname{Ker} \mathcal{E} \neq 0 .$$

2.5. <u>Algebraic fields of meromorphic functions.</u>

a) By an <u>algebraic field of transcendence degree</u> d we mean a finite algebraic extension of the field $\mathbb{C}(t_1, \dots, t_d)$ of all rational functions in d variables. Since the ground field $\mathbb{C}$ is of characteristic zero, any extension of this kind is primitive so is of the form $\mathbb{C}(t_1, \dots, t_d , \theta)$ with θ algebraic over $\mathbb{C}(t_1, \dots, t_d)$. Let $P(t_1, \dots, t_d, t) = 0$ be the minimal equation for $t = \theta$ over $\mathbb{C}(t_1, \dots, t_d)$. Chasing denominators and dividing off any factor in the variables $t_1, \dots, t_d$ only, we may assume that P is a polynomial in all the variables and that it is irreducible. If V is the algebraic variety defined in $\mathbb{C}^{d+1}$ where $t_1, \dots, t_d, t$ are are coordinates by the equation

$$P(t_1, \dots, t_d, t) = 0 ,$$

then V is an irreducible variety and the field $\mathbb{C}(t_1, \dots, t_d, \theta)$ is isomorphic to the field of rational functions on V. Moreover $d = \dim_{\mathbb{C}} V$.

We want to prove the following

Theorem (2.5.1). On a pseudoconcave manifold X of complex dimension n, the field $\mathcal{K}(X)$ of all meromorphic functions is an algebraic field of transcendence degree $d \leq n$.

That the transcendence degree of $\mathcal{K}(X)$ cannot exceed $\dim_{\mathbb{C}} X$ follows already from the fact that on pseudoconcave manifolds algebraic and analytic dependence are the same (theorem (2.4.1)). The remaining part of the theorem is a consequence of

Proposition (2.5.2). Let X be a pseudoconcave manifold and let $f_1, \ldots, f_k \in \mathcal{K}(X)$ be algebraically independent. There exists an integer $\nu = \nu(f_1, \ldots, f_k)$ such that any $f \in \mathcal{K}(X)$ which is algebraically dependent on $f_1, \ldots, f_k$, satisfies a non-trivial equation over $\mathbb{C}(f_1, \ldots, f_k)$ of degree $\leq \nu$.

Proof. We follow the proof of theorem (2.4.1).
First we find a holomorphic line bundle F and holomorphic sections $s_0 \not\equiv 0, s_1, \ldots, s_k$ of F such that

$$f_i = \frac{s_i}{s_0}, \quad 1 \leq i \leq k .$$

Secondly we find coordinate polycilinders $P_{a_i} \supset P'_{a_i}$, $1 \leq i \leq N$, such that

i) F restricted to a neighborhood of $\bar{P}_{a_i}$ is trivial

ii) $S(\bar{P}_{a_i}) \subset Y$

iii) $\cup P'_{a_i} \supset \bar{Y}$

iv) at each point a_i, $f_1 - f_1(a_i) = \xi_1^{(i)}, \ldots, f_k - f_k(a_i) = \xi_k^{(i)}$

are holomorphic and can be taken among a set of local holomorphic coordinates.

Thirdly, since the conditions i), ii), iii), iv) remain valid by small translations of P_{a_i} within its coordinate patch, we may determine for each a_i a small closed neighborhood $V(a_i)$ so that no matter how we translate the center a_i of P_{a_i} on a point of $V(a_i)$ the above four conditions remain valid. Let Q_i be the union of the translates of P_{a_i} just considered and let us compute $||F||$ with respect to the covering $\{Q_i\}_{1\leq i\leq N}$.

Finally, from the proof of theorem (2.2.1) we realize that there exists a holomorphic line bundle G and two holomorphic sections $\sigma_0 \not\equiv 0, \sigma_1$ of G such that $f = \frac{\sigma_1}{\sigma_0}$ and satisfying the following condition

v) $G|_{Q_i}$ is trivial.

We set

$$||F^k|| = ||F||^k = e^{k\rho}$$

and we choose

$$\nu+1 > Nk^k\rho^k .$$

Then ν depends only on $f_1,\ldots, f_k$ but not on f. We also define, with respect to the covering $\{Q_i\}$,

$$||G|| = e$$

and we choose the centers a_i of P_{a_i} in $V(a_i)$ so that at a_i also f is holomorphic, $1 \leq i \leq N$.

We can now proceed as in the proof of theorem (2.4.1) and we realize that if r is sufficiently large then f satisfies a non-trivial equation over $\mathbb{C}(f_1,\ldots, f_k)$ of degree $\leq \nu$.

<u>Proof of theorem (2.5.1)</u>. Let $f_1,\ldots, f_k$ be a maximal set of algebraically independent meromorphic functions. Let $f \in \mathcal{K}(X)$ be so chosen that its degree over $\mathbb{C}(f_1,\ldots, f_k)$ is maximal. This is possible by virtue of proposition (2.5.2). We claim that

$$\mathcal{K}(X) = \mathbb{C}(f_1,\ldots, f_k, f) .$$

Clearly $\mathbb{C}(f_1,\ldots, f_k, f) \subset \mathcal{K}(X)$. Let $h \in \mathcal{K}(X)$, we can find $\theta \in \mathcal{K}(X)$ such that

$$\mathbb{C}(f_1,\ldots, f_k, f, h) = \mathbb{C}(f_1,\ldots, f_k, \theta).$$

Then

$$\alpha \geq [\mathbb{C}(f_1,\ldots, f_k, \theta) : \mathbb{C}(f_1,\ldots, f_k) =$$

$$\mathbb{C}(f_1,\ldots, f_k, \theta) : \mathbb{C}(f_1,\ldots, f_k, f)].[\mathbb{C}(f1,\ldots,f_k, f) : \mathbb{C}(f_1,\ldots, f_k)].$$

But the second factor of this product equals α; therefore the first factor equals 1. This means that $h \in \mathbb{C}(f_1,\ldots, f_k, f)$ and thus our contention is proved.

Theorem (2.5.3). Let X be a pseudoconcave manifold and let $\tau : X \to \mathbb{P}_N(\mathbb{C})$ be a holomorphic map of rank $n = \dim_{\mathbb{C}} X$ at some point of X. Then $\operatorname{Im} \tau$ is contained in an irreducible algebraic variety Y of the same dimensions than X.

Proof. Let Y be the smallest algebraic subvariety of $\mathbb{P}_N(\mathbb{C})$ containing (X). Certainly Y exists and is irreducible, it is defined by the homogeneous prime ideal

$$\mathcal{P}_Y = \{p \in \mathbb{C}[z_0,\ldots, z_N] \mid p \circ \tau = 0\}$$

where $\mathbb{C}[z_0,\ldots, z_N]$ denotes the graded ring of homogeneous polynomials on $\mathbb{P}_N(\mathbb{C})$.

Let $\mathcal{R}(Y)$ be the field of rational functions on Y. Any element $f \in \mathcal{R}(Y)$ is represented as a quotient of two homogeneous polynomials $f = \frac{p}{q}$ with $q \notin \mathcal{P}_Y$. If $f = \frac{p}{q} = \frac{p'}{q'}$ then $pq' - p'q \in \mathcal{P}_Y$. This shows that $f \circ \tau$ is a well defined meromorphic function on X. We have therefore defined a, necessarily injective, homomorphism

$$\tau^* : \mathcal{R}(Y) \to \mathcal{K}(X).$$

Now

$\dim_{\mathbb{C}} Y$ = transendence degree of $\mathcal{R}(Y)$ _ transcendence degree of $\mathcal{K}(X) \leq \dim_{\mathbb{C}} X$. But $\dim_{\mathbb{C}} Y \geq \dim_{\mathbb{C}} \tau(X) = \dim_{\mathbb{C}} X$ by the assumption about the rank of the map τ.

In particular every connected complex submanifold of $\mathbb{P}_N(\mathbb{C})$ is a projective algebraic variety (Chow - theorem).

b) <u>Excercises.</u>

1. Let A be a pure-dimensional non-singular algebraic subvariety of $\mathbb{P}_n(\mathbb{C})$. Let $a = \dim_{\mathbb{C}} A$. Prove that A has a fundamental system of neighborhoods $V(A)$ in $\mathbb{P}_n(\mathbb{C})$ with a smooth boundary at which the Levi-form restricted to the analytic tangent plane has at least a negative eigenvalues. Let B be a connected complex submanifold of $V(A)$ with $\dim_{\mathbb{C}} B + a \geq n+1$. Prove that B is contained in an irreducible algebraic subvariety of $\mathbb{P}_n(\mathbb{C})$ of the same dimension then B. (cf. [15]).

2. Prove that any pseudoconcave complex Liegroup is a complex torus ([6]).

3. Let K be the canonical bundle of the pseudoconcave manifold X ; define the "canonical dimension of X" as the transcendence degree of $\mathcal{Q}(X, K)$. Prove that $0 \leq \text{can dim } X \leq \dim_{\mathbb{C}} X$. Prove by examples that any value in that range is permitted.

The proofs given in this chapter are inspired by an idea of Serre ([46]); the method of exposition follows very closely an improved version given by Siegel ([50]) for the case of a compact manifold. For the pseudoconcave case they were given first in [1] and [3].

A. Andreotti

Chapter III. Properly discontinuous pseudoconcave groups: the Siegel modular group.

3.1. Preliminaries.

a) The notions developed in the previous chapter can be slightly generalized with respect both of the notion of manifold and the notion of pseudoconcavity.

Let us consider first the following situation; X is a complex connected manifold and $\Gamma \subset \mathrm{Aut}(X)$ a group of automorphisms of X. We will say that Γ is properly discontinuous on X if for every compact set $K \subset X$ the set

$$\{\gamma \in \Gamma \mid \gamma K \cap K \neq \emptyset\}$$

is a finite set. In particular taking for K a point $\{x_o\}$ we have that the isotropy group of x_o

$$\Gamma_{x_o} = \{\gamma \in \Gamma \mid \gamma x_o = x_o\}$$

is a finite group.

For any point $x_o \in X$ there exists a Γ_{x_o}-invariant relatively compact neighborhood $U(x_o)$ such that

if $y \in \bar{U}(x_o)$ and $\gamma y \in U(x_o)$ for some $\gamma \in \Gamma$, then $\gamma \in \Gamma_{x_o}$.

In fact let us choose a coordinate patch around x_o and let $V(\varepsilon)$, for $\varepsilon > 0$ and sufficiently small, denote the coordinate ball with center x_o and radius ε.

Set $S(\varepsilon) = \{\gamma \in \Gamma \mid \gamma V(\varepsilon) \cap V(\varepsilon) \neq \emptyset$.

Choose ε_o small enough such that $V(\varepsilon)$ is relatively compact for $\varepsilon < \varepsilon_o$. Then $S(\varepsilon)$ is finite if $\varepsilon < \varepsilon_o$. Certainly $S(\varepsilon) \supset \Gamma_{x_o}$. If for any ε, $0 < \varepsilon < \varepsilon_o$, $S(\varepsilon) \supsetneqq \Gamma_{x_o}$ there exists a $\gamma \in \Gamma$, $\gamma \notin \Gamma_{x_o}$ with $\gamma \in \bigcap_{0<\varepsilon<\varepsilon_o} S(\varepsilon)$.

Therefore there exists a sequence $x_n \to x_o$ with $\gamma x_n \to x_o$. By continuity $\gamma x_o = x_o$ and thus $\gamma \in \Gamma_{x}$, a contradiction. Therefore there exists $\varepsilon_1 > 0$, $0 < \varepsilon_1 < \varepsilon_o$, such that $S(\varepsilon) = \Gamma_{x_o}$ for $0 < \varepsilon \leq \varepsilon_1$. It is then enough to take $U(x_o) = V(\varepsilon_1)$.

It follows then that the equivalence relation

$$R = \{(x, y) \in X \times X \mid x = \gamma y \text{ for some } \gamma \in \Gamma\}$$

is closed and therefore the quotient space $Z = X/\Gamma$ is a Hausdorff space. Let $p : X \to X/\Gamma$ denote the natural projection, then $\Gamma U(x_o)$ is a neighborhood or the orbit Γ_{x_o} of x_o and $p(\Gamma U(x_o)) = U(x_o)/\Gamma_{x_o}$.

If we stipulate that a function $f : U \to \mathbb{C}$ for U open in Z is <u>holomorphic</u> if $f \circ p$ is holomorphic on $p^{-1}(U)$ we can extend to this new type of spaces the notions considered in the previous chapter. This type of space is sometimes called a <u>generalized complex manifold</u> (or V-manifold).
In particular we can talk about the field $\mathcal{R}(Z)$ of meromorphic functions on Z and of (locally trivial) holomorphic line bundles on Z. Every statement about the generalized manifold Z can be restated in terms of X and the group Γ.

Note that if Γ has no fixed points[(1)] X/Γ has actually the structure of a complex manifold.

b) By a <u>Γ-automorphic holomorphic line bundle</u> over X we we mean a holomorphic line bundle (F, π, X) over X with a lifting of the action of Γ over X to a group of Bundle maps of F, i.e. for every $\gamma \in \Gamma$ a bundle map $\rho_\gamma : F \to F$ is given such that

i) $\gamma \circ \pi = \pi \circ \rho_\gamma$

ii) if $\gamma = \gamma_2\gamma_1$ then $\rho_\gamma = \rho_{\gamma_2} \circ \rho_{\gamma_1}$.

For instance if F is the trivial bundle $X \times \mathbb{C}$, we will have

$$\rho_\gamma(x, v) = (\gamma x, \rho_\gamma(x)v),$$

where $\rho_\gamma(x)$ is a never vanishing holomorphic function. Thus

---

(1) By this we mean that if $\gamma \in \Gamma$ and $x_o \in X$ are such that $\gamma x_o = x_o$ then γ is the identity.

$\{\rho_\gamma\}$ can be identified with a collection of holomorphic functions

$$\rho_\gamma : X \to \mathbb{C}^*$$

satisfying the consistency relation

$$\rho_{\gamma_2\gamma_1}(x) = \rho_{\gamma_2}(\gamma_1 x)\, \rho_{\gamma_1}(x).$$

<u>A system of functions of this sort is called a system of factors of automorphy</u>. For example if X is an open subset of $\mathbb{C}^n$ we obtain a system of factors of automorphy considering the jacobian determinants

$$\rho_\gamma = \det\left(\frac{d\,(x)}{d(x)}\right) \quad \text{for every } \gamma \in \Gamma.$$

Given two Γ-automorphic holomorphic line bundles $(F, \pi, X, \{\rho_\gamma\})$ and $(G, \omega, X, \{\sigma_\gamma\})$, a <u>bundle map</u> of the first bundle into the second will be a bundle map $\phi : F \to G$ such that

$$\phi \circ \rho_\gamma = \sigma_\gamma \circ \phi \quad \text{for every } \gamma \in \Gamma.$$

For instance, given a Γ-automorphic line bundle by a system of factors of automorphy $\{\rho_\gamma(x)\}$, this will be isomorphic to the trivial bundle (which corresponds to factors of automorphy identically equal to 1) if and only if there exists a never vanishing holomorphic function $\phi : X \to \mathbb{C}^*$ such that

$$\rho_\gamma(x) = \phi(x) \quad \phi(\gamma x)^{-1} \quad \text{(trivial factor of automorphy).}^{(1)}$$

Given a (locally trivial) holomorphic line bundle (f', π', Z) on Z its reciprocal image on X is a Γ-automorphic line bundle $(F, \pi, X, \{\rho_\gamma\})$ with the property that

$$\rho_\gamma(x_o) = 1 \quad \text{for every } \gamma \in \Gamma_{x_o}.$$

One can verify that this property characterizes the reciprocal images on X of locally trivial holomorphic line bundles on Z.

---

(1) Thus the classes of factors of automorphy $\{\rho_\gamma(x)\}$ modulo the equivalence relation $\{\rho_\gamma(x)\} \sim \{\sigma_\gamma(x)\}$ iff $\{\rho_\gamma^{-1}\sigma_\gamma(x)\}$ is a trivial factor of automorphy, correspond to classes of isomorphisms of Γ-automorphic line bundles $\{F, \pi, X, \{\rho_\gamma\}\}$ in which F is the trivial bundle.

c) Given a Γ-automorphic line bundle $\{F, \pi, X, \{\rho_\gamma\}\}$ we can consider the space of Γ-invariant sections fo F:

$\Gamma(X, F)^\Gamma = \{s \in \Gamma(X, F) \mid s(\gamma x) = \rho_\gamma(x)\, s(x)$ for every $x \in X$.

Given two Γ-invariant sections $s_0 \not\equiv 0$, s_1 of F, then

$\frac{s_1}{s_0}$ is a Γ-invariant meromorphic function on X i.e. an element of $p^* \mathcal{K}(Z)$.

Conversely for every Γ-invariant meromorphic function m $p^*\mathcal{K}(Z)$ we can find a Γ-automorphic holomorphic line bundle and two Γ-invariant sections of it: $s_0 \not\equiv 0$, and s_1 such that $m = \frac{s_1}{s_0}$.

The proof is straightforward and is thus omitted. We can also repeat the considerations of section 2.2 c) in this more general case.

3.2. Psuedoconcave properly discontinuous groups.

a) For practical reasons it is convenient to generalize the notion of pseudoconcavity as follows.

Let Ω be an open set in $\mathbb{C}^n$, and $z_0 \in \partial\Omega = \bar{\Omega} - \Omega$ a boundary point. We will say that z_0 is a pseudoconcave boundary point if we can find a complex 2-dimensional linear space E through z_0:

$z = z_0 + a_1 t_1 + a_2 t_2$; $a_1, a_2 \in \mathbb{C}^n$, linearly independent, (t_1, t_2) variable in $\mathbb{C}^2$ and a C^∞ function ϕ on a neighborhood V of z_0 in E, real-valued and such that

i) $V \cap \Omega \supseteq \{z \in V \mid \phi(z) < \phi(z_0)\}$

ii) $L(\phi)_{z_0} < 0$.

Let now Γ be a properly discontinuous group of automorphisms of a connected complex manifold X. We will say that Γ is a pseudoconcave group of automorphisms if we can find a non-empty open subset $\Omega \subset X$ such that

i) Ω is relatively compact in X

ii) for every point $z_0 \in \partial\Omega$ the orbit Γ_{z_0} contains either an interior point of Ω or a pseudoconcave boundary point of Ω.

Clearly for Γ = identity we obtain a generalization of the notion of pseudoconcave manifold given in the previous chapter. It is not a difficult exercise to carry over to this more general case all the theory developed there. In particular we obtain the theorem (3):

Let X be a connected complex manifold and Γ a pseudoconcave group of automorphisms of X. The field $\mathcal{K}(X)^{\Gamma}$ of Γ-invariant meromorphic functions on X is an algebraic field of transcendence degree $d \leq \dim_{\mathbb{C}} X$.

Note that $\mathcal{K}(X)^{\Gamma} \simeq p^{*}\mathcal{K}(Z)$ where $Z = X/\Gamma$.

Suppose for instance that there exists a Γ-automorphic line bundle F on X such that the quotient field $\mathcal{Q}(X, F)$ of the graded ring

$$\mathcal{A}(X, F)^{\Gamma} = \bigcup_{k=0}^{\infty} \Gamma(X, F^{k})^{\Gamma}$$

has transcendence degree equal to that of $\mathcal{K}(X)^{\Gamma}$ (for instance equal the dimension of X) then it follows, since $\mathcal{Q}(X, F)^{\Gamma}$ is algebraically closed in $\mathcal{K}(X)^{\Gamma}$ that we must have

$$\mathcal{Q}(X, F)^{\Gamma} = \mathcal{K}(X)^{\Gamma}.$$

b) The notion of pseudoconcave properly discontinuous group of automorphisms is stable by "commensurability"; two subgroups Γ_1, Γ_2 of Aut (X) are called commensurable if $\Gamma_1 \cap \Gamma_2$ is of finite index in Γ_1 and Γ_2.

Proposition (3.2.1). If Γ_1 and Γ_2 are commensurable and if one of them is properly discontinuous and pseudoconcave so is the other.

Proof. Let Γ_1 be properly discontinuous, then $G = \Gamma_1 \cap \Gamma_2$, as a subgroup of Γ_1 is properly discontinuous.

Set $\Gamma_2 = Ga_1 \cup \dots \cup Ga_k$ and let K be compact in X. If the set $\{\gamma \in \Gamma_2 \mid \gamma K \cap K \neq \emptyset$ is infinite, there exists an index

i with $1 \leq i \leq k$ and infinitely many g's in G such that $ga_1K \cap K \neq \emptyset$. But then for infinitely many g's in G
$g(K \cup a_1K) \cap (K \cup a_1K) \neq \emptyset$, which is a contradiction since $K \cup a_1K$ is compact.

Suppose that Γ_1 is pseudoconcave. It is enough to show that $G = \Gamma_1 \cap \Gamma_2$ is pseudoconcave. Let $\Omega \subset X$ satisfy the conditions i) and ii) specified above in a) for $\Gamma = \Gamma_1$. Set $\Gamma_1 = a_1G \cup \ldots \cup a_1G$.

Then $\Omega' = a_1^{-1}\Omega \cup \ldots \cup a_1^{-1}\Omega$ verifies the same conditions for $\Gamma = G$.

3.3 Siegel modular group.

We will apply the above considerations to the particular case of the Siegel upper half plane and Siegel modular group ([3]).

Let n be an integer, $n > 0$, and let $\mathcal{M}(n, \mathbb{C}) = \mathrm{Hom}(\mathbb{C}^n, \mathbb{C}^n)$ be the set of $n \times n$ matrices with complex entries. The generalized upper half plane is the set

$$H_n = \{ Z \in \mathcal{M}(n, \mathbb{C}) \quad {}^tZ = Z, \quad \mathrm{Im}\, Z > 0 \ .$$

This set can be identified with an open subset of $\mathbb{C}^{\frac{1}{2}n(n+1)}$. Note that H_1 is the usual Poincaré upper half plane.
Consider the simplectic group $S_p(n, \mathbb{R})$ i.e. the set of linear automorphisms of $\mathbb{R}^n \times \mathbb{R}^n$ which leave invariant the exterior form $\sum_{i=1}^{n} dx_i \wedge dy_i$, $(x, y) \in \mathbb{R}^n \times \mathbb{R}^n$.
in matrix notation

$$S_p(n, \mathbb{R}) = \{ g \in \mathcal{M}(2n, \mathbb{R}) \mid {}^tgJg = J \} \quad \text{where}$$

$$J = \begin{pmatrix} 0 & I \\ -I & 0 \end{pmatrix}, \quad \mathcal{M}(2n, \mathbb{R}) = \mathrm{Hom}(\mathbb{R}^{2n}, \mathbb{R}^{2n}).$$

We can let $S_p(n, \mathbb{R})$ operate on the generalized upper half plane as follows

$$\text{for } g = \begin{pmatrix} A & B \\ C & D \end{pmatrix} \in S_p(n, \mathbb{R}), \quad g : Z \to (AZ + B)(CZ + D)^{-1}.$$

<u>Proposition</u> (3.3.1). <u>This operation is a well defined automorphism of</u> H_n.

<u>Proof</u>. Let $Z \in H_n$ and let I denote the $n \times n$ identity matrix. The conditions defining H_n can be reformulated as follows:

$$({}^tZ, I) \; J\binom{Z}{I} = 0, \quad i({}^t\bar{Z}, I)\; J\binom{Z}{I} \quad 0.$$

Set

$$\binom{U}{V} = \begin{pmatrix} A & B \\ C & D \end{pmatrix} \binom{Z}{I} = \binom{AZ + B}{CZ + D}$$

Then, since g is symplectic, we must have

$$i({}^t\bar{U}, {}^t\bar{V}) \; J\binom{U}{V} > 0 ,$$

i.e.

$$i({}^t\bar{U}V - {}^t\bar{V}U) > 0 .$$

This shows that V must be non-singular, otherwise for a vector $w \neq 0$ $Vw = 0$ thus $i\,{}^t\bar{w}({}^t\bar{U}V - {}^t\bar{V}W)w = 0$ which is impossible. Then we can consider the matrix

$$\binom{Z_1}{I} = \binom{UV^{-1}}{I} = \binom{U}{V}V^{-1}.$$

From the first condition we derive

$$({}^tZ_1, I) \; J\binom{Z_1}{I} = 0 \quad \text{i.e.} \quad {}^tZ_1 = Z_1 .$$

From the second we derive

$$i({}^t\bar{Z}_1, I) \; J\binom{Z_1}{I} > 0 \quad \text{i.e.} \quad i({}^t\bar{Z}_1 - Z_1) > 0 .$$

This shows that $Z_1 = (AZ + B)(CZ + D)^{-1}$ is well defined and represents a point of H_n. Since the transformation is invertible it gives an automorphism of H_n.

By the map $Z \to (Z - iI)(Z + iI)^{-1}$ the generalized upper half plane is mapped into the generalized unit disc $\bar{Z}Z < I$ which is a bounded domain in $\mathbb{C}^{\frac{1}{2}n(n+1)}$ and $Sp(n, \mathbb{R})$ appears as a group of automorphisms of a bounded domain. Consider the discrete subgroup $Sp(n, \mathbb{Z})$ of $Sp(n, \mathbb{R})$ of those matrices with integer entrices. By the above remark it follows that $Sp(n, \mathbb{Z})$ acts in a properly discontinuous way on H_n

The transformations $g = \pm I$ are the only transformations which act as identity on H_n and are contained in $Sp(n, \mathbb{Z})$(1). The group $Sp(n, \mathbb{Z})$ (or more precisely the group $Sp(N, \mathbb{Z}) / \{\pm I\}$) is called the <u>modular group</u> of rank n. For $n = 1$ we get the usual modular group in one variable.

3.4 <u>Pseudoconcavity of the modular group.</u>

<u>Theorem</u> (3.4.1). <u>If</u> $n \geq 2$ <u>the modular group of rank</u> n <u>is a pseudoconcave properly discontinuous group of automorphisms of</u> H_n.

<u>Proof.</u> (α) Every psoitive definite symmetric matrix Y can be written in a unique way in the following Jacobi normal form:

$$Y = {}^tWDW \quad \text{with } D = \begin{pmatrix} d_1 & & 0 \\ & \ddots & \\ 0 & & d_n \end{pmatrix}, \; W = \begin{pmatrix} 1 & & w_{ij} \\ & \ddots & \\ 0 & & 1 \end{pmatrix}$$

For a positive real number μ, let Ω_μ be the open subset of H_n defined by $Z = X + iY \in \Omega_\mu$ if

i) $|x_{\alpha\beta}| < \mu$ where $X = (x_{\alpha\beta})$

ii) $|w_{ij}| < \mu$ for $i < j$

iii) $1 < \mu d_1 < \mu^2 d_2 < \ldots < \mu^n d_n$.

Note that if $X + iY \in \Omega_\mu$, then there exists a constant $c(\mu) > 0$ such that $Y > c(\mu)I$ $(c(\mu) = \min(\frac{1}{\mu}, \frac{1}{\mu^n}))$.

If $\mu \geq \mu_0$ is sifficiently large then Ω_μ is a "fundamental open set" for the modular group $\Gamma = Sp(n, \mathbb{Z})$, i.e.

---

(1) If $(AZ + B)(CZ + D)^{-1} = Z$ for every $Z \in H_n$, then taking $Z = \lambda iI$ we get that necessarily $B = C = 0$, $A = D$ and moreover $AZ = ZA$. Thus $A = \mu I$ and therefore $\mu = \pm 1$.

i) $\Gamma \Omega_\mu = H_n$

ii) $\{\gamma \in \Gamma \mid \gamma\Omega_\mu \cap \Omega_\mu \neq \emptyset\}$ is a finite set.

For the proof of this statement we refer to [48] and [49]. Notice that from this statement the proper discontinuity of Γ could be deduced by a more direct argument than the general one used before.

( ) We fix the parameter $\mu \geq \mu_0$ once and for all. We will say that a transformation $\gamma \in \Gamma$ is a <u>transformation at infinity</u> if

$$F = \gamma^{-1}(\Omega_\mu) \cap \Omega_\mu$$

contains a divergent sequence of points in H_n (i.e. a sequence with no accumulation point in H_n). In view of the definition of Ω_μ this is equivalent to say that

$$\sup_{Z \in F} d_n(Z) = \infty .$$

Note that the same is true for $\check{F} = \gamma(F) = \Omega_\mu \cap \gamma(\Omega_\mu)$.

We set

$$v(Z) = Y^{-1} \begin{bmatrix} 0 \\ \vdots \\ 0 \\ 1 \end{bmatrix}$$

A direct calculation shows that each component v_i of $v(Z)$ is of the form $v_i = d_n^{-1} \alpha_i$ where α_i is a polynomial in the w_{ij}'s. Therefore if $Z = X + iY$ describes F, $\| v(Z) \|$ takes a rbitrarily small values.

Set $r(Z) = \| v(Z) \|^{-1}$ so that $\sup_{Z \in F} r(Z) = \infty$, and moreover

$$\begin{bmatrix} 0 \\ \vdots \\ 0 \\ r(Z) \end{bmatrix} = Y\, w(Z) \quad \text{where} \quad w(Z) = \frac{v(Z)}{\| v(Z) \|} , \quad \| w(Z) \| = 1 .$$

<u>Lemma.</u> <u>If</u> $\gamma = \begin{pmatrix} A & B \\ C & D \end{pmatrix}$ <u>is a transformation at infinity, then in the matrix</u> C <u>the last row and column are zero</u>.

Indeed if $Z \in F$, there exists $\check{Z} \in \Omega_\mu$ such that $Z = \gamma^{-1}\check{Z}$, i.e.

$$\check{Z}(CZ + D) = AZ + B . \tag{1}$$

Multiplying (1) on the right by w(Z) and equating the real p arts we obtain

$$\check{Y}\ C\begin{bmatrix}0\\ \vdots\\ 0\\ r(Z)\end{bmatrix} = (\check{X}CX + \check{X}D - AX - B)\ w(Z).$$

When Z describes F, the right hand side represents a vector u(Z) of bounded norm. Since $\check{Z} \in \Omega_{\mu}$ there exists a constant $c > 0$, independent of $\check{Y}$ and x, such that

$$(\check{Y}x,\ x) \geq c\ \|x\|^2$$

where $(y, x) = {}^t yx$ denotes the euclidian scalar product. Therefore

$$c\left\|C\begin{bmatrix}0\\ \vdots\\ 0\\ r(Z)\end{bmatrix}\right\|^2 \leq \left(\check{Y}\ C\begin{bmatrix}0\\ \vdots\\ 0\\ r(Z)\end{bmatrix},\ C\begin{bmatrix}0\\ \vdots\\ 0\\ r(Z)\end{bmatrix}\right) \leq \|u(Z)\|\ \left\|C\begin{bmatrix}0\\ \vdots\\ 0\\ r(Z)\end{bmatrix}\right\|$$

i.e.

$$\left\|C\begin{bmatrix}0\\ \vdots\\ 0\\ r(Z)\end{bmatrix}\right\| \leq \frac{1}{c}\ \|u(Z)\|$$

which is bounded as Z describes F. This is possible only if the last column of C is zero.

As Z describes F, $\check{Z}$ describes $\check{F}$. Set now

$$v(\check{Z}) = \check{Y}^{-1}\begin{bmatrix}0\\ \vdots\\ 0\\ 1\end{bmatrix}$$

and proceed as before. Setting $w(\check{Z}) = \dfrac{v(\check{Z})}{\bar{v}(\check{Z})}$ we get

$$\begin{bmatrix}0\\ \vdots\\ 0\\ r(\check{Z})\end{bmatrix} = \check{Y}\ w(\check{Z}),\quad r(\check{Z}) = \|v(\check{Z})\|^{-1}$$

and

$$\sup_{\check{Z} \in \check{F}} r(\check{Z}) = \sigma.$$

Multiflying (1) on the left by ${}^t w(\check{Z})$ we get

$$(0,\dots, 0, \ r(\check{Z}))\, C\, Y = {}^t w(\check{Z}) \ (\check{X}\, C\, X + \check{X}D - AX - B).$$

Arguing as before we obtain that the last row of C must be zero.

As a consequence of this lemma we get that, for a transformation $\gamma = \begin{pmatrix} A & B \\ C & D \end{pmatrix}$ at infinity, det $(CZ + D)$ is independent of the last row and column of Z.

(γ) Consider on H_n the function $k(Z) = \det Y$ where $Z + X + iY$. A direct calculation shows that for every $= \begin{pmatrix} A & B \\ C & D \end{pmatrix})$ Sp(n, $\mathbb{Z}$) we have

$$k(\gamma Z) = k(Z) \ \det(CZ + D)^{-2} \quad .$$

Moreover one can prove the following property (cf. [49]) :

for every $Z_o \in H_n$, $\sup_{Z \in \Gamma Z_o} k(Z) < \infty$ and the maximum is attained at some point $Z_1 \in \Omega_\mu \cap \Gamma Z_o$.

If we set

$$\phi(Z) = \sup_{\gamma \in \Gamma} k(\gamma Z) \ ,$$

we obtain a Γ-invariant function on H_n.

Notice that the sets

$$B_c = \{ Z \in \Omega_\mu \mid \phi(Z) < c$$

for any real c, are relatively compact in H_n. In fact B_c is contained in the set $\{ Z \in \Omega_\mu \mid \det Y < c \}$ which is contained in a set of the form $\{ Z \in \Omega_\mu \mid d_n < \alpha$.

Choose $c > 0$ large enough to ensure that for

$$T = \{ Z \in \Omega_\mu \mid \phi(Z) > c \} \text{ every } \gamma \in \Gamma$$

with $\gamma(T) \cap T \neq \emptyset$ is a transformation at infinity. This can be done since the set $\{\gamma \in \Gamma \mid \gamma\Omega_\mu \cap \Omega_\mu \neq \emptyset$ is finite. Let $\gamma_1, \dots, \gamma_s$ be the set of these transformations so that, for $Z \in T$ we have

$$\phi(Z) = \sup_{\nu = 1,\dots,s} k(\gamma_\nu Z).$$

In fact for $\gamma \in \Gamma$ such that $\gamma(Z) \in \Omega_\mu$ either $K(\gamma(Z)) > c$ and thus $\gamma(Z) \in T$ so that $\gamma = \gamma_i$ for some i or else $k(\gamma(Z)) \leq c$ and therefore $k(\gamma(Z) < \phi(Z)$.

We now take for Ω the set B_{c+1}. We want to show that for any point $Z_o \in \partial\Omega$ there exists in ΓZ_o either an interior point of Ω or a pseudoconcave boundary point of Ω.

Let $p : H_n \to H_n/\Gamma$ be the natural projection on the quotient space. Since ϕ is Γ-invariant, $\Omega = p^{-1}p(\Omega)$. Therefore a point $Z_o \in \partial\Omega$ which is not equivalent to an interior point of Ω must be on the "surface" $\{\phi = c+1\}$. For such a point we must have

$k(Z_o) = k(\gamma_\nu Z_o)$ for those ν's, $1 \leq \nu \leq s$, for which $k(\gamma_\nu Z_o) = \phi(Z_o) = c+1$.

In particular for these γ's we must have

$$|\det (C Z_o + D)| = 1, \quad \gamma_\nu = \begin{pmatrix} A & B \\ C & D \end{pmatrix}.$$

Moreover $\det (C Z_o + D)$ is independent of the last row and column of Z_o.

Consider the linear space $L(Z_o)$ of complex dimension n, through Z_o, defined by the equations

$$Z_{\alpha\beta} = Z^o_{\alpha\beta} \quad \text{for} \quad 1 \leq \alpha \leq n-1, \quad 1 \leq \beta \leq n-1, \quad Z \text{ symmetric.}$$

Since on that space $|\det (C Z + D)| = 1$ we must have $\phi|_{L(Z_o)} = k(Z)$.

Therefore in a neighbrohood V of Z_o on $L(Z_o)$ the set $\Omega \cap L(Z_o)$ can be described as the set $\{Z \in V \mid k(Z) < c+1\}$.

The proof of the theorem will be complete if we show that the function $-\log k(Z)$ is strongly plurisubfarmonic in H_n (and thus the same is true for its restriction to $L(Z_o)$). To see this we remark that by the way $\det Y$ transforms under the action of $Sp(n, \mathbb{R})$ the form

$$K(Z)dv = (\det Y)^{-n+1)}dv,$$

(where dv is the euclidean volume element), is invariant.

Since $Sp(n, \mathbb{R})$ acts transitively on H_n that form, up to a positive factor must be the invariant Bergman form (cf. [51]). Therefore that quadratic hermitian form

$$-(n+1)\ \partial.\bar{\partial} \log k(Z)$$

is the Bergman metric of H_n and thus it is positive definite.

<u>**Remark.**</u> Let m be an integer. One can consider the subgroup Γ_m of $\Gamma = Sp(n, \mathbb{Z})$ defined by

$$\Gamma_m = \{ g \in \Gamma \mid g \equiv I \pmod{m} \} .$$

For every m, Γ_m and Γ are commensurable. Therefore for every m Γ_m is properly discontinuous and pseudoconcave.

3.5. <u>Poincaré series.</u>

a) Let D be a bounded domain in $\mathbb{C}^m$; for $\gamma \in \mathrm{Aut}(D)$ we set

$$p_\gamma(z) = \det \left(\frac{d(\gamma z)}{d(z)} \right) .$$

If $\Gamma \subset \mathrm{Aut}(D)$ is any subgroup of $\mathrm{Aut}(D)$ and if $l \in \mathbb{Z}$, then the set $\{ p_\gamma(Z)^l \}$ is a system of factors of automorphy for Γ and thus it defines a Γ-automorphic line bundle F^l. For any f holomorphic in D we can consider the "Poincaré series of weight k"

$$\Theta_k(f;\ z) = \sum_{\gamma \in \Gamma} f(\gamma z) \ \det \left(\frac{d(\gamma z)}{d(z)} \right)^k .$$

If Γ is properly discontinuous, if f is bounded in absolute value and if $k \geq 2$, then this series converges uniformly on compact sets. Indeed the convergence of that series reduced to the convergence of the series

$$\sum_{\gamma \in \Gamma} \left| \det \frac{d(\gamma z)}{d(z)} \right|^2$$

on compact sets.

Let $Q \subset\subset D$ be a polycilinder of radius R and let P be any polycilinder of radius $\rho < R$ contained in Q. Since $\det \frac{d(\gamma z)}{d(z)}$ is holomorphic we have at the center x_o of P

$$\left|\det \frac{d(\gamma z)}{d(z)}\right|^2_{z=x_o} \text{vol } P \leq {}_P\!\int \left|\det \frac{d(\gamma z)}{d(z)}\right|^2 dv = \text{vol } \gamma(P) \leq \text{vol } \gamma(Q).$$

Therefore if V is the concentric polycilinder of Q of radius $R-r$, for any point $y \in V$ we have

$$\left|\det \frac{d(\gamma z)}{d(z)}\right|^2_{z=y} \leq C \frac{\text{vol } \gamma(Q)}{\text{vol } (Q)} \quad (C = (\text{vol } P)^{-1} \cdot \text{vol } Q).$$

Let s be the number of transformations $\gamma \in \Gamma$ such that $\gamma Q \cap Q \neq \emptyset$. We have for $y \in V$

$$\begin{aligned} \sum \det \left| \frac{d(\gamma z)}{d(z)}\right|^2_{z=y} &\leq C \frac{\text{vol } \gamma(Q)}{\text{vol } (Q)} \\ &\leq \frac{C}{\text{vol } (Q)} \sum \text{vol } \gamma(Q) \\ &- \frac{C\, s}{\text{vol } (Q)} \text{vol } (D), \end{aligned}$$

since in $\bigcup_{\gamma \in \Gamma} \gamma(Q)$ any point is covered at most s times.

This proves the absolute uniform convergence of the given series in V. It follows then the absolute uniform convergence of that series on any compact subset of D.

Now if $\Theta_k(z)$ is a convergent Poincaré series of weight k, it satisfies the functional equation

$$\Theta_k(z) = \rho_\gamma(z)^{-k}\, \Theta_k(z) \quad \text{for every } \gamma \in \Gamma .$$

In fact

$$\begin{aligned} \Theta_k(\gamma z) &= \sum_{\sigma \in \Gamma} f(\sigma\gamma z) \left(\frac{d(\sigma\gamma z)}{d(\gamma z)}\right)^k \\ &= \sum_{\sigma \in \Gamma} f(\sigma\gamma z) \left(\frac{d(\sigma\gamma z)}{d(z)}\right)^k \left(\frac{d(z)}{d(\gamma z)}\right)^k \\ &= \rho_\gamma(z)^{-k} \sum_{\sigma \in \gamma} f(\sigma\gamma z) \left(\frac{d(\sigma\gamma z)}{d(z)}\right)^k \\ &= \rho_\gamma(z)^{-k}\, \Theta_k(z). \end{aligned}$$

This shows that any such Poincaré series represents an invariant section of the -automorphic line bundle F^{-k}.

One is led therefore to consider the ring $\mathcal{A}(D, F^{-2})^{\Gamma}$ and its quotient field $\mathcal{Q}(D, F^{-2})^{\Gamma}$.

Disposing of the freedom one has in the choice of the function f one can show (cf. [19]) that the field $\mathcal{Q}(D, F^{-2})^{\Gamma}$ contains m analytically independent meromorphic functions.

b) We can now apply the above consideration to H_n and to the modular group as H_n is isomorphic to a bounded domain of $\mathbb{C}^m$, $m = \frac{1}{2}n(n+1)$.

We then deduce the following conclusion.

<u>For</u> n 1 <u>the field of</u> Γ <u>-automorphic meromorphic functions on</u> H_n <u>coincides with the field</u> $\mathcal{Q}(H_n, F^{-2})^{\Gamma}$. <u>In particular</u>

(α) <u>the transcendence degree of that field equals</u> $\frac{1}{2}n(n+1)$;

(α) <u>every</u> Γ <u>-invariant meromorphic function can be written as the quotient of two Poincare series of the same weight</u>.

<u>Remark</u>. We have reached the above conclusion using the fact that H_n admits a bounded model. Using instead the usual unbounded model of H_n one can develop a simular argument replacing the factor of automorphy ρ_γ by

$$\sigma_\gamma(Z) = \det(CZ + D) \qquad \gamma = \begin{pmatrix} A & B \\ C & D \end{pmatrix} \in Sp(n, \mathbb{Z})$$

One is then led to the theory of Eisenstein series:

$$\sum_{\gamma \in \Gamma} f(\gamma Z) \frac{1}{\det (CZ+D)^k}$$

and reaches a similar conclusion but the proof of convergence of Eisenstein series is more difficult (cf. [49]).

The result proved in this chapter for the modular group was extended by A. Borel ([17]) to arithmetic groups acting on irreducible bounded domains of dimension ≥ 2 (cf. section 4.1 example 5).

Also one can show as in H. Cartan ([19]) that the quotient space X/Γ has the structure of an analytic space and is p seudoconcave if Γ . is in the sense of section 3.2.

A. Andreotti

Chapter IV. Projective imbeddings of pseudoconcave manifolds.[1]

4.1. Measure of pseudoconcavity.

a) Let U be an open subset of $\mathbb{C}^n$ and $\phi : U \to \mathbb{R}$ be a C^∞ function. We will say that ϕ is strongly q-pseudoconvex at the point $z_o \in U$ if the Levi-form

$$L(\phi)_{z_o} = \sum \left(\frac{\partial^2 \phi}{\partial z_\alpha \partial \bar{z}_\beta} \right)_{z_o} u_\alpha \bar{u}_\beta$$

(where $z_1, \ldots, z_n$ are local holomorphic coordinates at z_o) has at least $n-q$ positive eigenvalues. In particular a strongly 0-pseudoconvex function is a strongly plurisubharmonic function. As we have seen before (1.4) this notion is independent of the choice of local coordinates, and could also be formulated as follows:

there exists a $(n-q)$-complex-dimensional plane E through z_o such that $\phi|E$ is strongly plurisubharmonic.

Remark. For practical reasons it is sometimes convenient to release the assumption that ϕ is C^∞ and require only that

(i) in a neighborhood of z_o there is a finite number of C^∞ functions $\phi_1, \ldots, \phi_l$ such that $\phi = \sup(\phi_1, \ldots, \phi_l)$;

(ii) there exists an $(n-q)$-dimensional plane E through z_o such that $_{i\,E'}$ $1 \leq i \leq l$, is strongly plurisubharmonic in a neighborhood of z_o in E.

b) Let X be a connected complex manifold of complex dimension n. We will say that X is q-pseudoconcave if a C^∞ function $\phi : X \to \mathbb{R}$ is given such that

(1) In this chapter the notions of analytic set, complex space, normal complex space, Stein space will occasionally be used although our main concern are complex manifolds. For the basic definitions we refer to the following Chapter 5.

(i) for every $c > \inf \phi$ the sets

$$B_c = \{x \in X \mid \phi(x) > c\}$$

are relatively compact in X;

(ii) there exists a compact subset K of X such that at every point $z_o \in X - K$ ϕ is strongly q-pseudoconvex.

Remarks.

1. For all $c > \inf \phi$, except for a set of measure 0, the boundary $\partial B_c = \bar{B}_c - B_c$ is smooth (By Sard's theorem). The Levi-form restricted to the analytic tangent plane at $z_o \in \partial B_c$ to ∂B_c, $c < \inf_K \phi$, has at least n-q-1 positive eigenvalues. Therefore if $n-q-1 \geq 1$ these manifolds are a special case of the pseudoconcave manifolds studied in chapter 1.

For this reason when we speak of a q-pseudoconcave manifold X, if X is not compact, we will assume that $0 \leq q \leq n-2$, hence $n \geq 2$. Indeed only the eigenvalues of the Levi-form in the direction of the level surfaces of ϕ do have a geometric meaning; taking a rapidly increasing convex function of ϕ the remaining eigenvalue can be forced to be positive on any given compact set.

2. The notion of q-pseudoconcave manifold could be generalized in the sense of the remark made in a), above.

Examples.

1. Every compact connected manifold X is q-pseudoconcave for any q. Take $K = X$, $\phi = 1$.

2. Let Z be a connected compact complex manifold of complex dimension $n \geq 2$. Let $\{x_1, \ldots, x_k\}$ be a finite subset of Z, then $X = Z - \{x_1, \ldots, x_k\}$ is 0-pseudoconcave. Indeed if $z_j^{(i)}$, $1 \leq j \leq n$, are local complex coordinates at x_i, vanishing at x_i we can select $\varepsilon > 0$ such that the coordinate balls

$$B_i = \{ \sum_{j=1}^{n} |z_j^{(i)}|^2 < \varepsilon \}, \quad 1 \leq i \leq k ,$$

are relatively compact in their coordinate patch and are disjoint. Take $K = Z - \bigcup_{i=1}^{k} B_i$ and for ϕ a C^∞ function such that $\phi \geq \varepsilon$ on K,

$$\phi |_{B_i} = \Sigma | z_j^{(i)} |^2 , \quad 1 \leq i \leq k .$$

3. Now generally let Z be compact connected manifold of complex dimension $n \geq 2$. Let Y be a complex submanifold of Z with $\dim_{\mathbb{C}} Y = q \leq n-2$.
Then $X = Z - Y$ is q-pseudoconcave.
Indeed, at every point $y \in Y$ we can choose a coordinate neighborhood $U(y)$ with coordinates $z_1^U, \ldots, z_n^U$ such that

$$U(y) \cap Y = \{ z \in U \mid z_{q+1}^U = \ldots = z_n^U = 0 \}.$$

Let $\phi^U = \sum_{j=q+1}^{n} |z_j^U|^2$. Cover Y with a finite number $U_1, \ldots, U_l$ of these neighborhoods and let $U_0 = Z - Y$ and $\phi^{U_0} = 1$. Let $\{\rho_i\}$ be a C^∞ partition of unity subordinate to the covering $\{U_i\}_{0 \leq i \leq l}$. Set $\phi = \sum_0^l \rho_i \phi^{U_i}$.

Then $\phi > 0$ on X and for $\varepsilon > 0$ $\{\phi > \varepsilon\}$ is relatively compact in X. Moreover at each point of Y $L(\phi)$ has n-q positive eigenvalues. Therefore if $\varepsilon > 0$ is sufficiently small on $X - \{\phi \geq \varepsilon\}$ $L(\phi)$ has n-q positive eigenvalues.

4. Let H_n be the generalized upper half plane with $n \geq 2$, and let Γ be the modular group. From (3.4, α)) it follows that there is an integer $l > 0$ such that for every point $Z_0 \in H_n$ the isotropy group Γ_{Z_0} has an order $\leq l$.

Let p be a prime number with $p > 1$. Then the group $\Gamma_p = \{ g \in \Gamma \mid g \equiv I \pmod p \}$ cannot have fixed points[1]. Therefore $X = H_n/\Gamma_p$ is a complex manifold; From the proof of theorem 3.4.1 it follows that X is q-pseudoconcave with

$$q = \tfrac{1}{2}n(n-1) = \dim_{\mathbb{C}} H_{n-1} .$$

Here the concavity is to be understood in the general sense of remark 2 above.

5. More generally let D be an irreducible symmetric bounded domain in $\mathbb{C}^n$ with $n \geq 2$. Let Γ be an arithmetic group of automorphisms of D without fixed points. A result of A. Borel shows that $X = D/\Gamma$ is q-pseudoconcave for some q, $0 \leq q \leq n-2$ ([17]).

4.2 The problem of projective imbedding of pseudoconcave manifolds.

a) We have already remarked that if X is a pseudoconcave manifold and $\tau : X \to \mathbb{P}_N(\mathbb{C})$ a one-to-one holomorphic map of X onto $\tau(X)$, then $\tau(X)$ is contained in an irreducible algebraic variety Y of the same dimension than X (theorem (2.5.3)). If $Y^* \xrightarrow{\pi} Y$ is the normalization of Y (cf. chapter 5), then the map $\tau : X \to Y$ factors through π. Moreover Y^* is again a projective algebraic variety. Thus X is isomorphic to an open subset of the projective algebraic variety Y^*.

---

(1) Indeed if $A \in \Gamma_p$ has a fixed point, for some $l_1 \leq l$ we must have $A^{l_1} = I$. Now $A = I + pB$ thus

$$\binom{l_1}{1} pB + \binom{l_1}{2} p^2B^2 + \dots + p^{l_1}B^{l_1} = 0.$$

Dividing this relation by p we see that each entry b_{ij} of B must be divisible by p. Thus $A = I + p^2B_1$. Arguing in the same way we see that each entry of B_1 is devisible by p^2, thus $A = I + p^4B_2$. And so on. It follows that we must have $B = 0$ and thus $A = I$.

The problem of projective imbedding of speudoconcave manifolds can be loosely formulated as follows: to find some useful criterion to ensure that a pseudoconcave manifold X is isomorphic to an open set of some projective algebraic variety. We will then say that X admits a <u>projective imbedding</u>.

b) If X is compact such criterion is provided by the theorem of Kodaira [33]:

<u>Theorem</u> (4.2.1). <u>Let</u> X <u>be a compact complex manifold, the following are equivalent conditions:</u>

A. X <u>admits a projective imbedding.</u>

B. X <u>carries a Kähler metric whose associated exterior form has integral periods.</u>

C. <u>There exists over</u> X <u>a holomorphic line bundle</u> $F \xrightarrow{\pi} X$ <u>whose total space</u> F <u>is</u> <u>0-pseudoconcave.</u>

D. <u>There exist over</u> X <u>a holomorphic line bundle</u> F <u>such that the graded ring</u> $\mathcal{A}(X, F) ; \bigcup_{l=0}^{\infty} \Gamma(X, F^l)$ <u>separates points and gives local coordinates everywhere on</u> X.

Let $n = \dim_{\mathbb{C}} X$; to say that $\mathcal{A}(X, F)$ separates points and gives local coordinates everywhere on X means the following:

(α) given $x \neq y$ X there exist an integer l, $l > 0$, and $s_0, s_1 \in \Gamma(X, F^l)$ such that

$$\det \begin{bmatrix} s_0(x) & s_0(y) \\ s_1(x) & s_1(y) \end{bmatrix} \neq 0 :$$

(β) for every x X there exists an integer $l > 0$ and $s_0, \ldots, s_n \in \Gamma(X, F^l)$ such that

$$(\Sigma (-1)^i s_i \, d s_0 \wedge \ldots \wedge \widehat{ds_i} \wedge \ldots \wedge ds_n)_x \neq 0.$$

In other words the meromorphic function $\frac{s_1}{s_0}$ (or $\frac{s_0}{s_1}$) takes different "values" at x and y while the meromorphic functions $\frac{s_1}{s_0}, \ldots, \frac{s_n}{s_0}$ (if $s_0(x) \neq 0$, otherwise renumber the sections s) provide local coordinates at x.

We will prove here only the implications $D \Leftrightarrow A$. The other implications will be proved in chapter VI.

Proof.

$D \Rightarrow A$. For any two points $x \neq y$ X we can find an integer $l = l(x, y)$ and two sections $\alpha, \beta \in \Gamma(X, F^l)$ such that

$$\det \begin{bmatrix} \alpha(x) & \alpha(y) \\ \beta(x) & \beta(y) \end{bmatrix} \neq 0 .$$

Replacing α and β by convenient linear combinations, we may assume that $\alpha(y) = \beta(x) = 0$ and thus $\alpha(x)\beta(y) \neq 0$. Therefore also

$$\det \begin{bmatrix} \alpha^k(x) & \alpha^k(y) \\ \beta^k(x) & \beta^k(y) \end{bmatrix} \neq 0 .$$

Set

$$A_l = \{(x,y) \in X \times X \mid \forall (\alpha,\beta) \in \Gamma(X, F^l) \times \Gamma(X, F^l);\ \alpha(x)\beta(y) - \alpha(y)\beta(x) = 0 \}.$$

By the above remark for any integer $k > 0$ $A_{kl} \subset A_l$. Now for each l A_l is an analytic subset of $X \times X$ containing the diagonal Δ of $X \times X$. If $A_l \supsetneq \Delta$ then for some $k > 0 : A_{kl} \subsetneq A_l$. Since a decreasing sequence of analytic subsets of a compact manifold must be stationary there exists an $l_o > 0$ such that $A_{l_o} = \Delta$. Therefore

(i) the sections of $\Gamma(X, F^{l_o})$ do separate any couple of points $x \neq y$ in X. Similarly by a simpler argument one proves that l_o can be chosen such that also

(ii) the sections of $\Gamma(X, F^{l_o})$ have no common zeros,

(iii) they give local coordinates everywhere on X.

Let $\{s_0, \ldots, s_k\}$ be a basis of $\Gamma(X, F^{l_0})$ and consider the map $\tau : X \to \mathbb{P}_k(\mathbb{C})$ given by

$$x \to (s_0(x), \ldots, s_k(x)).$$

This map is holomorphic by (ii), is one-to-one (by(i)) and biholomorphic (by(iii)). Therefore τ is an isomorphism of X onto $\tau(X)$ which is a projective algebraic manifold by theorem (2.5.3).

$A \Rightarrow D$. It is enough to prove that implication for $X = \mathbb{P}_n(\mathbb{C})$. Consider on $\mathbb{P}_n(\mathbb{C})$ the line bundle of the hyperplane sections. this is given by

$$\pi : \mathbb{P}_{n+1}(\mathbb{C}) - \{0, \ldots, 0, 1\} \to \mathbb{P}_n(\mathbb{C}),$$

$$(z_0, \ldots, z_{n+1}) \to (z_0, \ldots, z_n)$$

On $U_i = z_i \neq 0$ on $\mathbb{P}_n$, $\pi^{-1}(U_i) \simeq U_i \times \mathbb{C}$ as we can write for $z = (z_0, \ldots, z_{n+1})$ on $\pi^{-1}(U_i)$

$$(z_0, \ldots, z_{n+1}) = (\frac{z_0}{z_i}, \ldots, 1, \ldots, \frac{z_n}{z_i}, \frac{z_{n+1}}{z_i}).$$ It is then

immediate to verify that this is a holomorphic line bundle F. A section on $\mathbb{P}_n$ is given by $(z_0, \ldots, z_n) \to (z_0, \ldots, z_{n+1})$ where $z_{n+1} = a_0 z_0 + \ldots + a_n z_n$. Then $\Gamma(\mathbb{P}_n(\mathbb{C}), F)$ separates points and gives local coordinates everywhere.

Remark. If X is not compact the proof of the implication $D \Rightarrow A$ breaks down, although one has reasonable conditions to ensure that condition D is fulfilled. For instance:

) if $X = H/\Gamma$ is a quotient of a bounded domain $H \subset \mathbb{C}^n$ by a properly discontinuous group (without fixed points) then one can show directly, by means of Poincare series, that condition D holds ;

) if X is a complex manifold and on X there exists a holomorphic line bundle such that F is "positive" and $F \otimes K^{-1}$ "positive and complete", K denoting the canonical bundle of X.

We refer for these statements to [13].

We thus formulate the problem of projective imbedding in the following manner:

Let X be q-pseudoconcave manifold with $0 \leq q \leq n-2$, $n = \dim_{\mathbb{C}} X$. Assume that there exists on X a holomorphic line bundle F such that $\mathcal{A}(X, F)$ separates points and gives local coordinates everywhere on X. Does X admit a projective imbedding?

4.3 Solution of the problem for 0-pseudoconcave manifolds (cf. [10], [12]).

Theorem (4.3.1). Let X be a 0-pseudoconcave manifold. Suppose that there exists on X a holomorphic line bundle F such that $\mathcal{A}(X, F)$ separates points and gives local coordinates everywhere on X. Then X admits a projective imbedding.

Proof (following an idea of H. Grauert).

(α) We choose $c_0 \in \mathbb{R}$, $\inf_X \phi < c_0 < \inf_K \phi$ such that $B_{c_0} = \{x \in X \mid \phi(x) > c_0\}$ has a smooth boundary. Because $\overline{B}_{c_0}$ is compact, we can choose an l sufficiently large so that the sections of $\Gamma(X, F^l)$ have no common zeros, separate points and give local coordinates everywhere on $\overline{B}_{c_0}$. This is done by the same argument developed in the previous section.

(β) Let $N+1 = \dim_{\mathbb{C}} \Gamma(X, F^l)$ and let $s_0, \ldots, s_N$ be a basis of $\Gamma(X, F^l)$. We consider the map

$$\tau : B_{c_0} \to \mathbb{P}_N(\mathbb{C})$$

defined by $x \mapsto (s_0(x), \ldots, s_N(x))$. This map is holomorphic and one-to-one.(1) By theorem (2.5.3) $\sigma(B_{c_0})$ is contained in an irreducible algebraic variety Z of the same dimension than X.

(γ) Because X is 0-pseudoconcave any analytic subset A of X contained in $X - \bar{B}_{c_0}$ must be 0-dimensional.

Assume if possible that A has an irreducible component A_0 of dimension ≥ 1. Then $\phi|_{A_0}$ has a maximum at some point $z_0 \in A_0$. Since A_ρ is of positive dimension, we can find a one-to-one holomorphic map $\mu : D \to X$ where

$D^1 = \{t \in \mathbb{C} \mid |t| < 1\}$ such that

$$\mu(0) = z_0, \qquad \mu(D^1) \subset A_0 .$$

Then $\phi \circ \mu$ is a subharmonic function on D^1, non-constant and having a maximum at $t = 0$. This is a contradiction.

( ) Let $n = \dim_{\mathbb{C}} X$. Consider the subset of X :

$$A = \{x \in X \mid (\Sigma(-1)^h s_{i_h} \, ds_{i_0} \wedge \ldots \wedge \widehat{ds_{i_h}} \wedge \ldots \wedge ds_{i_n})_x = 0$$

$$\text{for all } \{i_0, \ldots, i_n\} \subset \{0, \ldots, N\} \} .$$

This is an analytic subset of X contained in $X - \bar{B}_{c_0}$. Thus it is a discrete set by (γ), thus at most countable. Obviously σ extends to a holomorphic map

$$\tilde{\sigma} : X - A \to Z$$

which is of rank n at each point. We agree to delete from A all points over which σ may be extended by a holomorphic map of rank n.

---

(1)

We can also assume $\mathcal{K}(X) = \mathbb{C}(\frac{s_1}{s_0}, \ldots, \frac{s_N}{s_0})$.

We will call this set A again.

(ε) If we can show that A is a finite set then one completes the proof as follows. We can replace l by a convenient multiple of l so that the corresponding new map is of rank n at the points of A. Let $Z' \xrightarrow{\omega} Z$ be the normalization of Z. It is known that Z^* is also a projective algebraic variety (cf. [52], [53]). Since X is normal the new map $\tilde{\tau}: X \quad Z$ factors through the normalization Z^* of Z by a holomorphic map

$$\tilde{\tau}^* : X \to Z^* .$$

Since $\tilde{\tau}^*$ is of rank n everywhere, and since Z^* being normal is locally irreducible, $\tilde{\tau}^*$ must be one-to-one and therefore an isomorphism.(1)

(ξ) We are thus reduced to show that A is a finite set. Let $\hat{T}$ be the closure of the graph T of $\tilde{\tau}$ in $X \times \mathbb{P}_N(\mathbb{C})$. The set $\hat{T}$ must be an analytic set as it is the irreducible component containing T of the analytic set

$$\{(x, t) \in X \times \mathbb{P}_N(\mathbb{C}) \mid s_i(x)\, t_j - s_j(x)\, t_i = 0 \quad \text{for } 0 \leq i, j \leq N .$$

Let $\tilde{\tau}^* : X-A \to Z^*$ be the factorization of $\tilde{\tau}$ through the normalization $Z^* \xrightarrow{\omega} Z$ of Z. Let $\hat{T}^*$ be the closure of the graph of $\tilde{\tau}^*$. This is also an analytic set as it is contained in the analytic set $(I \times \omega)^{-1}(\hat{T})$ which has the same dimension n than $\hat{T}^*$. Let $\alpha : \hat{T}^* \to X$, $\beta : \hat{T}^* \to Z^*$ be the natural projections. Now let $a \in A$ and set $\tilde{\tau}^*(a) = \beta\alpha^{-1}(a)$. Let U be a coordinate ball centered at a and not containing any other point of A.

---

(1) $\tilde{\tau}^*$ is of rank n everywhere and generally one-to-one as

$$\mathcal{K}(X) = \mathbb{C}\left(\frac{s_1}{s_0}, \dots, \frac{s_N}{s_0}\right) .$$

Now $\tilde{\tau}^*$ must be one-to-one as Z is normal; so $\beta|_{\alpha^{-1}(U)}$ must be one-to-one as it is one-to-one on $\alpha^{-1}(U-a)$ and since $\beta|_{\alpha^{-1}(a)}$ must also be one-to-one. Therefore $\Omega = \beta\alpha^{-1}(U)$ is open and thus a neighborhood in Z^* of $\tau^*(a)$. Since Ω is normal, the holomorphic map $\lambda = \alpha\beta^{-1} : \Omega - \tilde{\tau}^*(a) \to U$ extends to a holomorphic map $\lambda : \Omega \to U$.

We claim that each component B of $\tilde{\tau}^*(a)$ which is not of co-cimension 1 must be an irreducible component of the singular set of Z^*. In fact if B contains a non-singular point $b \in Z^*$ and if B is of codimension ≥ 2, then λ must have a non-vanishing Jacobian at b. Thus λ^{-1} is defined near a and $\lambda^{-1}(a) = b$ i.e. $\tilde{\tau}^*$ must be a local isomorphism at a. Hence $a \notin A$.

(η) Now we remark that α is a proper map because Z^* is compact. Thus for each $a \in A$ $\tilde{\tau}^*(a)$ is a union of finitely many compact analytic subsets of $Z^* - \tau^*(\bar{B}_{c_0}) = Y$ of codimension one and finitely many irreducible components of the singular set of Z^* (which must also be contained in Y.) Moreover, from the previous argument it follows that if a_1, a_2 are in A and $a_1 \neq a_2$ then $\tilde{\tau}^*(a_1) \cap \tilde{\tau}^*(a_2) = \emptyset$.

We will prove that A is a finite set if we show that Y contains at most finitely many irreducible compact analytic subsets of codimension 1 (here we use that the singular set of Z^* has a finite number of irreducible components).

Let $\varepsilon > 0$ be so small that $c_0 - \varepsilon > \inf_X \phi$ and that τ extends biholomorphically to $B_{c_0-\varepsilon} = \{\phi > c_0 - \varepsilon\}$. Set

$$Y_{c_0-\varepsilon} = Z^* - \tilde{\tau}^*(B_{c_0-\varepsilon}).$$

Then $Y_{c_0-\varepsilon}$ is compact in Y and we can, by the isomorphism given by τ, transpose the function ϕ on $Y - \bar{Y}_{c_0-\varepsilon}$. Let us extend this function by putting it equal to $c_0 - \varepsilon$ on $\bar{Y}_{c_0-\varepsilon}$ and call ψ the continuous function thus obtained. The function $\psi : Y \to \mathbb{R}$, just defined, has the following properties

i) ψ is continuous;

ii) $Y_c = \{\psi < c\} \subset\subset Y$ for any $c < c_o = \sup_Y \psi$;

iii) on $Y - \bar{Y}_{c_o - \varepsilon}$ is C^∞ and strongly plurisbharmonic.

Then every compact irreducible analutic subset of Y of dimension ≥ 1 must be contained in $Y_{c_o - \varepsilon}$ by the same argument given in (γ).
Moreover Y is holomorphically convex as it follows from i), ii), iii) and the solution of Levi problem (cf. chapter VI).
It then follows by the reduction theory of holomorphically convex spaces (cf. [20]) that there is a compact analytic subset $C \subsetneq Y$ of dimension 1 at each one of its points and such that any irreducible compact analytic subset of Y of dimension ≥ 1 must be contained in C (see section 6.3 : remarks 3 and 4). Since C has finitely many irreducible components of codimension 1 the theorem is proved.

<u>Remark</u>: only if $\dim_{\mathbb{C}} X = 2$ this result can be considered satisfactory as the only type of pseudoconcavity available is 0-pseudoconcavity.

4.4. <u>The case of</u> $\dim_{\mathbb{C}} X \geq 3$.

If $\dim_{\mathbb{C}} X \geq 3$ in the previous theorem the assumption that the ring $\mathcal{A}(X, F)$ separates points can be dropped; one has in fact the following

<u>Theorem</u> (4.4.1). <u>Let</u> X <u>be a 0-pseudoconcave manifold with</u> $\dim_{\mathbb{C}} X \geq 3$. <u>Suppose there exists on</u> X <u>a holomorphic line bundle</u> F <u>such that the ring</u> $\mathcal{A}(X, F)$ <u>gives local coordinates everywhere on</u> X. <u>Then</u> $\mathcal{A}(X, F)$ <u>does also separate points of</u> X <u>so that</u> X <u>admits a projective imbedding</u>.

The proof of this theorem is rather complicated and will be omitted. However the interest of the theorem resides in the fact that it does not hold for $\dim_{\mathbb{C}} X = 2$.

We will indicate how to construct a counterexample (cf. [10]) based on the next

Lemma. Let V be a connected compact manifold. Let $W \subset\subset U$ be Stein[(1)] open subsets of V. Let X be a connected complex manifold and let

$$\pi : X \to V - \bar{W}$$

be an holomorphic map making X into a λ-sheeted non-ramified covering of $V - \bar{W}$. If $\dim_{\mathbb{C}} V \geq 2$ and if V is simple connected either $\lambda = 1$, or else X cannot be compactified (i.e. X cannot be isomorphic to an open set of a compact complex space).

The reason for the validity of this lemma can be said briefly as follows. If X would compactify into $\tilde{X}$ then one can show that the holomorphic map π extends to a holomorphic map $\tilde{\pi} : \tilde{X} \to V$.

Since V is simply connected the map π must be ramified, if $\lambda > 1$, somewhere in V. But the ramification set must be analytic and contained in U i.e. it is a compact analytic subset of a Stein manifold. Therefore the ramification set consists of a finite set of points E.

Since V is simply connected, non-singular, of dimension ≥ 2, also $V-E$ is simply connected. Therefore $\tilde{X} - \tilde{\pi}^{-1}(E)$ is an irreducible non-ramified covering of $V-E$. But this is possible only if $\lambda = 1$.

Remark. One could also assume that X is ramified over V along an analytic set $F \subset V-U$ such that $V-F$ remains simply connected.

(1) A Stein manifold is a holomorphic convex manifold on which holomorphic functions separate points.

Construction of the example. Let $T = \mathbb{C}^2/\Gamma$ be an algebraic torus of complex dimension two defined as the quotient of $\mathbb{C}^2$ by the group Γ of translations generated by the vector columns of the matrix

$$\begin{bmatrix} 1 & 0 & z_{11} & z_{12} \\ 0 & 1 & z_{21} & z_{22} \end{bmatrix} \text{ where } Z = \begin{pmatrix} z_{11} & z_{12} \\ z_{21} & z_{22} \end{pmatrix}$$

is a point of Siegel half plane H_2, ${}^tZ = Z$, $\operatorname{Im} Z > 0$.

Let $\tau : T \to T$ the map which associates to each point of the group T its inverse. This map is involutive, $\tau^2 = I$, and has 16 fixed points. Let $K = T/\tau$; this is a complex analytic space of dimension 2 with 16 conical, non-degenerate double points. A theorem of Kummer shows that K is isomorphic to an algebraic surface ϕ_o of fourth order in $\mathbb{P}_3(\mathbb{C})$ with 16 isolated, non-degenerate, conical singular points. Select 16 disjoint small balls U_α in $\mathbb{P}_3$, $1 \leq \alpha \leq 16$, centered at the singular points of ϕ_o and let $U = \bigcup_{\alpha=1}^{16} U_\alpha$. Consider a surface ϕ_ε of fourth order in $\mathbb{P}_3(\mathbb{C})$ close to ϕ_o and non-singular. It is known that ϕ_ε, as a non-singular surface of $\mathbb{P}_3(\mathbb{C})$, is simply connected. Moreover

i) $U_\varepsilon = \phi_\varepsilon \cap U$ is a Stein open subset of

ii) $\phi_\varepsilon - U_\varepsilon$ is diffeomorphic to $\phi_o - U$.

Therefore $\phi_\varepsilon - U_\varepsilon$ admits a non-ramified 2-sheeted covering, the torus T where 16 small neighborhoods of the 16 fixed points of τ have been removed. Let X be this double covering. We give X the complex structure that makes the natural map $\pi : X \to \phi_\varepsilon - U_\varepsilon$ holomorphic.

Since $\phi_\varepsilon - U_\varepsilon$ is 0-pseudoconcave X is also 0-pseudoconcave. Moreover if F denotes the holomorphic line bundle of the hyperplane section of ϕ_ε, $\mathcal{A}(\phi_\varepsilon - U_\varepsilon, F)$ gives local coordinates, everywhere on $\phi_\varepsilon - U_\varepsilon$. The same is therefore true for $\mathcal{A}(X, \pi^* F)$. But the preceeding lemma tells us that X cannot be compactified, in particular X cannot admit a projective imbedding.

Remarks.

1. We have constructed an example of a 0-pseudoconcave manifold that cannot be compactified; actually if we let ε vary a little on a small disc Δ around ε, we have constructed a family $U = \{X_t\}_{t \in \Delta}$ of 2-dimensional manifolds, all 0-pseudoconcave and no one of which can be compactified.
2. Making use of the remark made after the lemma one can show that example 3 of (2.3) gives also a non-compactifiable complex structure on $\mathbb{P}_2(\mathbb{C}) - \{0\}$.

A. Andreotti

Chapter V.. Meromorphic functions on complex spaces.

5.1. Preliminaries

a) Let Ω be an open set in $\mathbb{C}^n$. A closed subset $A \subset \Omega$ is called an analytic set if A is locally the zero set of a finite number of holomorphic functions i.e. for every $z_o \in A$ there exi ts a neighborhood U of z_o in Ω and $f_1, \dots, f_k \in \mathcal{H}(U)$ such that

$$A \cap U = \{ z \in U \mid f_1(z) = f_2(z) = \dots = f_k(z) = 0 \}.$$

On A we define as the sheaf O_A of holomorphic functions, the sheaf of germs of restrictions to A of holomorphic functions in some open set of Ω. A morphism (or holomorphic map) from the analytic set $(A, \mathcal{O}_A)$ to the analytic set $(B, \mathcal{O}_B)$ is a continuous map $\tau : A \to B$ such that

$$\tau^* \mathcal{O}_{B,\tau(x)} \subset \mathcal{O}_{A,x} \quad \text{for every } x \in A.$$

The notion of isomorphism is then defined and then the notion of complex space is obtained replacing in the definition of complex manifold open sets of a numerical space and isomorphisms between them by analytic sets and isomorphisms between them. The "structural sheaf" will be denoted by $\mathcal{O}$.
A point $x \in X$ of a complex space X is called a simple point if some neighborhood U of x in X is isomorphic to some open subset of some $\mathbb{C}^n$. Let $S(X)$ be the set of non-simple points of X, this is called the singular set of X. This is an analytic subset of X as one can prove.
A complex space X is called irreducible if $X-S(X)$ is a connected manifold; its dimension is called the dimension of X. The dimension of an arbitrary complex space X is, by definition, the maximal dimension of its components. If $Y \subset X$ are complex spaces then $\dim Y \leq \dim X$. If $x \in X$ is a point of the complex space X we define $\dim_x X$ as inf. $\dim U(x)$ when $U(x)$ describes a fundamental system of neighborhoods of x in X.
Always holds: $\dim S(X) < \dim X$.

b) The notion of meromorphic function on a complex space is defined in the same way as for complex manifolds. Given a complex space X we can then define two rings of meromorphic functions:

$\mathcal{Q}(X)$, the ring of meromorphic functions which are quotient of two global holomorphic functions on X, the second of which is not a divisor of zero.

$\mathcal{K}(X)$, the ring of all meromorphic functions (which are locally the quotient of two holomorphic functions, the second of which is not a zero divisor).

If X is irreducible, as we will always assume in the sequel, $\mathcal{Q}(X)$ and $\mathcal{K}(X)$ are fields.

c) A complex space $(X, \mathcal{O})$ is called normal if for every $x \in X$ the ring $\mathcal{O}_x$ is integrally closed in its quotient ring. A complex space $(X, \mathcal{O})$ is called weakly normal if every continuous meromorphic function $f : U \to \mathbb{C}$ defined on an open set $U \subset X$ is holomorphic on U. Given a complex space $(X, \mathcal{O}_X)$ there exists a (normal) complex space $(\hat{X}, \mathcal{O}_{\hat{X}})$ and a morphism $\rho : \hat{X} \to X$ which is surjective and such that for any normal complex space $(Y, \mathcal{O}_Y)$ any morphism $Y \to X$ such that the image of any irreducible component of Y is not contained in S(X) factors through $\hat{X}$:

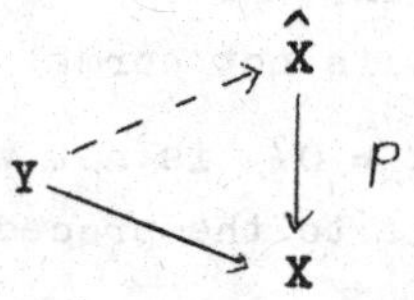

The space $(\hat{X}, \mathcal{O}_{\hat{X}})$ is uniquely defined up to isomorphisms and is called the normalization of $(X, \mathcal{O})$.
Analogous statement holds replacing the word "normal" by "weakly-normal". We thus obtain the notion of weak-normalization space of $(X, \mathcal{O})$. A space X and its weak normalization are homeomorphic. A normal space is weakly normal.

Examples.

1. The space $X = \{(x, y) \in \mathbb{C}^2 \mid x^3 = y^2\}$ is neither normal nor weakly-normal as the function $\frac{y}{x}$ is continuous on X and integral over the local ring of X at the origin as it satisfies the equation $(\frac{y}{x})^2 - x = 0$ on X.

Its normalization and weak normalization is the complex space $\mathbb{C}$ with $\mathbb{C} \xrightarrow{\sigma} X$ defined by the map $t \to (t^2, t^3)$.

2. The space $X = \{(x, y) \in \mathbb{C}^2 \mid x^2 - y^2 + y^3 = 0\}$ is not normal as $\frac{x}{y}$ is integral over the local ring at the origin but not holomorphic; it satisfies the equation $(\frac{x}{y})^2 + y - 1 = 0$ on X.
Its normalization is $\mathbb{C}$ with $\mathbb{C} \xrightarrow{\rho} X$ defined by $t \to (t(1-t^2), 1-t^2)$. However X is weakly normal. In fact if g is holomorphic and continuous near the origin on X $g \circ \rho$ must be holomorphic in the neighborhoods of $t = 1$, $t = -1$ and assume the same value α at $t = 1$ and $t = -1$. One verifies that any holomorphic function of this type can be written as a series of the form $\alpha + \sum_{r+s=1}^{\infty} a_{rs} t^r (1-t^2)^{r+s}$, but this is the pull back by ρ of the holomorphic function near the origin in $\mathbb{C}^2$ $\alpha + \sum_{r+s=1}^{\infty} a_{rs} x^r y^s$.

3. Simpler examples can be obtained by taking into consideratin reducible spaces. So for instance

$\{(x, y) \in \mathbb{C}^2 \mid xy = 0\}$ is not normal but weakly normal;

$\{(x, Y) \in \mathbb{C}^2 \mid (x-y^2)\, x = 0\}$ is not weakly normal, its weak normalization is isomorphic to the preceding set.

As reference one can consult [29], [42] and [8] for the weak normalization.

5.2 Pseudoconcavity for complex spaces.

The notion of pseudoconcave manifold can be extended to complex spaces with the following definition.

The complex space $(X, \mathcal{O})$ is called pseudoconcave if an open sutset $Y \neq \emptyset$ of X is given such that

i) $Y \subset\subset X$;

ii) For every $z_o \in \partial Y = \bar{Y} - Y$ there exists a fundamental sequence of neighborhoods $\{U_\nu\}$ of z_o in X such that

$$\widehat{(U_\nu \cap Y)}_{U_\nu} \text{ is a neighborhood of } z_o.$$

Here

$$\widehat{(U_\nu \cap Y)}_{U_\nu} = \{z \in U_\nu \mid |f(z)| \leq \sup_{U_\nu \cap Y} |f| \text{ for all } f \in \mathcal{H}(U_\nu)\}.$$

This is generalization of the notion given for complex manifolds. in fact if X is such a manifold and $Y \subset\subset X$ has a smooth boundary and if $\forall z_o \in \partial Y$ the Levi-form restricted to the analytic tangent plane has one negative eigenvalue, one can construct, using the disc theorem, a fundamental sequence of coordinate polycilinders P_ν centered at z_o such that every holomorphic function on $P_\nu \cap Y$ extends to P. Taking $U_\nu = P_\nu$ the above condition ii) is then fulfilled.

For a pseudoconcave space one has the equivalent of theorem (2.5.1) (cf. [1]).

Theorem (5.2.1). If X is a pseudoconcave space of dimension n, then the field $\mathcal{K}(X)$ of meromorphic functions on X is an algebraic field of transcendence degree $d \leq n$.

We remark that from condition ii) it follows again that every holomorphic function on a pseudoconcave space X is constant. Therefore $\mathcal{Q}(X) = \mathbb{C}$.

5.3 The Poincaré problem.

a) If we drop the assumption of pseudoconcavity there is no hope to obtain a statement of the nature of theorem (5.2.1); we have already remarked that for $X = \mathbb{C}$, the functions $e^z, e^{z^2}, \dots$ are algebraically independent and, therefore, the transcendence degree of $\mathcal{K}(\mathbb{C})$ is infinite. However if we take as ground field $\mathcal{Q}(X)$ instead of $\mathbb{C}$ the situation is more hopeful. First of all one has the following useful fact

Theorem (5.3.1). If X is a normal space, then $\mathcal{Q}(X)$ is algebraically closed in $\mathcal{K}(X)$.

The proof is the same as the proof given for manifolds (theorem (2.2.2)). Moreover the previous counter-example disappears as one has:

Theorem (5.3.2).

(a) If X is a Stein space[1] then $\mathcal{Q}(X) = \mathcal{K}(X)$.

(b) If X is an open connected subset of a Stein manifold then $\mathcal{Q}(X) = \mathcal{K}(X)$.

---

(1) A Stein space (or holomorphically complete space) is a complex space X (with countable topology) satisfying the following conditions

(i) $\mathcal{H}(X)$ separates points i.e. if $x \neq y$, $x, y \in X$, there exists an $f \in \mathcal{H}(X)$ with $f(x) \neq f(y)$.

(ii) for any divergent sequence $\{x_i\} \subset X$ there exists an $f \in \mathcal{H}(X)$ such that

$$\sup |f(x_i)| = \infty.$$

A space X in which (ii) is satisfied is called holomorphically convex. A sheaf $\mathcal{F}$ of $\mathcal{O}$-modules on a complex space X is called coherent (cf. [28]) if for each $x \in X$ there is a neighborhood U of x and an exact sequence of the form

$$\mathcal{O}^p_U \to \mathcal{O}^q_U \to \mathcal{F}_U \to 0 .$$

In particular the sheafs $\mathcal{O}^p$, $p \geq 1$, are all coherent. Kernels, cokernels, images, coimages of homomorphisms of coherent sheaves are coherent. For a coherent sheaf $\mathcal{F}$ on a Stein space X the space $\Gamma(X, \mathcal{F})$ generates $\mathcal{F}_x$ over $\mathcal{O}_x$

Proof. (a) Given $h \in \mathcal{K}(X)$ consider the sheaf
$$\mathcal{J} = \{f \in \mathcal{O} \mid f_x h_x \in \mathcal{O}_x \text{ for } x \in X .$$
Locally $h = \frac{p}{q}$ with p and q holomorphic and q not a zero divisor. Thus locally
$$\mathcal{J} = \{f \in \mathcal{O} \mid fp \equiv 0 \pmod{q}\}$$
$$= \operatorname{Ker}\{\mathcal{O} \xrightarrow{p} \mathcal{O}/q\mathcal{O}\}$$
Since $\mathcal{O}$, $q\mathcal{O}$ and thus $\mathcal{O}/q_{\mathcal{O}}$ are coherent, it follows that $\mathcal{J}$ is coherent. If $x_0 \in X$ is a point where h is holomorphic we have $\mathcal{J}_{x_0} = \mathcal{O}_{x_0}$. Therefore by theorem A there exists $s \in \Gamma(X, \mathcal{J})$ with $s(x_0) \neq 0$. But, by the definition of $\mathcal{J}$, $sh = t \in \Gamma(X, \mathcal{O})$. Thus $h = \frac{t}{s}$ and $s \not\equiv 0$ (since X is assumed irreducible), hence $h \in \mathcal{O}(X)$.

(b)$_1$ Let us assume first that X is an open set of $\mathbb{C}^n$ (a particular Stein manifold). Given $h \in \mathcal{K}(X)$ and denoting by $\mathcal{M}$ the sheaf of germs of meromorphic functions on $\mathbb{C}^n$, h defines a section $h : X \to \mathcal{M}$ of the stacked space $\mathcal{M}$ over the open set X.

Let $\hat{X}$ be the connected component of $h(X)$ in $\mathcal{M}$. Because the natural projection $\pi : \mathcal{M} \to \mathbb{C}^n$ is a local homeomorphism, X acquires a natural complex structure of an n-dimensional complex manifold. This manifold is Hausdorff as the stacked space $\mathcal{M}$ is Hausdorff. By the very meaning of $\hat{X}$, there

---

for every $x \in X$ (theorem A of H. Cartan). Every complex subspace of the numberical space $\mathbb{C}^N$ is a Stein space. Every Stein space (of finite dimension) admits a proper one-to-one holomorphic map into a numerical space $\mathbb{C}^N$. For a Stein manifold that map can be chosen to be an isomorphism onto its image.

exists a meromorphic function $\hat{h}$ on $\hat{X}$ such that $\hat{h}\,|_{h(X)} = \pi^* h$; at each point of X the germ of $\hat{h}_{\hat{x}}$ is the germ represented by $\hat{x}$.

Now $\hat{X}$ is the largest domain spread over $\mathbb{C}^n$ on which $\hat{h}$ can be continued. Therefore $-\log \delta(\hat{x})$, $\delta(\hat{x})$ polycilindrical distance of $\hat{x}$ from the boundary of $\hat{X}$ is a plurisubharmonic function (cf. [31]).
Hence $\hat{X}$ is a Stein manifold. We can then apply to $\hat{h}$ the argument developed in (a) and write $\hat{h} = \frac{\hat{p}}{\hat{q}}$ with $\hat{p}$, $\hat{q}$ holomorphic on $\hat{X}$ and $\hat{q} \not\equiv 0$. Then $\frac{\hat{p}}{\hat{q}}\,\Big|_{h(X)} = \pi^* h$ and since π^* is an isomorphism between X and $h(X)$ we get the desired result.

$(b)_2$ Let us now consider the general case, in which X is an open subset of a Stein manifold Y. By the imbedding theorem for Stein manifolds (cf. [28], [31]) we may assume that Y is a submanifold of some numerical space $\mathbb{C}^N$. By a theorem of Coquier-Grauert ([23]) one can find a connected neighborhood U of Y in $\mathbb{C}^N$ and a holomorphic retraction $\tau: U \to Y$, $\tau(y) = y$ for all $y \in Y$. Consider the meromorphic function $\tau^* h$ on $\tau^{-1}(X)$. By the special case $(b)_1$ we can find two holomorphic functions p, q with $q \not\equiv 0$ on $\tau^{-1}(X)$ such that $q\,\tau^* h = p$. Since $q \not\equiv 0$ there exists a multi index $\alpha \in \mathbb{N}^n$ such that $D^\alpha q_{|X} \not\equiv 0$ while $D^\beta q_{|X} \equiv 0$ if $|\beta| < |\alpha|$. Then $D^\alpha(q\,\tau^* h) = D^\alpha p$. Restricting to X we get

$$D^\alpha a_{|X}\, h = D^\alpha p_{|X} \quad \text{with} \quad D^\alpha q_X \not\equiv 0$$

as we wanted.

The previous theorem gives the solution, in some particular cases of the so-called <u>Poincaré problem</u>: <u>when, for a complex space</u> X, <u>do we have</u> $\mathcal{Q}(X) = \mathcal{K}(X)$? Of course this problem is not always solvable, for instance for compact projective irreducible algebraic varieties we have $\mathcal{Q}(X) = \mathbb{C}$ but

$\mathcal{K}(X) \neq \mathbb{C}$ if the variety is not a point. One could hope that $Q(X) = \mathcal{K}(X)$ if $H(X)$ separates points of X, but this in general is still an open question.

b) For a complex space X which is not "holomorphically separated" one is lead naturally to consider the equivalence relation $\mathcal{R}$ on X

$$x \underset{\mathcal{R}}{\sim} y \quad \text{iff} \quad f(x) = f(y) \quad \text{for all} \quad f \in H(X).$$

Let $Y = X/\mathcal{R}$ be the topological quotient space, and let $p : X \to Y$ be the natural projection. A set $U \subset Y$ is open iff $p^{-1}(U)$ is open in X, thus Y is a Hausdorff space. We define for U open in Y

$$\mathcal{O}_Y(U) = \{f : U \to \mathbb{C} \mid f \text{ continuous, } p^*f : p^{-1}(U) \to \mathbb{C} \text{ holomorphic}\}.$$

We give in this way to Y the structure of a ringed space (a space with a sheaf of rings of germs of continuous functions) and the problem arises to see when $(Y, \mathcal{O})$ is a complex space. If this is the case, then $(Y, \mathcal{O})$ is a holomorphically separated space and $p : X \to Y$ is a holomorphic surjection. Moreover one can then consider $\mathcal{K}(X)$ as an extension of the field $p^*\mathcal{K}(Y)$ which contains $Q(X)$ and may be equal to $Q(X)$ if on Y the Poincaré problem is solvable, as one obviously has $Q(X) = p^*Q(Y)$.

To decide when $(Y, \mathcal{O})$ has the structure of a complex space we have the following sufficient criterion (of topological nature):

<u>Proposition</u> (5.3.3). <u>Let</u> X <u>be an irreducible weakly-normal complex space</u> <u>If</u>

(i) $Y = X/\mathcal{R}$ <u>is locally compact,</u>

(ii) $p : X \to Y$ <u>is semiproper,</u> <u>i.e. for every compact set</u> $K \subset Y$ <u>there exists a compact set</u> $K' \subset X$ <u>such that</u> $p(K') = K$.

<u>Then</u> $(Y, \mathcal{O})$ <u>is a weakly normalcomplex space.</u>

<u>Example</u>. Let X be a weakly-normal, holomorphically convex space. Then p is a proper map. Indeed if $K \subset Y$ is compact, $p^{-1}(K)$ must be compact, otherwise we can select a divergent sequence $\{x_i\} \subset p^{-1}(K)$ and thus find an $f \in \mathcal{H}(X)$ with $\sup_i |f(x_i)| = \infty$. But $\sup_{p^{-1}(K)} |f| = \sup_K p_* f$ where $p_* f$ is the function on $Y = X/\mathcal{R}$ whose pull back by p is f. Since $p_* f$ is continuous this is impossible.

Moreover Y is locally compact. Let $y \in Y$ and $x \in X$ such that $p(x) = y$. Since X is holomorphically convex, we can find a finite set of functions $f_1, \dots, f_k$ in $\mathcal{H}(X)$ such that the set

$$U = \{z \in X \mid |f_i(z) - f_i(x)| < 1 \quad \text{for } 1 \le i \le k\}$$

has a compact closure K'. Then $p(U)$ is a neighborhood of y and is contained in $p(K')$ which is compact.
We satisfy thus the conditions stated by the previous proposition and therefore $(Y, \mathcal{O})$ is an analytic space which is also weakly-normal.
But on Y holomorphic functions separate points and holomorphic convexity is inherited by Y from X. Thus $(Y, \mathcal{O})$ is a weakly normal Stein space. In this case we thus have

$$p^* \mathcal{Q}(Y) = p^* \mathcal{K}(Y) = \mathcal{G}(X).$$

<u>Proof of proposition</u> (5.3.3).

(α) First we remark that, given a compact set $K \subset X$ we can find finitely many holomorphic functions $f_1, \dots, f_k$ in $\mathcal{H}(X)$ such that:

$x_1 \underset{\mathcal{R}}{\sim} x_2$, x_1, x_2 in K, iff $f_1(x_1) = f_1(x_2), \dots f_k(x_1) =$

Indeed the equivalence relation $\mathcal{R}$ is represented in $X \times X$ by the analytic set

$$\mathcal{R} = \bigcap_{f \in \mathcal{H}(X)} \{(x_1, x_2) \in X \times X \mid f(x_1) - f(x_2) = 0\}.$$

Locally $\mathcal{R}$ can be given by finitely many of its defining equations. Since $K \times K$ is compact finitely many of these defining equations suffice to define $\mathcal{R} \cap K \times K$.

(β) Secondly we show the following fact:

Let X, Y be complex spaces and let $\phi : X \to Y$ be holomorphic Let $g : Y \to \mathbb{C}$ be continuous. Then

(i) if g is meromorphic so is $\phi^* g$

(ii) if ϕ is semiproper and surjective then if $\phi^* g$ is meromorphic so is g.

This fact is proved making use of the following remark.

A continuous function $h : X \to \mathbb{C}$, defined on a complex space X is meromorphic iff the graph of h is an analytic subset of $X \times \mathbb{C}$.

This is a particular case of a general theorem (cf. [11], prop. 3.12).

In this instance if X is weakly normal there is almost nothing to prove; in general a space and its weak normalization have the "same" meromorphic functions. Making use of this remark, statement (i) then follows from the fact that the graph of $\phi^* g$ equals $(\phi \times \mathrm{id})^{-1}$ (graph of g) while $\phi \times \mathrm{id}: X \times \mathbb{C} \to Y \times \mathbb{C}$ is holomorphic and the graph of g is analytic.

Denote by $\tilde{\Gamma}$ and Γ the graphs of $\phi^* g$ and g respectively. We have a commutative diagram

$$\begin{array}{ccc} \tilde{\Gamma} & \xrightarrow{\chi} & \Gamma \\ \downarrow{\scriptstyle \tilde{\pi}} & & \downarrow{\scriptstyle \pi} \\ X & \xrightarrow{\phi} & Y \end{array}$$

where $\tilde{\pi}$ and π are the natural projections and where $\chi = (\phi \times \mathrm{id})|_{\tilde{\Gamma}}$.

Since $\tilde{\pi}$ and π are homeomorphisms χ is semiproper because ϕ is semiproper. Moreover Γ is closed in $Y \times \mathbb{C}$. Therefore the composed map

$$\tilde{\chi} : \tilde{\Gamma} \xrightarrow{\chi} \Gamma \xrightarrow{i} Y \times \mathbb{C}$$

is semiproper and holomorphic. By assumption $\phi^* g$ is meromorphic so that $\tilde{\Gamma}$ is analytic.

A theorem of Kuhlmann (cf. [40], se also [11], a generalization of a theorem of Remmert [45]) states that the image of a holomorphic semiproper map between complex spaces is an analytic set. By virtue of that theorem $\operatorname{Im} \tilde{\chi} = \Gamma$ must be analytic and therefore g must be meromorphic.

(γ) Let $b \in Y$ and let V be a neighborhood of b with a compact closure $\bar{V}$.

Choose $K \subset X$ compact such that $p(K) = \bar{V}$. Select $f_1, \ldots, f_k \in \mathcal{H}(X)$ such that

$$x_1 \tilde{\mathcal{R}} x_2 ; \quad x_1, x_2 \text{ in } K \quad \text{iff}$$

$$f_1(x_1) = f_1(x_2), \ldots, f_k(x_1) = f_k(x_2).$$

Let $g_i : Y \to \mathbb{C}$ be such that $p^* g_i = f_i$ for $1 \leq i \leq k$. Consider the map

$$\chi : \bar{V} \to \mathbb{C}^k$$

defined by $\chi(x) = (g_1(x), \ldots, g_k(x))$. Let $L = (\bar{V})$. Then, being continuous, L must be compact as $\bar{V}$ is compact. Since χ is bijective χ must be a homeomorphism of $\bar{V}$ onto L. Therefore $Z = \chi(V)$ is open in L and there exists an open set $\Omega \subset \mathbb{C}^k$ (open in $\mathbb{C}^k$) such that $Z = L \cap \Omega$. In particular Z is closed in Ω.

Set $W = p^{-1}(V)$ and consider the composite map

$$\phi : W \xrightarrow{p} V \xrightarrow{\chi} Z \xrightarrow{i} \Omega .$$

This is given by $x \mapsto (f_1(x), \ldots, f_k(x))$, therefore ϕ is a holomorphic map which is also semiproper as p is semiproper. By the Kuhlmann theorem $\phi(W) = Z$ is an analytic set.

Let $(\hat{Z}, \hat{\mathcal{O}})$ be the weak-normalization of Z. We get a bijective map $\psi : V \to \hat{Z}$.

We claim that ψ is an isomorphism of $(V, \mathcal{O}_V)$ onto $(\hat{Z}, \hat{\mathcal{O}})$. This follows directly by application of statement (β) and the fact that on a weakly normal space continuous meromorphic functions are holomorphic.

5.4. Relative theorems.

The previous considerations lead us naturally to the following situation.

Let X, Y be irreducible complex spaces and let $p : X \to Y$ a holomorphic surjection. This gives an injective map

$$p^* : \mathcal{K}(Y) \to \mathcal{K}(X)$$

and we want to investigate when $\mathcal{K}(X)$ considered as an extension of $p^*\mathcal{K}(Y)$ is an algebraic field.

For instance Y could be the separation space of X and in favorable conditions we would have $\mathcal{K}(Y) = \mathcal{Q}(Y) \simeq \mathcal{Q}(X)$.

In this connection one can prove the following:

Theorem (5.4.1). If $p : X \to Y$ is a proper holomorphic map then $\mathcal{K}(X)$ is an algebraic field over $p^*\mathcal{K}(Y)$, of transcendence degree $\leq \dim_{\mathbb{C}} X - \dim_{\mathbb{C}} Y$.

Example. Let X be a weakly-normal holomorphically convex space. Let Y be the separation space of X. Since Y is a Stein space, $\mathcal{K}(Y) = \mathcal{Q}(Y)$ and $p^*\mathcal{Q}(Y) \simeq \mathcal{Q}(X)$. We thus obtain the following result as $p : X \to Y$ is proper:

For a weakly-normal holomorphically convex space X the field $\mathcal{K}(X)$ is an algebraic field over $\mathcal{Q}(X)$ of transcendence degree $\leq \dim_{\mathbb{C}} X - \dim_{\mathbb{C}} Y$, where Y is the separation space of X.

If in theorem (5.4.1) Y is a point we get back theorem (5.2.1) for a compact space. One is lead to conjecture an analogue of theorem (5.4.1) also for a pseudoconcave situation. This actually is the case. For this purpose we introduce the following definition.

Let X, Y be irreducible complex spaces and let $f : X \to Y$ be a holomorphic surjection. We say that f is a pseudoconcave map if an open, non-empty, subset $\Omega \subset X$ is given with the following properties

(i) $f : \bar{\Omega} \to Y$ is a proper map

(ii) for every $y \in Y$ every irreducible component of $f^{-1}(y)$ intersects Ω

(iii) for every point $z_o \in \partial\Omega = \bar{\Omega} - \Omega$ there exist two fundamental sequences of neighborhoods $\{U_\nu\}$, $\{V_\nu\}$ with $V_\nu \subset U_\nu$, such that

(α) $f(U_\nu \cap \Omega) = f(U_\nu)$

(β) for every $y \in f(U_\nu)$ we have

$$\widehat{(U_\nu \cap \Omega \cap f^{-1}(y))}_{U_\nu \cap f^{-1}(y)} \supset V_\nu \cap f^{-1}(y) .$$

__Theorem__ (5.4.2). __Let__ X, Y __be irreducible complex spaces and let__ X __be normal__.

__Let__ $p : X \to Y$ __be a pseudoconcave map__. __Then__ $\mathcal{K}(X)$ __is an algebraic field over__ $p^*\mathcal{K}(Y)$ __of transcendence degree__ $\leq \dim_{\mathbb{C}} X - \dim_{\mathbb{C}} Y$.

__Remark__. For the validity of this theorem actually one needs only to suppose that (α) p is pseudoconcave over some open set $A \subset Y$ with $A \neq \emptyset$ and such that $p^{-1}(A)$ is normal,
(β) for every compact subset $K \subset Y$ we can find a compact set $K' \subset X$ such that for every $y \in K$, every irreducible component of $p^{-1}(y)$ intersects K'.

The proof of this theorem is rather complicated. It can be found in [11] with all the material considered in this chapter.

__Remarks__.

1. Theorem (5.4.2) is more general then theorem (5.4.1). For example the family $V = \{X_t\}_{t \in \Delta}$, where $\Delta = \{t \in \mathbb{C} \mid |t| < 1\}$ constructed in 4.4, remark 1, gives an example of a pseudo-

concave map in which no one of the fibers X_t can be compactified.

2. The assumption of pseudoconcavity in theorem (5.4.2) cannot be relaxed. Here is a counter example due to Kodaira and Kas (cf. [11]) :
Take countably many copies Δ_k of the unit disc Δ in $\mathbb{C}$, and countably many copies $\mathbb{C}^*_k$ of $\mathbb{C}^*$.
Consider the sets

$$M = \bigcup_{k\in\mathbb{Z}} \Delta_k \times \mathbb{C}^*_k$$

$$S = \bigcup_{k\in\mathbb{Z}} \{0\} \times \mathbb{C}^*_k \ .$$

On M-S consider the cyclic group Γ generated by the automorphism

$$g : (t_k, w_k) \longmapsto (t_{k+1} = t_k,\ w_{k+1} = t_k w_k)$$

where $t_k \in \Delta_k$, $w_k \in \mathbb{C}^*_k$. Define the action of Γ on S to be the identity. One verifies that $X = M/\Gamma$ is a two-dimensional manifold. The manifold X looks like $\Delta \times \mathbb{C}^*$ in which $\{0\}\times \mathbb{C}^*$ is replaced by S (cf. [11]). We have a natural holomorphic map $p : X \to \Delta$ and one has

$$\mathcal{H}(X) = p^* \mathcal{H}(\Delta) \ .$$

Indeed, let $f \in \mathcal{H}(X)$, then $f | \Delta_k \times \mathbb{C}^*_k$ is given by a holomorphic function $f_k = \sum_{n=-\infty} a_{k,n}(t)\, w^n$. in the Laurent expansion of f_k $a_{k,n}(t) \in \mathcal{H}(\Delta)$. Now

$f_{k+1}(t, w) = f_k(t, t^{-1}w)$, for $t \neq 0$, therefore

$a_{k,n}(t) = a_{0,n}(t)\, t^{-nk}$. This means that if $n \neq 0$ $a_{0,n}$ has

a zero of infinite order at $t = 0$, thus $a_{0,n} = 0$ if $n \neq 0$ and therefore in the expansion of f only the term independent of w appears i.e. $f \in \mathcal{H}(\Delta)$.
It follows then that $\mathcal{Q}(X) = p^*\mathcal{Q}(\Delta)$ and that Δ is the reduction space of X.

On X consider the functions given on $\Delta_0 \times \mathbb{C}_0^*$ by

$$f = w \; ; \qquad g = \sum_{n \in \mathbb{Z}} t^{n^2} w^n .$$

One verifies easily that these define meromorphic functions on X.
Set $t = e^{\pi i \tau}$ with $\mathrm{Im} > 0$ since $|t| < 1$. Set also $w = e^{2\pi i z}$, then g is transformed into the Jacobi theta series

$$\theta(\tau ; z) = \sum_{n \in \mathbb{Z}} e^{\pi i (n^2 + 2nz)} .$$

For any value of τ, $\theta(\tau, z)$ has an infinite number of zeros of the form $z_0(\tau) + l + k\tau$; $l, k \in \mathbb{Z}$.

We show that f and g are algebraically independent over $\mathcal{Q}(X)$. If, on the contrary, f and g were algebraically dependent over $\mathcal{Q}(X)$ there would exist an irreducible polynomial in X, Y, $F(t;X,Y) \not\equiv 0$ with coefficients that can be assumed to be holomorphic in $t \in \Delta$, such that

$$F(t;\, f,\, g) \equiv 0.$$

Now for some $t = t_0 \neq 0$ $F(t_0, X, 0) \not\equiv 0$, but this equation $f(t_0, X, 0) = 0$ should be satisfied for all values $X = e^{2\pi i z_0} t_0^{2k}$, $k \in \mathbb{Z}$. This is impossible since a polynomial in one variable has only finitely many zeros. The conclusion is that the transendence degree of $\mathcal{K}(X)$ over $\mathcal{Q}(X)$ is ≥ 2, while if theorem (5.4.2) is holding for X, one should have a transcendence degree ≤ 1.

A Andreotti

Chapter VI. E. E. Levi problem.

6.1 Preliminaries.

a) Čech cohomology with values in a sheaf. Let $\mathcal{F}$ be a sheaf of abelian groups over a topological space X. Let $\mathcal{U} = U_{i\ i\in I}$ be an open covering of X; we set

$$C^q(\mathcal{U},\mathcal{F}) = \prod_{(i_0,\ldots,i_q)} \Gamma(U_{i_0\ldots i_q},\mathcal{F}) \quad \text{for } q \geq 0,$$

where $(i_0,\ldots,i_q)$ runs over all $(q+1)$-tuples extracted from I and where we have set $U_{i_0\ldots i_q} = U_{i_0} \cap U_{i_1} \cap \ldots \cap U_{i_q}$.

Thus an element $f \in C^q(\mathcal{U},\mathcal{F})$ is given by a collection $\{f_{i_0\ldots i_q}\}$ where $f_{i_0\ldots i_q} \in \Gamma(U_{i_0\ldots i_q},\mathcal{F})$.

Obviously $C^q(\mathcal{U},\mathcal{F})$ is an abelian group. We define

$$\delta_q : C^q(\mathcal{U},\mathcal{F}) \to C^{q+1}(\mathcal{U},\mathcal{F})$$

by the formula

$$(\delta_q f)_{i_0\ldots i_{q+1}} = \sum_{j=0}^{q+1} (-1)^j r^{U_{i_0\ldots\hat{i}_j\ldots i_{q+1}}}_{U_{i_0\ldots i_{q+1}}} f_{i_0\ldots\hat{i}_j\ldots i_{q+1}}$$

where $r^{U_{i_0\ldots\hat{i}_j\ldots i_{q+1}}}_{U_{i_0\ldots i_{q+1}}} \Gamma(U_{i_0\ldots\hat{i}_j\ldots i_{q+1}},\mathcal{F}) \to \Gamma(U_{i_0\ldots i_{q+1}},\mathcal{F})$

is the natural restriction homomorphism.

One verifies that $\delta_{q+1} \circ \delta_q = 0$, therefore the sequence of abelian groups and maps

$$C^0(\mathcal{U},\mathcal{F}) \xrightarrow{\delta_0} C^1(\mathcal{U},\mathcal{F}) \xrightarrow{\delta_1} C^2(\mathcal{U},\mathcal{F}) \xrightarrow{\delta_2} \ldots$$

is a cochain comples. Its cohomology in dimension q will be denoted by $H^q(\mathcal{U},\mathcal{F})$.

Note that, since $\mathcal{F}$ is a sheaf,

$$H^0(\mathcal{U},\mathcal{F}) = \operatorname{Ker}\delta_0 = \Gamma(X,\mathcal{F})$$

so that the above complex can be completed with the augmentation map

$$0 \to \Gamma(X,\mathcal{F}) \xrightarrow{\varepsilon} C^0(\mathcal{U},\mathcal{F}) \quad \dots$$

The groups $H^q(\mathcal{U},\mathcal{F})$ are called the <u>Čech cohomology groups of the covering $\mathcal{U}$ with values in $\mathcal{F}$</u>.

If $\mathcal{V} = \{V_j\}_{j\in J}$ is another covering of X by open sets which is a refinement of $\mathcal{U} = \{U_i\}_{i\in I}$ then one can define a refinement function $\tau : J \to I$ with the property $V_j \subset U_{\tau(j)}$.

By means of τ one defines a homomorphism of complexes $\tau^* : C^*(\mathcal{U},\mathcal{F}) \to C^*(\mathcal{V},\mathcal{F})$ by setting

$$(\tau^* f)_{j_0\dots j_q} = r^{U_{\tau(j_0)\dots\tau(j_q)}}_{V_{j_0\dots j_q}} f_{\tau(j_0)\dots\tau(j_q)}$$

for $f \in C^q(\mathcal{U},\mathcal{F})$.

This defines a homomorphism of cohomology groups

$$\tau^{\mathcal{U}}_{\mathcal{V}} : H^*(\mathcal{U},\mathcal{F}) \to H^*(\mathcal{V},\mathcal{F}).$$

One verifies that $\tau^{\mathcal{U}}_{\mathcal{V}}$ depends only on the coverings $\mathcal{U},\mathcal{V}$ but not on the choice of the refinement function. One can then define

$$H^q(X,\mathcal{F}) = \varinjlim_{\mathcal{U}} H^q(\mathcal{U},\mathcal{F}), \qquad q \geq 0,$$

the direct limit taken over all coverings $\mathcal{U}$ of X. This is called the q-th <u>Čech cohomology group of</u> X <u>with values in</u> $\mathcal{F}$.

b) <u>Homomorphisms of sheaves.</u> Given two sheaves of abelian groups $(\mathcal{F},\pi,X)$, $(\mathcal{G},\omega,X)$ over X a <u>homomorphism</u> of $\mathcal{F}$ into $\mathcal{G}$ is a continuous map $\phi : \mathcal{F} \to \mathcal{G}$ such that

i) $\omega \circ \phi = \pi$

ii) for every $x \in X$ the induced map $\phi_x : \mathcal{F}_x \to \mathcal{G}_x$ is a group homomorphism.

If ϕ is injective $\mathcal{F}$ can be considered as a subsheaf of $\mathcal{G}$ and one can then define a <u>quotient sheaf</u> $\mathcal{G}/\mathcal{F}$, as the quotient space of $\mathcal{G}$ by the equivalence relation

$$g_x \sim g'_x \qquad \text{iff} \qquad g_x - g'_x \in \phi(\mathcal{F}_x),$$

the topology being the quotient topology. One has thus $(\mathcal{G}/\mathcal{F})_x = \mathcal{G}_x/\mathcal{F}_x$.

A sequence of sheavesand homomorphisms

(1) $\quad 0 \to \mathcal{F} \xrightarrow{\alpha} \mathcal{G} \xrightarrow{\beta} \mathcal{H} \to 0$

is called <u>exact</u> if for each $x \in X$ the sequence

$$0 \to \mathcal{F}_x \xrightarrow{\alpha_x} \mathcal{G}_x \xrightarrow{\beta_x} \mathcal{H}_x \to 0$$

is an exact sequence. In particular if $\mathcal{F}$ is a subsheaf of $\mathcal{G}$ the sequence

$$0 \to \mathcal{F} \to \mathcal{G} \to \mathcal{G}/\mathcal{F} \to 0 \text{ is exact.}$$

Analogously one defines the notion of long exact sequence of sheaves.

Given a sheaf homomorphism $\phi : \mathcal{F} \to \mathcal{G}$, this induces a homomorphism of cohomology groups

$$\phi^* : H^q(X, \mathcal{F}) \to H^q(X, \mathcal{G}).$$

One has the following theorem of Serre:

<u>Given an exact sequence</u> (1) <u>of sheaves over a paracompact space</u> X <u>then one has an exact cohomology sequence</u>

$$0 \to H^0(X, \mathcal{F}) \xrightarrow{\alpha^*} H^0(X, \mathcal{G}) \xrightarrow{\beta^*} H^0(X, \mathcal{H}) \xrightarrow{\delta}$$

$$\to H^1(X, \mathcal{F}) \xrightarrow{\alpha^*} H^1(X, \mathcal{G}) \xrightarrow{\beta^*} H^1(X, \mathcal{H}) \xrightarrow{\delta}$$

$$\to H^2(X, \mathcal{F}) \xrightarrow{\alpha^*} \cdots$$

where δ is a "connecting homomorphism" which is defined in the usual way.

Čech cohomology is particularly useful if the topological space X admits a system of coverings $\mathcal{U}^a$ cofinal to all coverings of X such that for any

$U_{i_0 \dots i_q}$ $H^r(U_{i_0} \dots {}_{i_q}, \mathcal{F}) = 0$ for al $r > 0$.

One has in fact the following important theorem of Leray:

Let $\mathcal{F}$ be a sheaf of abelian groups on the paracompact space X. Let $\mathcal{U}$ be an open covering of X such that

$$H^r(U_{i_0 \dots i_q}, \mathcal{F}) = 0 \ \text{for all} \ i_0, \dots, i_q, \ \text{all} \ q \ \text{and all} \ r > 0.$$

Then the natural homomorphism

$$H^q(\mathcal{U}, \mathcal{F}) \to H^q(X, \mathcal{F}) \ \text{is an isomorphism for all} \ q \geq 0.$$

Example.

Let X be a complex manifold and $\mathcal{U} = \{U_i\}$ be an open covering of X by open set of holomorphy. Then any $U_{i_0 \dots i_q}$ is also an open set of holomorphy. A theorem of H. Cartan and J. P. Serre ("Theorem B") statesthat for any coherent sheaf $\mathcal{F}$ on an open set U of holomorphy (or more general on a Stein space) one has $H^r(U, \mathcal{F}) = 0$ if $r > 0$. Thus coverings by open set of holomorphy are Leray-coverings.

c) Acyclic resolutions. Let $\mathcal{F}$ be a sheaf of abelian groups on the paracompact space X. By a resolution of $\mathcal{F}$ we mean an exact sequence of sheaves

$$(2) \quad 0 \to \mathcal{F} \to \mathcal{F}_0 \xrightarrow{d_0} \mathcal{F}_1 \xrightarrow{d_1} \mathcal{F}_2 \xrightarrow{d_2} \dots$$

the resolution is called acyclic if

$$H^q(X, \mathcal{F}_i) = 0 \ \text{for all} \ q > 0 \ \text{and all} \ i \geq 0.$$

De Rham theorem. **If** (2) **is an acyclic resolution of** $\mathcal{F}$ **then Cech cohomology of** X **with values in** $\mathcal{F}$ **is naturally isomorphic to the coholomogy of the complex**

$$\Gamma(X,\mathcal{F}_0)\overset{d_0^*}{\to}\Gamma(X,\mathcal{F}_1)\overset{d_1^*}{\to}\Gamma(X,\mathcal{F}_2)\overset{d_2^*}{\to}\dots$$

i.e.

$$H^q(X,\mathcal{F})\simeq\frac{\operatorname{Ker}\{\Gamma(X,\mathcal{F}_q)\overset{d_q^*}{\to}\Gamma(X,\mathcal{F}_{q+1})\}}{\operatorname{Im}\{\Gamma(X,\mathcal{F}_{q-1})\overset{d_{q-1}^*}{\to}\Gamma(X,\mathcal{F}_q)\}}$$

Examples.

1. A sheaf of abelian groups $\mathcal{G}$ is called **flabby** if for every open set $U\subset X$ the restriction map $\Gamma(X,\mathcal{G})\to\Gamma(U,\ \ )$ is surjective. For a flabby sheaf $\mathcal{G}$ one has that $H^q(X,\mathcal{G})=0$ for all $q>0$. Every sheaf $\mathcal{F}$ of abelian groups can be considered as a subsheaf of the flabby sheaf $\mathcal{C}^0$ = sheaf of germs of arbitrary (not necessarily continuous) sections of $\mathcal{F}$. Making use of this fact one realizes that any sheaf $\mathcal{F}$ admits a flabby (thus acyclic) resolution

$$0\to\mathcal{F}\to\mathcal{C}^0\to\mathcal{C}^1\to\mathcal{C}^2\to\dots$$

2. A sheaf $\mathcal{G}$ is called **soft** if for every closed subset $A\subset X$ the restriction map $\Gamma(X,\mathcal{G})\to\Gamma(A,\mathcal{G})$ is surjective. For a soft sheaf $\mathcal{G}$ one has again $H^q(X,\mathcal{G})=0$ for all $q>0$. Therefore soft resolutions of a sheaf $\mathcal{F}$ are acyclic.

3. Let X be a differentiable manifold of dimension n. Let $\mathcal{A}^r$ denote the sheaf of germs of real-valued differentiable C^∞ forms of type r $(r=0,1,\dots)$ and let $d:\mathcal{A}^r\to\mathcal{A}^{r+1}$ be the sheaf homomorphism induced by exterior differentiation. The kernel of the homomorphism $d:\mathcal{A}^0\to\mathcal{A}^1$ is the sheaf of germs of constant functions $\mathbb{R}$ Since for each r the sheaf $\mathcal{A}^r$ is a soft sheaf we obtain a sequence

$$0\to\mathbb{R}\overset{\varepsilon}{\to}\mathcal{A}^0\overset{d}{\to}\mathcal{A}^1\overset{d}{\to}\mathcal{A}^2\overset{d}{\to}\dots\overset{d}{\to}\mathcal{A}^n\to 0.$$

One has $d \cdot d = 0$. Moreover the sequence is an exadt sequence (Poincaré lemma). We have thus obtained a soft resolution of the sheaf $\mathbb{R}$ The De Rham theorem tells us that the cohomology group $H^*(X, \mathbb{R})$ can be computed as the cohomology of the complex $\{\mathcal{A}^*(X), d\}$ of global differential forms.

4. Let X be a a complex manifold of complex dimension n. Let $\mathcal{A}^{r,s}$ denote the sheaf of germs of complex-valued C^∞ differentiable forms of type (r, s) i.e. of those forms ϕ that can be written in a system of local holomorphic coordinates as

$$\phi = \sum_{\substack{\alpha_0 < \dots < \alpha_r \\ \beta_0 < \dots < \beta_s}} a_{\alpha_0, \dots, \alpha_r \beta_0, \dots, \beta_s}(z)$$

$$dz_{\alpha_0} \wedge \dots \wedge dz_{\alpha_r} \quad d\bar{z}_{\beta_0} \wedge \dots \wedge d\bar{z}_{\beta_s} \quad .$$

Let $\bar{\partial}$ be the operator induced by exterior differentiation with respect to antiholomorphic coordinates. We obtain in this way a sequence

$$0 \to \Omega^r \xrightarrow{\varepsilon} \mathcal{A}^{r,0} \xrightarrow{\bar{\partial}} \mathcal{A}^{r,1} \xrightarrow{\bar{\partial}} \dots \xrightarrow{\bar{\partial}} \mathcal{A}^{r,n} \to 0$$

in which $\Omega^r = \operatorname{Ker}\{\mathcal{A}^{r,0} \to \mathcal{A}^{r,1}\}$ is the sheaf of germs of holomorphic forms of degree r. The sequence is a complex as $\bar{\partial}\bar{\partial} = 0$ and exact (Dolbeault lemma). We obtain in this way for any $r \geq 0$ a soft resolution of the sheaf Ω^r.
In particular for $r = 0$ $\Omega^0 = \mathcal{O}$ and we get a soft resolution of the structure sheaf $\mathcal{O}$ of X. We have thus

$$H^k(X, \mathcal{O}) = \frac{\operatorname{Ker}\{\Gamma(X, \mathcal{A}^{0,k}) \xrightarrow{\bar{\partial}} \Gamma(X, \mathcal{A}^{0,k+1})\}}{\operatorname{Im}\{\Gamma(X, \mathcal{A}^{0,k-1}) \xrightarrow{\bar{\partial}} \Gamma(X, \mathcal{A}^{0,k})\}} \quad \text{for any } k \geq 0.$$

References: [25], [30].

6.2 E.E. Levi problem.

a) Let X be an n-dimensional complex manifold. To clarify the exposition we will restrict our attention to open relatively compact subsets D of X with a smooth boundary. We can thus assume that D is defined as follows. We have given a C^∞ function $\phi : X \to \mathbb{R}$ with $d\phi \neq 0$ on ∂D and such that $D = \{x \in X \mid \phi(x) < 0\}$ (1).

As for the case $X = \mathbb{C}^n$ we define D to be an open set of holomorphy if for every $z_o \in \partial D$ there exists an $f \in \mathcal{H}(D)$ which is not holomorphically extendable over z_o. In chapter I we have also proved that if D is an open set of holomorphy, then (*) the Levi-form at each point of ∂D restricted to the analytic tangent plane to ∂D has no negative eigenvalues (theorem (1.4.2)). The Levi problem consists in asking if an open set D with a smooth boundary and relatively compact in X which verifies the condition (*) is a domain of holomorphy. When no additional conditions are put on the nature of the manifold X the answer to this question is negative as the following example of Grauert shows.

Example. Let $n \geq 2$; consider the real torus

$$X = \mathbb{R}^{2n} / \mathbb{Z}^{2n}$$

and let $\tau : \mathbb{R}^{2n} \to X$ be the natural projection. Let

$$Q = \{x = (x_1, \dots, x_{2n}) \in \mathbb{R}^{2n} \mid |x_1| \leq \tfrac{1}{2} \text{ for } 1 \leq i \leq 2n\}$$

be the unit cube in $\mathbb{R}^{2n}$, so that $\tau(Q) = X$.

Consider the following set

$$Q = \{x \in Q \mid |x_1| < \varepsilon\} \quad \text{where } 0 < \varepsilon < \tfrac{1}{2} ,$$

---

(1) It is easy to verify that if D has a smooth boundary ∂D this can be also defined by a global equation.

and let $D = \tau(Q_\varepsilon)$. Then D is an open subset of X, relatively compact (as X is compact) with a smooth boundary given by $\tau(\{x_1 = \pm \varepsilon\})$.

Consider now an $\mathbb{R}$-linear surjective map $\lambda : \mathbb{R}^{2n} \to \mathbb{C}^n$ so that $\mathbb{R}^{2n}$ inherits through λ the complex structure of $\mathbb{C}^n$. Since $\mathbb{Z}^{2n}$ is a group of translations, these are holomorphic for every λ we consider. Therefore X, and thus D, is provided with a complex structure. Moreover D in any of these structures is Levi-flat i.e. $L(\phi)_{z_0}\Big|_{T_{z_0}(\partial D)}$ for every $z_0 \in \partial D$ is identically zero. In fact this is a local property since τ is a local isomorphism it is enough to verify this condition on $\tau^{-1}(\partial D) = \{x_1 = \pm \varepsilon\}$. But now we can take for ϕ the function $\phi = x_1 \mp \varepsilon$ which has identically zero Levi-form.

The set $Y = \tau(\{x_1 = 0\})$ is a real compact submanifold of D of real dimension $2n-1$. Let λ^{-1} be given by

$$x = (A, \bar{A}) \begin{pmatrix} z \\ \bar{z} \end{pmatrix} \qquad z \in \mathbb{C}^n,\ A = (a_{ij})_{\substack{1 \le i \le 2n \\ 1 \le j \le n}} .$$

In particular $x_1 = 2 \operatorname{Re} \ a_{11}z_1 + \dots + a_{1n}z_n$, therefore $\{x_1 = 0\}$ is the union of a cone - parameter family of complex linear subspaces

$$\mathbb{C}^{n-1}_t = \{a_{11}z_1 + \dots + a_{1n}z_n = it\} \quad , \quad t \in \mathbb{R}.$$

Let $c = {}^t(0, c_2, \dots, c_{2n})$ be a real vector in $\{x_1 = 0\}$ such that the numbers $c_2, \dots, c_{2n}$ are linearly independent over the rationals. Since the matrix A has to satisfy the only condition $\det (A, \bar{A}) \neq 0$, we can choose A so that

$c \in \{a_{11}z_1 + \dots + a_{1n}z_n = 0\}$. Then for every $t \in \mathbb{R}$ $\tau(\mathbb{C}^{n-1}_t)$ must be everywhere dense in Y. Moreover

$Y = \bigcup_{t \in \mathbb{R}} (\mathbb{C}^{n-1}_t)$.

We can now show that $\mathcal{H}(D) = \mathbb{C}$. Indeed, let $f \in \mathcal{H}(D)$ and let z_o be so chosen in Y so that $|f(z_o)| = \max_Y |f|$. There exists a $t_o \in \mathbb{R}$ such that $z_o \in \tau(\mathbb{C}^{n-1}_{t_o})$.

By the maximum principle f must be constant on $\tau(\mathbb{C}^{n-1}_{t_o})$ as $f \circ \tau$ must be constant on $\mathbb{C}^{n-1}_{t_o}$. But $\tau(\mathbb{C}^{n-1}_{t_o})$ being dense in Y, f must be constant on Y because f is continuous. The set of zeros of a non-constant holomorphic function on a complex connected manifold is of real dimension $\leq 2n-2$. Therefore f must be constant on D as Y has real dimension $= 2n-1$. Since $\mathcal{H}(D) = \mathbb{C}$ every holomorphic function is estendable at each boundary point of D and therefore D cannot be an open set of holomorphy.

b) If we wish to obtain an affirmative answer to the Levi-problem, one need to reinforce the Levi condition (*). For instance one could assume that the Levi form restricted to the analytic tangent plane is nowhere degenerate on ∂D. Then we have the following

<u>Theorem</u> (6.2.1). (H. Grauert, [26]). <u>If D satisfies condition (*) and if the Leviform restricted to the analytic tangent plane to ∂D is non-degenerate (thus is positive definite everywhere on ∂D) then D is an open set of holomorphy.</u>

When the conditions of this theorem are satisfied we say that D has a strongly Levi-comvex boundary.

The proof of this theorem will occupy the next section. We divide the proof in several steps.

6.3 <u>Proof of Grauert's theorem.</u>

<u>Step 1</u>. <u>A criterion</u> for <u>finiteness for cohomology</u>.

Let X be a complex manifold and let $\mathcal{F}$ be the sheaf of germs of holomorphic sections of some holomorphic vector bundle F.

For every coordinate patch U where $F|_U$ is trivial, we have $\Gamma(U, \mathcal{F}) \simeq \mathcal{H}(U)^r$ where r is the fiber dimension of F. Now $\mathcal{H}(U)$, and therefore $\mathcal{H}(U)^r$ has the structure of a Frechet space. We can transpose this structure on $\Gamma(U, \mathcal{F})$. Banach's open mapping theorem ensures us that this structure is independent of the trivialization chosen for F over U.

If V is open and $V \subset U$ the natural restriction map

$$r^U_V : \Gamma(U, \mathcal{F}) \to (V, \mathcal{F})$$

is continuous, and, if $V \subset\subset U$, then r^U_V is a compact map by Vitali's theorem. Similar considerations could be repeated for any coherent sheaf even if it is not locally free. Since for the sake of the proof we need only to consider the locally free case the corresponding argument is omitted.

Let $\mathcal{W} = \{W_i\}_{i \in I}$ be a countable covering of X by coordinate patches, then the Čech cochain groups

$$C^q(\mathcal{W}, \mathcal{F}) = \prod \Gamma(W_{i_0 \dots i_q}, \mathcal{F})$$

as a countable product of Fréchet spaces have the structure of a Fréchet space. Since the restriction maps are continuous, the coboundary operator δ is continuous.

Therefore in the complex

$$C^0(\mathcal{W}, \mathcal{F}) \xrightarrow{\delta} C^1(\mathcal{W}, \mathcal{F}) \xrightarrow{\delta} C^2(\mathcal{W}, \mathcal{F}) \xrightarrow{\delta} \dots$$

the spaces are Fréchet spaces and the maps are continuous i.e. it is a "topological complex of Fréchet spaces".

<u>Lemma</u> (6.3.1). <u>Let</u> A , B <u>be open sets of</u> X <u>such that</u>

i) $B \subset\subset A$

ii) $r^A_{B*} : H^q(A, \mathcal{F}) \to H^q(B, \mathcal{F})$ <u>is surjective</u>

<u>then</u> $\dim_{\mathbb{C}} H^q(B, \mathcal{F}) < \infty$.

Proof.

(α) Choose a countable covering $\mathcal{U} = \{U_i\}_{i \in IN}$ of A by open sets of holomorphy contained in coordinate patches on which F is trivial. We may assume that $B \cap U_i \neq \emptyset$ for only finitely many U_i's, say $B \cap U_i \neq \emptyset$ for $1 \leq i \leq k$.
In each of the open sets $\hat{U}_i$, $1 \quad i \quad k$, we can take an open set U_i such that

$$\hat{U}_i \subset\subset U_i, \quad B \subset \bigcup_{i=1}^{k} \hat{U}_i .$$

We now choose a countable covering $\mathcal{V} = \{V_j\}_{j \in J}$ by open sets of jolomorphy of the set B and such that $\mathcal{V}$ is a refinement of the covering $\{\hat{U}_i \cap B\}_{1 \leq i \leq k}$.
For each V_j let $\tau(j)$ be such that $V_j \subset U_{\tau(j)}$.

(β) We now consider the cochain groups

$$C^q(\mathcal{U}, \mathcal{F}) = \prod \Gamma(U_{i_0 \dots i_q}, \mathcal{F}) \qquad \text{for } A ,$$

$$C^q(\mathcal{V}, \mathcal{F}) = \prod \Gamma(V_{j_0 \dots j_q}, \mathcal{F}) \qquad \text{for } B ,$$

and the restriction map $\tau^* : C^q(\mathcal{U}, \mathcal{F}) \to C^q(\mathcal{V}, \mathcal{F})$ which is defined as follows:

$$(\tau^* f)_{j_0 \dots j_q} = f_{\tau(j_0) \dots \tau(j_q)} \mid V_{j_0 \dots j_q} \qquad \text{for } f \in C^q(\mathcal{U}, \mathcal{F}).$$

If we consider the Frechet structure on the cochain groups, then τ^* is a compact map since $V_j \subset \hat{U}_{\tau(j)} \subset\subset U_{\tau(j)}$ and since there is only a finite number of groups $\Gamma(U_{\tau(j_0) \dots \tau(j_q)}, \mathcal{F})$.

(γ) Consider the map

$$\omega : Z^q(\mathcal{U}, \mathcal{F}) \oplus C^{q-1}(\mathcal{V}, \mathcal{F}) \to Z^q(\mathcal{V}, \mathcal{F})$$

defined by

$$\omega(\alpha \oplus \beta) = \tau^* \alpha + \delta \beta ,$$

where Z denotes the space of cocycles. By the Leray theorem we have $H^q(\mathcal{U}, \mathcal{F}) = H^q(A, \mathcal{F})$ and $H^q(\mathcal{V}, \mathcal{F}) = H^q(B, \mathcal{F})$. By the assumption the map r^A_B induced by τ^* among these groups is surjective; hence ω is. Since τ^* is compact, $\delta = \omega - \tau^*$ has a closed image of finite codimension by a theorem of L. Schwartz. (1) Hence $\dim_{\mathbb{C}} H^q(B, \mathcal{F}) < \infty$.

<u>Step 2</u>. <u>Mayer-Victoris sequence.</u> <u>Let</u> X <u>be a paracompact topological space and let</u> X_1, X_2 <u>be two open subsets of</u> X <u>such that</u> $X = X_1 \quad X_2$. <u>Let</u> $\mathcal{F}$ <u>be a sheaf of abelian groups on</u> X. <u>One has the following exact sequence</u>

$$0 \to H^0(X, \mathcal{F}) \to H^0(X_1, \mathcal{F}) \oplus H^0(X_2, \mathcal{F}) \to H^0(X_1 \cap X_2, \mathcal{F}) \to$$

$$\to H^1(X, \mathcal{F}) \to H^1(X_1, \mathcal{F}) \oplus H^1(X_2, \mathcal{F}) \to H^1(X_1 \cap X_2, \mathcal{F}) \to \dots$$

<u>Proof</u>. Choose a flabby resolution of $\mathcal{F}$ on X:

$$0 \to \mathcal{F} \to \mathcal{C}^0 \to \mathcal{C}^1 \to \dots$$

One has the following exact sequence for every $q \geq 0$:

$$0 \to \Gamma(X, \mathcal{C}^q) \xrightarrow{\alpha} \Gamma(X_1, \mathcal{C}^q) \oplus \Gamma(X_2, \mathcal{C}^q) \xrightarrow{\beta} \Gamma(X_1 \cap X_2, \mathcal{C}^q) \to 0$$

where $\alpha(a) = a|_{X_1} \oplus a|_{X_2}$, $\beta(a \oplus b) = a|_{X_1 \cap X_2} - b|_{X_1 \cap X_2}$.

Taking the firect sum over all $q \geq 0$ we obtain an exact sequence of complexes. Writing down the corresponding exact cohomology sequence we get the desired result.

<u>Step 3</u>. <u>The bumps lemma.</u>

<u>Lemma</u> (6.3.2). <u>Let</u> B <u>be an open relatively compact subset of the complex manifold</u> X <u>with a smooth, strongly Levi-convex boundary.</u>

(1)

Let E, F be Fréchet spaces, let $u : E \to F$ be a continuous linear surjection and let $k : E \to F$ be a compact linear map. Then $u+k$ has a closed image of finite codimension, (cf. [28].)

<u>Then there exist arbitrarily fine finite coverings</u> $\mathcal{U} = \{U_i\}_{1 \le i \le t}$ <u>of</u> ∂B <u>and for each</u> $\mathcal{U}$ <u>a sequence of open sets</u> B_j, $0 \le j \le t$, <u>with smooth strongly Levi-complex boundaries such that</u>

i) $B = B_0 \subset B_1 \subset \dots \subset B_t$

ii) $B_0 \subset\subset B_t$

iii) $B_i - B_{i-1} \subset\subset U_i$ <u>for</u> $1 \le i \le t$

iv) $H^r(U_i \cap B_j, \mathcal{F}) = 0$ <u>for any</u> i <u>and</u> j, <u>any coherent sheaf</u> $\mathcal{F}$ <u>and for all</u> $r > 0$.

<u>Proof</u>. Without loss of generality, we may assume that on a neighborhood U of B there is a C^∞ function $\phi : U \to \mathbb{R}$, such that $B \cap U = \{x \in U \mid \phi(x) < 0\}$, $(d\phi)_x \neq 0$ for every $x \in U$, $L(\phi)_x > 0$ for every $x \in U$. The covering $\mathcal{U} = \{U_i\}_{1 \le i \le t}$ will be chosen with sufficiently small coordinate balls $U_i \subset U$ such that $U_i \cap B$ in the coordinates of U_i is elementary convex and thus a domain of holomorphy.

We now select C^∞ functions $\rho_i \ge 0$ on X, $1 \le i \le t$, with $\operatorname{supp} \rho_i \subset\subset U_i$ and such that $\Sigma \rho_i(x) > 0$ for all $x \in \partial B$.

Choose $\varepsilon_1, \dots, \varepsilon_t$ successively with $\varepsilon_i > 0$ and sufficiently small so that the functions

$$\phi_1 = \phi - \varepsilon_1 \rho_1, \quad \phi_2 = \phi - \varepsilon_1 \rho_1 - \varepsilon_2 \rho_2, \dots, \phi_t = \phi - \varepsilon_1 \rho_1 - \dots - \varepsilon_t \rho_t$$

have all their Levi forms $L(\phi_i)_x > 0$ for all $x \in U$.

Define

$$B_i = B \cup \{x \in U \mid \phi_i(x) < 0\}, \quad \text{for } 1 \le i \le t,$$

$$B_0 = B.$$

If the ε_i are chosen sufficiently small then

∂B_i will be smooth for $1 \le i \le t$,

$B_i \cap U_j$ will be convex in the coordinates of U_j and thus an open set of holomorphy.

We then have

(i) $B = B_0 \subset B_1 \subset \dots \subset B_t$ since $\phi = \phi_0 \geq \phi_1 \geq \dots \geq \phi_t$.

(ii) $B_0 \subset\subset B_t$ since $\phi_0 > \phi_t$ on ∂B

(iii) $B_i - B_{i-1} \subset\subset U_i$ since $\phi_i - \phi_{i-1} = \varepsilon_i \rho_i$ has support $\subset\subset U_i$

(iv) $H^r(B_i \cap U_j, \mathcal{F}) = 0$ for $r > 0$, all i,j and all coherent sheaves $\mathcal{F}$ since $B_i \cap U_j$ is an open set of holomorphy, by the theorem of H. Cartan and J.P. Serre.

Step 4.

Lemma (6.3.3). Let B a relatively compact open subset of the complex manifold X with smooth strongly Levi-convex boundary. There exists an open set A in X such that

(i) $B \subset\subset A$

(ii) for any coherent sheaf $\mathcal{F}$ on X and any $r > 0$, the restriction map $H^r(A, \mathcal{F})$ $H^r(B, \mathcal{F})$ is surjective.

Proof. Applying the previous lemma, taking $A = B_t$, it suffices to show that for $r > 0$ the restriction maps

$$H^r(B_i, \mathcal{F}) \quad H^r(B_{i-1}, \mathcal{F})$$

are surjective for each k, $1 \leq i \leq t$.

Write $B_i = B_{i-1} \cup (U_i \cap B_i)$ and apply the Mayer-Vietoris sequence:

$$H^r(B_i, \mathcal{F}) \quad H^r(B_{i-1}, \mathcal{F}) \oplus H^r(U_i \cap B_i, \mathcal{F}) \to H^r(B_{i-1} \cap U_i, \mathcal{F})$$

Since $H^r(U_i \cap B_i, \mathcal{F}) = 0 = H^r(B_{i-1} \cap U_i, \mathcal{F})$, the conclusion follows from the exactness of the sequence.

As a corollary of the above considerations we get the following

Theorem (6.3.4) Let $B \subset\subset X$ be open with strongly Levi-convex boundary. Then for any locally free sheaf $\mathcal{F}$ on X we have

$$\dim_{\mathbb{C}} H^r(B, \mathcal{F}) < \infty \quad \text{for every } r > 0.$$

Note. This theorem is true for any coherent sheaf as lemma (6.3.1) can be carried over to this more general case with the same proof.

Step 5.

Solution of the Levi problem. It is enough to show that given a divergent sequence $\{x_\nu\} \subset D$ we can find $f \in \mathcal{H}(D)$ such that

$$\sup_\nu f(x_\nu) = \infty \quad \text{(condition D of (1.3.1))}.$$

Since $D \subset\subset X$ it is not restrictive to assume that the sequence x_ν converges to a boundary point $z_0 \in \partial D$.

Choose a coordinate patch U around z_0 with coordinates $z_1, \dots z_n$, zero at z_0, such that

$$D \cap U = \{z \in U \mid \phi(z) < 0\}$$

where $\phi : U \to \mathbb{R}$ is a C^∞ function with $d\phi \neq 0$ in U and $L(\phi) > 0$ in U. We can write

$$\phi(z) = \operatorname{Re} f(z) + L(\phi)_{z_0}(z) + O(\|z\|^3)$$

with $f \in \mathcal{H}(U)$.

If U is sufficiently small, then $\{f = 0\} \cap D = \emptyset$. Choose a C^∞ function ρ in U with $\rho \geq 0$, $\operatorname{supp} \rho \subset\subset U$, $\rho(0) > 0$, and set $\phi_1 = \phi - \varepsilon\rho$. Take $\varepsilon > 0$ so small that $d\phi_1 \neq 0$ on U and $L(\phi_1) > 0$ on U.

Set

$$A = D \cup \{x \in U \mid \phi_1(x) < 0\}.$$

Then A has a smooth strongly Levi-convex boundary and f is holomorphic on $A \cap U$. In particular by theorem (6.3.4) $\dim_{\mathbb{C}} H^1(A, \mathcal{O}) = r < \infty$.

Write $A = D \cup (A \cap U)$ and apply the Mayer-Vietoris sequence for the sheaf $\mathcal{O}$:

$$0 \to H^0(A, \mathcal{O}) \to H^0(D, \mathcal{O}) \oplus H^0(A \cap U, \mathcal{O}) \to H^0(D \cap U, \mathcal{O}) \xrightarrow{\delta}$$
$$\to H^1(A, \mathcal{O}) \to \dots$$

Consider the sequence of functions $\frac{1}{f^\nu}$ for $\nu = 1, 2, \ldots$ in $D \cap U$. These are holomorphic and linearly independent over $\mathbb{C}$. Since $\dim_{\mathbb{C}} H^1(A,\mathcal{O}) = r$ we can find constants $c_1, \ldots, c_{r+1} \in \mathbb{C}$ not all zero such that

$$\delta(c_1 \frac{1}{f} + \ldots + c_{r+1} \frac{1}{f^{r+1}}) = 0.$$

Therefore there must exist holomorphic functions $h_d \in \mathcal{H}(D)$ and $h_{A \cap U} \in \mathcal{H}(A \quad U)$ such that

$$\sum_1^{r+1} c_i \frac{1}{f^i} + h_{A \cap U} = h_D = m.$$

Hence m is a meromorphic function in A, holomorphic in D with principal part at z_o $\sum_1^{r+1} c_i \frac{1}{f^i}$. For the function $g = h_D$ we thus have

$$\lim_{\substack{z \in D \\ z \to z_o}} |g(x)| = \infty. \qquad \text{Q.E.D}$$

<u>Remarks.</u>

1. Let us call a manifold X <u>0-pseudoconvex</u> if there exist a C^∞ function $\phi : X \to \mathbb{R}$ and a compact set $K \subset X$ such that

   (i) for every $c \in \mathbb{R}$ the set
   $$B_c = \{x \in X \mid \phi(x) < c\}$$
   is relatively compact in X

   (ii) the Levi form $L(\phi)_x$ of ϕ at every point $x \in X - K$ is positive definite.

A 0-pseudoconvex manifold for which we can take K empty is also called 0-<u>complete</u>.

It can be shown that theorem (6.3.4) extends to 0-pseudoconvex manifolds as follows:

<u>Theorem</u> (6.3.5). <u>For a 0-pseudoconvex manifold</u> X, <u>and any coherent sheaf</u> $\mathcal{F}$ <u>on</u> X <u>one has</u>

$$\dim_{\mathbb{C}} H^r(X, \mathcal{F}) < \infty \quad \text{for all } r > 0.$$

<u>If moreover</u> X <u>is 0-complete then</u>

$$H^r(X, \mathcal{F}) = 0 \quad \text{for all } r > 0. \quad (\text{cf. } [2]).$$

The proof is obtained from theorem (6.3.4) by the Mittag-Leffler procedure making use of the fact that for $\sup_K \phi < c_1 < c_2$ the restriction map

$$H^0(B_{c_2}, \mathcal{F}) \to H^0(B_{c_2}, \mathcal{F})$$

has a dense image.

2. Let D be an open relatively compact subset of $\mathbb{C}^n$ with a smooth boundary and such that there exists a C^∞ function $\phi : \mathbb{C}^n \to \mathbb{R}$ with the properties

$$D = \{z \in \mathbb{C}^n \mid \phi(z) < 0\},$$

$d\phi \neq 0$ on ∂D, $L(\phi) \geq 0$ on some neighborhood of ∂D. Then D is an open set of holomorphy.

Indeed, replacing ϕ by an increasing convex function of ϕ we may assume that there exists on D a function $\phi : D \to \mathbb{R}$ with $\{\phi < \text{constant}\} \subset\subset D$ and with $L(\phi) \geq 0$ outside a compact set. Then if c is a large positive constant the function $\psi = \phi + c \sum_1^n |z_i|^2$ has the property $\{\psi < \text{constant}\} \subset\subset D$, $L(\psi) > 0$ on D. It follows then that D is an increasing union of open sets of holomorphy and thus a set of holomorphy.

One could also apply the previous theorem of vanishing for cohomology to derive the fact that D is an open set of holomorphy.

3. Let $D = \{x \in X \mid \phi(x) < 0\}$ be a strongly Levi-convex relatively compact subset of the complex manifold X. We have proved that for any coherent sheaf $H^1(D, \mathcal{F})$ is finite-dimensional. If $\varepsilon > 0$ is sufficiently small then $D_\varepsilon = \{x \in X \mid \phi(x) < -\varepsilon\}$ has the same property. Moreover one can show that the restriction map

$$H^1(D, \mathcal{F}) \to H^1(D_\varepsilon, \mathcal{F})$$

is an isomorphism, (2). It follows in particular that $\mathcal{H}(D) = H^0(D, \mathcal{O})$ separates points on $D - D_\varepsilon$.

In fact let us consider the sheaf $\mathcal{F}$ of germs of holomorphic functions vanishing at the points $x, y \in D - D_\varepsilon$. $\mathcal{F}$ is a coherent sheaf and we have the exact sequence

$$0 \to \mathcal{F} \to \mathcal{O} \to \mathbb{C}_x \oplus \mathbb{C}_y \to 0.$$

From this we deduce the commutative diagram with exact rows:

$$\begin{array}{ccccc} H^0(D,\Theta) & \xrightarrow{\alpha} & \mathbb{C}_x \oplus \mathbb{C}_y & \xrightarrow{\beta} & H^1(D,\mathcal{F}) \\ \downarrow & & \downarrow & & \downarrow \gamma\ \wr \\ H^0(D,\Theta) & \rightarrow & 0 & \rightarrow & H^1(D,;\mathcal{F}) \end{array}$$

Since γ is an isomorphism, $\operatorname{Im}\beta = 0$ thus α is surjective.

4. Let D be as in 3. Consider the analytic set

$$C = \{(x, y) \in D \times D \mid f(x) = f(y) \text{ for every } f \in \mathcal{H}(D)\}.$$

Clearly $C \supset$ the diagonal Δ of $D \times D$. Consider the closure $S = \overline{C-\Delta}$ of $C-\Delta$. This is an analytic set. Therefore $S \cap \Delta$ is an analytic set. Any compact irreducible analytic subset of Δ of dimension ≥ 1 must be contained in $S \cap \Delta$. Moreover since $\mathcal{H}(D)$ separates points in $D-D_\varepsilon$ the set $S \cap \Delta$ must be compast.

Therefore:

<u>if</u> D <u>is a strongly Levi-convex relatively compact open subset of a complex manifold</u> X <u>there exi ts in</u> D <u>a compact analytic subset</u> A <u>of dimension</u> ≥ 1 <u>at each point such that any irreducible compact analytic subset of</u> D <u>of dimension</u> ≥ 1 <u>is contained in</u> A.

This argument can be carried through for complex spaces.

6.4. Characterization of projective algebraic manifolds, Kodaira's theorem ([3?]).

As an application of the previous considerations we give here a proof of theorem (4.2.1) of Kodaira.

We have already given the proof of the implications $A \Leftrightarrow D$. We will give here the proof of the implications $A \Rightarrow B \Rightarrow C \Rightarrow D$

$A \Rightarrow B$. It is enough to prove this for $X = \mathbb{P}_n(\mathbb{C})$. Let $z_0, \dots, z_n$ be homogeneous coordinates in $\mathbb{P}_n(\mathbb{C})$ and let $U_i = z_i \neq 0$ where we assume as coordinates

$$y_1 = \frac{z_0}{z_i}, \dots, y_{i-1} = \frac{z_{i-2}}{z_i}, y_i = \frac{z_{i-1}}{z_i}, y_{i+1} = \frac{z_{i+1}}{z_i}, \dots, y_n = \frac{z_n}{z_i} .$$

Consider the hermitian form $ds^2_\lambda = \lambda\, \partial . \bar\partial \log \sum_0^n z_i \bar z_i$ where $\lambda \in \mathbb{R}$ and $\lambda > 0$. Then $ds^2_\lambda = \lambda\, \partial . \bar\partial \log (1 + \sum_{j=1}^n y_j \bar y_j)$ on U_i and by direct calculation we get

$$\frac{1}{\lambda} ds^2_\lambda = (1 + \Sigma y_j \bar y_j)^{-2} \{(1 + \Sigma y_j \bar y_j) \Sigma dy_j d\bar y_j - (\Sigma y_j dy_j)(\Sigma y_j d\bar y_j)\}$$

$$\geq (1 + \Sigma y_j \bar y_j)^{-2} \Sigma dy_j d\bar y_j .$$

Thus ds^2_λ defines a hermitian metric on $\mathbb{P}_n(\mathbb{C})$. Its exterior form is given by the 2-form

$$w_\lambda = \frac{i\lambda}{2} \partial\bar\partial \log \sum_0^n z_i \bar z_i = \frac{i\lambda}{2} \partial\bar\partial \log (1 + \sum_0^n y_j \bar y_j).$$

By the way it is constructed $\partial\omega_\lambda = \bar\partial\omega_\lambda = 0$ thus $d\omega_\lambda = 0$ and therefore the hermitian metric ds^2_λ is a Kahler metric (cf. [51]). Now $H_2(\mathbb{P}_n(\mathbb{C}), \mathbb{Z}) = \mathbb{Z} g$ where g is a projective line $\mathbb{P}_1(\mathbb{C})$ in $\mathbb{P}_n(\mathbb{C})$. Taking for g the projective line $g \equiv \{z_2 = \dots = z_n = 0\}$ we get $\omega_\lambda|_g = \dfrac{\lambda dy d\bar y}{(1+y\bar y)^2}$ where $y = \dfrac{z_1}{z_0}$.

Therefore $\int_g \omega_\lambda = \lambda \int_g \omega_1$. But $\int_g \omega_1 > 0$ thus one can choose λ to make that period an integer (1).

B => C.

(α) The assumption B can be replaced by the assumption that there exists on X a Kahler metric whose exterior form has rational periods as we can always multiply the metric by a positive integer to make the periods of the corresponding exterior form integral.

Singular homology or cohomology based on differentiable singular simplices will be denoted by a suffix "s", Čech cohomology by the suffix "v" and the cohomology of the de Rham complex of <u>differentiable forms</u> by the suffix "dR".

(1) We can take $\lambda = \frac{1}{\pi}$ as $\int_g \omega_1 = \pi$.

We then have the following commutative diagram of isomorphisms and inclusions:

$$\begin{array}{ccc}
\mathrm{Hom}_{\mathbb{Q}}({}_sH_2(X,\mathbb{Z})\otimes\mathbb{Q},\mathbb{Q}) & \overset{i}{\hookrightarrow} & \mathrm{Hom}_{\mathbb{C}}({}_2H_2(X,\mathbb{Z})\otimes\mathbb{C},\mathbb{C}) \\
\beta\uparrow s & & \beta'\uparrow s \\
{}_sH^2(X,\mathbb{Q}) & \overset{j}{\hookrightarrow} & {}_sH^2(X,\mathbb{C}) \\
\alpha\uparrow s & & \alpha\uparrow s \\
{}_vH^2(X,\mathbb{Q}) & \overset{k}{\hookrightarrow} & {}_vH^2(X,\mathbb{C}) \\
 & & \gamma\uparrow s \\
 & & {}_{dR}H^2(X,\mathbb{C})
\end{array}$$

If $_1, _2, \dots$ is a basis of ${}_sH_2(X,\mathbb{Z})$ modulo torsion, a cohomology class of ${}_sH^2(X,\mathbb{C})$ is caracterized, via β', by its values on $\gamma_1, \gamma_2, \dots$ and it is in the image of j if its values on $\gamma_1, \gamma_2, \dots$ are rational numbers. A direct calculation shows that the isomorphism $\beta'\circ\alpha'\circ\gamma$ is induced by the map which associates to a differential 2-form ϕ the singular cochain

$$\sigma \to \int_\sigma \phi .$$

It follows that if a closed 2-form ω has rational periods i.e.

$$\int_{\gamma_i} \omega \in \mathbb{Q}$$

then it is represented in Čech cohomology by a 2-cocycle $\{c_{ijk}\}$ with $c_{ijk} \in \mathbb{Q}$ (modulo coboundaries with values in $\mathbb{C}$).
In this argument $\mathbb{C}$ can be replaced by $\mathbb{R}$ everywhere.

(β) Define the sheaf $\mathcal{H}$ of germs of complex-valued pluriharmonic functions on X by the exact sequence of sheaves

$$0 \to \mathbb{C} \overset{\alpha}{\to} \mathcal{O} \oplus \overline{\mathcal{O}} \overset{\beta}{\to} \mathcal{H} \to 0$$

where $\alpha(c) = c \oplus c$ and $\beta(f \oplus \bar{g}) = f - \bar{g}$.
Let $\mathcal{A}^{r,s}$ denote the sheaf of germs of C^∞ complex-valued forms of type (r, s). Set for U open in X $\Gamma(U,\mathcal{A}^{r,s}) = \mathcal{A}^{r,s}(U)$. we have the following

Lemma (6.4.1) **Let** U **be a contractible domain of holomorphy in** $\mathbb{C}^n$. **One has the exact sequence**

$$0 \to \Gamma(U,\mathcal{H}) \to \mathcal{A}^{0,0}(U) \overset{\partial\bar\partial}{\to} \mathcal{A}^{1,1}(U) \overset{d}{\to} \mathcal{A}^{1,2}(U) \oplus \mathcal{A}^{2,1}(U),$$

where $d = \partial + \bar\partial$ **is the exterior differentiation.**

Proof. Obviously the composition of any two consecutive maps in the sequence is zero. Since U is a domain of holomorphy the resolution

$$0 \to \mathbb{C} \to \Omega^0 \overset{d}{\to} \Omega^1 \overset{d}{\to} \Omega^2 \overset{d}{\to} \dots$$

is acyclic, Ω^i denoting the sheaf of germs of holomorphic i-forms. Since U is contractible the complex of sections on U

$$0 \to \mathbb{C} \to \Omega^0(U) \overset{d}{\to} \Omega^1(U) \overset{d}{\to} \Omega^2(U) \overset{d}{\to} \dots$$

is exact.

Let $f^{0,0} \in \mathcal{A}^{0,0}(U)$ with $\partial\bar\partial f^{0,0} = 0$. Then $\bar\partial f^{0,0} = f^{0,1}$ satisfies $\partial f^{0,1} = 0$ i.e. $f^{0,1}$ is antiholomorphic. Moreover $df^{0,1} = 0$ thus $f^{0,1} = dg^{0,0}$ with $g^{0,0} \in \bar\Omega(U)$ i.e. antiholomorphic in U. Hence $\bar\partial f^{0,0} = \bar\partial g^{0,0}$, thus $\bar\partial(f^{0,0} - g^{0,0}) = 0$ i.e. $f^{0,0} - g^{0,0} = h^{0,0}$ is holomorphic in U. We have then $f^{0,0} = g^{0,0} + h^{0,0}$, a sum of an antiholomorphic and a holomorphic function on U. Hence $f^{0,0} \in \Gamma(U,\mathcal{H})$. This proves exactness at $\mathcal{A}^{0,0}(U)$.

Let $g^{1,1} \in \mathcal{A}^{1,1}(U)$ with $dg^{1,1} = 0$ i.e. $\partial g^{1,1} = 0 = \bar\partial g^{1,1}$. Now $\bar\partial g^{1,1} = 0$, thus there exists a (1,0) form $h^{1,0}$ on U such that $g^{1,1} = \bar\partial h^{1,0}$, because U is a domain of holomorphy. Now $\partial\bar\partial h^{1,0} = 0$ thus $\partial h^{1,0} = h^{2,0}$ is holomorphic and closed. Hence $h^{2,0} = dk^{1,0}$ with $k^{1,0}$ holomorphic in U. Thus $\partial h^{1,0} = \partial k^{1,0}$ and $\partial(h^{1,0} - k^{1,0}) = 0$. Consequently $h^{1,0} - k^{1,0} = \partial k^{0,0}$ for some C^∞ function on U. Now $g^{1,1} = \bar\partial h^{1,0} = \bar\partial(h^{1,0} - k^{1,0}) = \bar\partial\partial k^{0,0}$. This shows exactness at $\mathcal{A}^{1,1}(U)$.

(γ) Let $\omega = i \sum g_{\alpha\bar\beta}\, dz_\alpha \wedge d\bar z_\beta$ be the exterior form of the Kahler metric. Let $\mathcal{U} = \{U_i\}$ be an open covering of X by coordinate balls. Since $d\omega = 0$ by the previous lemma we can write

$$\omega|_{U_i} = \partial\bar\partial W_i \ .$$

Since $(g_{\alpha\bar\beta})$ is hermitian, ω is real, so that $\omega|_U = \bar\partial\,\partial\bar W_i$, and therefore $\omega|_{U_i} = \frac{1}{2}\,\partial\bar\partial(W_i - \bar W_i)$. We can thus assume W_i to be purely imaginary.

Then on $U_i \cap U_j$

$$W_i - W_j = 2i \quad \operatorname{Im} p_{ij} = p_{ij} - \bar p_{ij} ,$$

where $p_{ij} : U_i \cap U_j \to \mathbb{C}$ is holomorphic and determined up to addition by a real constant.

On $U_i \cap U_j \cap U_k$ we must have

$$p_{ij} + p_{jk} + p_{ki} = c_{ijk} \in \mathbb{R} .$$

Moreover $c_{ijk} - c_{ljk} + c_{lik} - c_{lij} = 0$ on $U_i \cap U_j \cap U_k \cap U_l$. Hence $\{c_{ijk}\}$ represents an element of $_vH^2(\mathcal{U}, \mathbb{R})$.

Now the cocycle $\{c_{ijk}\}$ is the Čech representative of the de Rham cohomology class represented by ω (up to a sign). Indeed we have $\omega|_{U_i} = d(\bar\partial W_i)$,

$$\bar\partial W_i - \bar\partial W_j = -\bar\partial\bar p_{ij} = -d\bar p_{ij}, \quad -(\bar p_{ij} + \bar p_{jk} + \bar p_{ki}) = - c_{ijk},$$

which is the string of homomorphisms that make explicit the isomorphism of the de Rham theorem:

By the assumption that ω has integral periods, by (α) we derive that we can assume that the real numbers c_{ijk} are integers:

(1) $p_{ij} + p_{jk} + p_{ki} = c_{ijk} \in \mathbb{Z}$

(δ) Let $g_{ij} = e^{2\pi i p_{ij}}$ so that $g_{ij} : U_i \cap U_j \to \mathbb{C}^*$ is holomorphic. Because of (1) we get

$$g_{ij}\, g_{jk} = g_{ik} \quad \text{on} \quad U_i \cap U_j \cap U_k \, .$$

Therefore the collection $\{g_{ij}\}$ is a set of transition functions for a holomorphic line bundle $F \xrightarrow{\pi} X$.

Now

$$|g_{ij}|^2 = e^{-2\pi i W_j}\, e^{2\pi i W_i} = k_j k_i^{-1} , \text{ where } k_i = e^{-2\pi i W_i} : U_i \to \mathbb{R}^+ .$$

Let $V_i = \pi^{-1}(U_i) \simeq U_i \times \mathbb{C}$ and let z be the base coordinate and ξ_i the fiber coordinate. One verifies that the function

$$\chi(z, \xi) = k_i(z) |\xi_i|^2$$

is a well-defined function on F as on $V_i \cap V_j$ we have

$$k_i(z) |\xi_i|^2 = k_j(z) |\xi_j|^2 \, .$$

Indeed $\chi(z, \xi)$ defined a hermitian metric on the fibers of F. Consider the tube of radius r around the 0-section of F:

$$B_r = \{ (z, \xi) \in F \mid \chi(z, \xi) < r \} .$$

For every $r > 0$ B_r is relatively compact as the base space X is compact. Moreover the Levi-form of $\log \chi(z, \xi)$ restricted to the analytic tangent space to ∂B_r reduces to

$$\partial . \bar\partial \log \chi(z, \xi) = \partial . \bar\partial \log k_i = 2\pi \sum g_{\alpha\bar\beta}\, dz_\alpha\, d\bar z_\beta ,$$

as the analytic tangent plane to ∂B_r is given by

$$\frac{\partial \log k_i}{\partial z_\alpha} dz_\alpha + \frac{1}{\xi_i} d\xi_i = 0 , \quad \text{we can take } dz_1, \ldots, dz_n \text{ as}$$

coordinates along that plane. Hence the Levi-form restricted to the analytic tangent plane to B_r is strictly positive definite: This shows that the complex space F is 0-pseudoconvex. If we consider instead of the bundle F the bundle F^{-1} which is given by transition functions $\{g_{ij}^{-1}\}$, then by a simple verification we see that the space F^{-1} is 0-pseudoconcave.

C ⟹ D.

Let $\mathcal{J}$ be a sheaf of ideals on X, for instance the sheaf of germs of holomorphic functions vanishing on some finite set $S \subset X$. Consider on F the sheaf $\mathcal{F} = \pi^{*}\mathcal{J} \otimes_{\mathcal{O}_X} \mathcal{O}_F$. In the previous instance, this is the sheaf of germs of holomorphic functions on F vanishing on the finite set of fibers $\pi^{-1}(S)$. Every $f \in \Gamma(V_i, \mathcal{F})$ has a power series expansion

$$f = \sum_{\alpha=0}^{\infty} f_\alpha(z)\, \xi_i^\alpha .$$

Set

$$\Gamma_k(V_i, \mathcal{F}) = \{ f \in \Gamma(V_i, \mathcal{F}) \mid f_\alpha \equiv 0 \quad \text{for } \alpha < k \}$$

so

$$\Gamma_0(V_i, \mathcal{F}) = \Gamma(V_i, \mathcal{F}) \supset \Gamma_1(V_i, \mathcal{F}) \supset \Gamma_2(V_i, \mathcal{F}) \supset \dots$$

This gives a filtration of $\Gamma(V_i, \mathcal{F})$ which is independent of the choice of the fiber coordinate ξ_i.
Similarly one filters the groups $\Gamma(V_{i_0 \dots i_q}, \mathcal{F})$, therefore we get a filtration of the Čech complex $C^*(\mathcal{V}, \mathcal{F})$:

$$C_0^*(\mathcal{V}, \mathcal{F}) = C^*(\mathcal{V}, \mathcal{F}) \supset C_1^*(\mathcal{V}, \mathcal{F}) \supset C_2^*(\mathcal{V}, \mathcal{F}) \supset \dots ,$$

Which is compatible with the coboundary operator.

If $\xi \in C^q(\mathcal{V}, \mathcal{F})$ and $\xi = \{ \sum_{\alpha=0}^{\infty} f^{(\alpha)}_{i_0 \dots i_q}(z) \quad \xi_{i_q}^\alpha \}$

one verifies that, for given α,

$$\{ f^{(\alpha)}_{i_0 \dots i_q} \} \in C^q(\mathcal{U}, \mathcal{J} \otimes \underset{\sim}{F}^{-\alpha})$$

where $\underset{\sim}{F}^{-}$ denotes the sheaf of germs of holomorphic sections of the bundle $F^{-\alpha}$. Therefore we get a (split) exact sequence

(2) $0 \to C^q_{k+1}(\mathcal{V}, \mathcal{F}) \to C^q_k(\mathcal{V}, \mathcal{F}) \to C^q(\mathcal{U}, \mathcal{J} \otimes \underset{\sim}{F}^{-k}) \to 0 .$

Since the filtration of $C^*(\mathcal{V},\mathcal{F})$ is compactible with the coboundary operator, we obtain a filtration of cohomology

$$H^q_0(\mathcal{V},\mathcal{F}) = H^q(\mathcal{V},\mathcal{F}) \supset H^q_1(\mathcal{V},\mathcal{F}) \supset H^q_2(\mathcal{V},\mathcal{F}) \supset \dots$$

while from (2) we get

$$H^q_k(\mathcal{V},\mathcal{F}) / H^q_{k+1}(\mathcal{V},\mathcal{F}) \simeq H^q(\mathcal{U}, \mathcal{G} \otimes \underset{w}{F}^{-k}).$$

Therefore

$$(3) \quad \mathop{\mathcal{L}}_{k=0} H^q_k(\mathcal{V},\mathcal{F}) / H^q_{k+1}(\mathcal{V},\mathcal{F}) \simeq \mathop{\widehat{\bigcup}}_{k=0} H^q(\mathcal{U}, \mathcal{G} \otimes \underset{w}{F}^{-k}) .$$

From theorem (6.3.5), if $q > 0$, $H^q(\mathcal{V},\mathcal{F}) = H^q(F,\mathcal{F})$ is finite-dimensional as $\mathcal{V}$ is a Leray covering of F. Thus if $q > 0$ in (3) the left hand side is finite-dimensional. The same must be true for the right hand side of (3), therefore

<u>for</u> $q > 0$ <u>there exist</u> $k_0 > 0$ <u>such that if</u>

$$k \geq k_0, \quad H^q(X, \mathcal{G} \otimes \underset{w}{F}^{-k}) = 0.$$

Now we apply this result to the exact sequence

$$0 \to \mathcal{G} \otimes \underset{w}{F}^{-k} \to \underset{w}{F}^{-k} \to Q \to 0$$

where $\mathcal{G}$ is the sheaf of germs of holomorphic functions vanishing at two distinct points or vanishing of second order at a given point. Writing down the corresponding cohomology sequence one realizes that the ring $\mathcal{A}(X, F^{-1}) = \bigcup_{k=0}^{\infty} \Gamma(X, F^{-k})$ separates points and gives local coordinates everywhere on X.

<u>Remark</u>. One could apply theorem (6.3.4) instead of theorem (6.3.5) replacing in the argument F by a tube B_r of radius r around the 0-section of F, thus avoiding the use of the Mittag-Leffler argument to go from the first to the second of these theorems.

The extension to complex spaces is given by Brauert in [27] .

A. Andreotti

Chapter VII. Generalizations of the Levi-problem.

7.1. d-open sets of holomorphy.

Let X be a complex manifold, let D be an open relatively compact subset of X with a smooth boundary and let d be an integer, $d \geq 0$. Let $\mathcal{F}$ be a locally free sheaf on X, for instance $\mathcal{F} = \mathcal{O}$ or $\mathcal{F} = \Omega^d$. We repeat the considerations of the previous chapter replacing the space $\mathcal{H}(D) = H^0(D, \mathcal{O})$ by the cohomology group $H^d(D, \mathcal{F})$.

Let $z_o \in \partial D$, an element $\xi \in H^d(D, \mathcal{F})$ is said to be <u>extendable over</u> z_o if there exists a neighborhood V of z_o in X and an element $\hat{\xi} \in H^d(D \cup V, \mathcal{F})$ such that $\;_D = \;$.

<u>**Remark:** if $d > 0$ the necessary and sufficient condition for an element $\xi \in H^d(D, \;)$ to be extendable over $z_o \in \partial D$ is that there exists a neighborhood W of z_o in X such that</u> $\xi|_{W \cap D} = 0$.

In fact, if ξ is extendable over z_o, we can find a neighborhood W of z_o in X which is an open set of holomorphy with $W \subset V$, Then $\hat{\xi}|_W = 0$ thus $\xi|_{W \cap D} = 0$.

Conversely suppose there exists a neighborhood W of z_o such that $\xi|_{W \cap D} = 0$. We may assume that W is an open set of holomorphy. By the Mayer-Vietoris sequence we get the exact sequence :

$$H^d(D \cup W, \mathcal{F}) \to H^d(D, \mathcal{F}) \oplus H^d(W, \mathcal{F}) \to H^d(D \cap W, \mathcal{F}).$$

Since $H^d(W, \mathcal{F}) = 0$, because $d > 0$, and since the image of ξ in $H^d(D \cap W, \mathcal{F})$ is zero, it follows that there exists $\hat{\xi} \in H^d(D \cap W, \mathcal{F})$ such that $\hat{\xi}|_D = \xi$.

We say that D is a <u>d-open set of holomorphy with respect to the sheaf</u> $\mathcal{F}$ if for every $z_o \in \partial D$ we can find $\xi \in H^d(D, \mathcal{F})$ which is not extendable over z_o.

For $d = 0$ we realized (Theorem (1.4.2) of E.E. Levi) that the boundary of D has to have a particular shape. In view of that result we are lead to consider the following situation:

Let U be an open set in $\mathbb{C}^n$ and let $\phi : U \to \mathbb{R}$ be a C^∞ function on U. We set

$$\theta^- = \{x \in U \mid \phi(x) < 0\},$$

and we assume that on $S = \{x \in U \mid \phi(x) = 0\}$ $d\phi \neq 0$, so that S is a smooth hypersurface, which constitutes the boundary of θ^- in U.

Let us consider the Levi-form of ϕ restricted to the analytic tangent plane to S at every point $z_o \in S$, and let us assume that at a point $z_o \in S$ it has p positive and q negative eigenvalues $(p+q \leq n-1)$. Then one can prove the following

<u>Theorem</u> (7.1.1). <u>There exists a fundamental sequence of neighborhoods</u> U_ν <u>of</u> z_o <u>in</u> U <u>such that, for any locally free sheaf</u> $\mathcal{F}$ <u>on</u> U <u>we have</u>

$$H^s(\theta^-_\nu, \mathcal{F}) = 0 \text{ if } \begin{cases} s > n-p-1 \\ \text{or} \\ 0 < s < q. \end{cases}$$

<u>These neighborhoods can be chosen to be open sets of holomorphy.</u> (cf. [2]). The proof of this theorem is rather tedious and will be omitted. As a corollary we get the analogue of E.E. Levi's theorem :

<u>Theorem</u> (7.1.2). (<u>E.E. Levi</u>). <u>If</u> D <u>is a</u> d<u>-open set of holomorphy for a locally free sheaf</u> $\mathcal{F}$ <u>then at any point</u> $z_o \in \partial D$

$$(*)\begin{cases} \text{The number of negative eigenvalues of } L(\phi)_{z_o}|_{T_{z_o}} \text{ is } \leq d \\ \text{the number of positive eigenvalues of } L(\phi)_{z_o}|_{T_{z_o}} \text{ is } \leq n-d-1. \end{cases}$$

One can then formulate the analogue of Levi-problem for $H^d(D, \mathcal{F})$. The example of Grauert shows that condition (*) is not sufficient to ensure that it is solvable as in the example any point $z_o \in \partial D$ has a fundamental sequence of neighborhoods U_ν such that $D \cap V_\nu$ is an open set of holomorphy. We have thus to reinforce condition (*) assuming for instance that $L(\phi)_{z_o}|T_{z_o}$ for every $z_o \in \partial D$ is non-degenerate. An open set D of this sort will be called <u>strictly Levi d-convex</u>. Then one has the following

<u>Theorem</u> (7.1.3). <u>(Grauert)$_d$</u>. <u>If</u> D <u>is strictly Levi d-convex then</u> D <u>is a d-open set of holomorphy for any locally free sheaf</u> $\mathcal{F}$ <u>on</u> X.

7.2. <u>Proof of theorem</u> <u>(Grauert)$_d$</u>. (cf. [8]). The proof follows the same pattern given for $d = 0$. It is based on the following remarks.

(α) As a substitute for theorem B of H. Cartan and J.P. Serre, we need only one half of the vanishing theorem (7.1.1)

$$H^s(\mathcal{O}_\nu^-, \mathcal{F}) = 0, \text{ if } s > d,$$

since $(n-p-1) = n-(n-d-1)-1 = d$. Here $\mathcal{F}$ is any locally free sheaf (although this part of the vanishing theorem holds for any coherent sheaf $\mathcal{F}$ on X). Moreover one has to realize that this statement is stable by small deformations of the boundary, precisely :

given any sufficiently small neighborhood U_ν of $z_o \in \partial D$ in X which is an open set of holomorphy and given any C^∞ function $\rho : U \to \mathbb{R}$ with $\rho \geq 0$, $\rho(z_o) > 0$, $\operatorname{supp} \rho \subset\subset U_\nu$, we can find $\varepsilon_0 > 0$ such that if we set $\phi_1 = \phi - \varepsilon\rho$ with

$$0 < \varepsilon < \varepsilon_0, \quad U_\nu(\varepsilon) = \{x \in U_\nu | \phi_1(x) < 0\}$$

then we still have

$$H^s(\mathcal{O}_\nu^-(\varepsilon), \mathcal{F}) = 0 \quad \text{for} \quad s > d .$$

(β) This enables us to repeat the proof of the bumps lemma with condition iv) replaced by

iv) $H^r(U_i \cap B_j, \mathcal{F}) = 0$ for any i, j, any locally free sheaf $\mathcal{F}$, and any $r < d$.

Then the analogue of lemma (6.3.3) and the criterion of finiteness gives

<u>Theorem</u> (7.2.1). <u>Let</u> B <u>open relatively compact in</u> X <u>with a smooth boundary on which the Levi-form restricted to the analytic tangent space has at least</u> $n-d-1$ <u>positive eigenvalues. Then for any locally free (or coherent) sheaf</u> $\mathcal{F}$ <u>on</u> X <u>we have</u>

$$\dim_{\mathbb{C}} H^r(B, \mathcal{F}) < \infty \text{ if } r > d.$$

(γ) We now have to construct for every $z_0 \in \partial D$ a sequence of cohomology classes $\xi_\nu \in H^d(W \cap D, \mathcal{F})$ defined in a fixed region $W \cap D$ where W is a neighborhood of z_0 in X, such that

i) they are linearly independent over $\mathbb{C}$

ii) every non-tirvial linear combination of them is not extendable over z_0.

We postpone the proof of this fact to the end. We remark that to achieve this purpose we will make use of the assumptions that $\mathcal{F}$ is locally free and that the Levi form on the analytic tangent plane at the boundary of D is non-degenerate.

(δ) Let us suppose for the moment that point (γ) has been settled. Then the proof proceeds as in the case $d = 0$.

Let A be constructed as in the solution of (Levi-problem)$_0$ making a bump on D to swallow the point $z_0 \in \partial D$,

$$A = D \cup (A \cap U).$$

We then apply Mayer-Vietoris

$$\to H^d(A, \mathcal{F}) \to H^d(D, \mathcal{F}) \oplus H^d(A \cap U, \mathcal{F}) \to H^d(D \cap U, \mathcal{F}) \xrightarrow{\delta}$$
$$\to H^{d+1}(A, \mathcal{F}) \to \cdots$$

We consider the classes $\xi \in H^d(D \cap V, \mathcal{F})$ constructed in (γ). Let $r = \dim_{\mathbb{C}} H^{d+1}(A, \mathcal{F})$ (which is finite by theorem (7.2.1)). We can find constants $c_1, \ldots, c_{r+1} \in \mathbb{C}$, not all zero, such that

$$\delta(c_1 \xi_1 + \ldots + c_{r+1} \xi_{r+1}) = 0.$$

Therefore there must exist cohomology classes $h_D \in H^d(D, \mathcal{F})$ and $h_{A\cap U} \in H^d(A \cap U, \mathcal{F})$ such that

$$h_D = \sum_{i=1}^{r+1} c_i \xi_i + h_{A\cap U} \quad \text{on } D \cap U.$$

We set $g = h_D \in H^d(D, \mathcal{F})$, then in any sufficiently small neighborhood W of z_o which is an open set of holomorphy $h_{A\cap U|W} = 0$ thus

$$h_D|_W = \sum_{i=1}^{r+1} c_i \xi_i|_W .$$

is not extendable over z_O because of property ii) of (γ).

(ε) It remains to prove statement (γ) :

Since $\mathcal{F}$ is locally free, in a small neighborhood W of z_O in X, $\mathcal{F} \simeq \mathcal{O}^p$ for some p. Thus it is enough to prove the statement for the sheaf $\mathcal{F} = \mathcal{O}$. The proof can be based on the following

<u>Lemma</u> (7.2.2). <u>Consider in</u> $\mathbb{C}^n - \{0\}$ $(n \geq 2)$ <u>the differential forms of type</u> $(0, n-1)$

$$\psi_\alpha = \frac{\Sigma(-1)^j \bar{z}_j^{\alpha_j+1} \, d\bar{z}_1^{\alpha_1+1} \wedge \ldots \wedge \widehat{d\bar{z}_j^{\alpha_j+1}} \wedge \ldots \wedge d\bar{z}_n^{\alpha_n+1}}{(\Sigma z_j^{\alpha_j+1} \bar{z}_j^{\alpha_j+1})^n}$$

<u>for</u> $\alpha \in \mathbb{N}^n$. <u>These forms are</u> $\bar{\partial}$<u>-closed therefore for any open set</u> U <u>in</u> $\mathbb{C}^n - \{0\}$ <u>they represent cohomology classes</u> $\xi_\alpha \in H^{n-1}(U, \mathcal{O})$. <u>If</u> U <u>contains the closed half sphere</u>

$$\Sigma = \{z \in \mathbb{C}^n \mid \sum_{j=1}^n |z_j|^2 = \varepsilon, \ \operatorname{Re} z_1 \geq 0\}$$

<u>Then the classes</u> $\xi_\alpha \in H^{n-1}(U, \mathcal{O})$ <u>are linearly independent</u> (cf. [5]).

Proof. The fact that the forms ψ_α are $\bar\partial$-closed can be verified by direct computation.

Let $\eta = \Sigma c_\alpha \psi_\alpha$ be a finite linear combination of those forms with coefficients $c_\alpha \in \mathbb{C}$. Let us assume that $\eta = \bar\partial \mu$ for some C^∞ form μ of type $(0, n-2)^\alpha$ on U.
If $\omega = dz_1 \wedge \ldots \wedge dz_n$, then for any holomorphic function f on $\mathbb{C}^n$ we have the generalized Cauchy formula:

$$\int_S f\, \omega \wedge \eta = \frac{(2\pi i)^n}{(n-1)!} \Sigma c_\alpha \frac{\partial^{|\alpha|} f}{\partial z^\alpha}(0) \quad \text{where} \quad S = \{\sum_{j=1}^n |z_j|^2 = \varepsilon\}.$$

Let $\sigma > 0$ be so small that the part

$S_\sigma = \{\Sigma |z_j|^2 = \varepsilon, \ \mathrm{Re}\, z_1 \geq -\sigma\}$ of S is contained in U. We have

$$\int_{S_\sigma} f\, \omega \wedge \eta = \int_{S_\sigma} f\, \omega \wedge \bar\partial \mu = \int_{S_\sigma} d(f\, \omega \wedge \mu) = \int_{\partial S_\sigma} f \omega \wedge \mu .$$

Thus

$$\frac{(2\pi i)^n}{(n-1)!} \Sigma c_\alpha \frac{\partial^{|\alpha|} f}{\partial z^\alpha}(0) = \int_{\partial S_\sigma} f\, \omega \wedge \mu + \int_{S - S_\sigma} f\, \omega \wedge \eta .$$

Let $A = \{\sum_{j=1}^n |z_j|^2 < \varepsilon + \delta\} \cap \{\mathrm{Re}\, z_1 < -\sigma/2\}$ and

$$B = \{\sum_{j=1}^n |z_j|^2 < \sigma/4\} .$$

Then A and B are Runge domains in $\mathbb{C}^n$. Moreover $\mathrm{Re}(z_1 + \sigma/2)$ is <0 on A and >0 on B. Therefore $A \cup B$ is also a Runge domain in $\mathbb{C}^n$. Let K be a compact subset of B containing the origin in the interior and let g be any holomorphic function on B. We can find a sequence $\{f_\nu\}$ of holomorphic functions in $\mathbb{C}^n$ such that

$f_\nu \to g$ uniformly on K, $f_\nu \to 0$ uniformly on $\partial S_\sigma \cup (S - S_\sigma)$

as $K \subset B$ and $\partial S_\sigma \cup (S - S_\sigma) = \overline{(S - S_\sigma)} \subset A$. Therefore for any g we must have

$\frac{(2\pi i)^n}{(n-1)!} \sum c_\alpha \frac{\partial^{|\alpha|} g}{\partial z^\alpha}(0) = 0$. Hence all c_α's are zero.

Let us now consider the case $d = n-1$. By suitable choice of local coordinates at z_o, ∂D near z_o is given by a local equation $h = 0$ and we may assume that z_o is at the origin and that

$$h = \operatorname{Re} z_1 + \sum a_{ij} z_i \bar{z}_j + O(\|z\|^3)$$

where $\sum a_{ij} z_i \bar{z}_j > 0$ and where $h > 0$ corresponds to the side of D. If $\varepsilon > 0$ is sufficiently small then

$\{\sum |z_j|^2 = \sigma\} \{h > 0\}$ for any σ with $0 < \sigma < \varepsilon$ contains the closed half sphere $\{\sum |z_j|^2 = \sigma\} \cap \{\operatorname{Re} z_1 \geq 0$. By the lemma above, in the coordinate ball W of radius ε centered at the origin the forms ψ_α represent linearly independent cohomology classes of $H^d(W, \mathcal{O})$, not extendable over z_o.

In the general case $d \leq n-1$ the local equation of ∂D near z_o can be written as $h = 0$ with

$$h = \operatorname{Re} z_1 + a_{11} |z_1|^2 + 2\operatorname{Re}\,(z_1 \sum_2^n a_j \bar{z}_j) + \sum_2^{d+1} \lambda_j |z_j|^2 -$$

$$- \sum_{d+2}^{n} \mu_j |z_j|^2 + O(\|z\|^2) \quad \text{where}$$

$\lambda_j > 0$, $\mu_j > 0$; $a_{11}|z_1|^2 + 2\operatorname{Re}\,(z_1 \sum_2^{d+1} a_j \bar{z}_j) + \sum_2^{d+1} \lambda_j |z_j|^2$

is positive definite and where the side of D corresponds to the side $h > 0$.

If W is a sufficiently small polycilindrical neighborhood of $z_o = 0$ in those coordinates, then

$$W \cap \{z_1 = \ldots = z_{d+1} = 0\} \cap \bar{D} = \{z_o\}.$$

Consider the forms of the lemma in the coordinates $z_1, \dots, z_{d+1}$, as forms defined in W. These have singularities outside of D. Moreover their classes again satisfy property ii) of (γ) as this is true for their restrictions to the space $\{z_{d+2} = \dots = z_n = 0\}$.

This achieves the proof of statement (γ).

7.3 <u>Finiteness theorems.</u> In theorem (7.2.1) we emphasized only the positive eigenvalues of the Levi form. With the same procedure one can prove the following more precise theorem, cf. [2]:

<u>Theorem</u> (7.3.1). <u>Let</u> D <u>be a relatively compact open subset of the complex manifold</u> X <u>with a smooth, strictly d-Levi-convex boundary. Then for any locally free sheaf</u> $\mathcal{F}$ <u>on</u> X <u>we have</u>

$$\dim_{\mathbb{C}} H^r(D, \mathcal{F}) < \infty \quad \text{if} \quad r \neq d.$$

The group $H^d(D, \mathcal{F})$ not considered in the theorem, is actually infinite - dimensional and moreover has a natural structure of a Fréchet space. Theorem (7.3.1) can be considered as the "intersection" of the following general theorems of finiteness.

A complex manifold X is called <u>q-pseudoconvex</u> if there exists on X a C^∞ function $\phi : X \to \mathbb{R}$ and a compact set K such that

i) the sets
$$B_c = \{x \in X \mid \phi(x) < c\}$$
are relatively compact in X for every $c \in \mathbb{R}$

ii) on X-K the Levi-form $L(\phi)$ of ϕ has at least n-q positive eigenvalues.

If we can choose $K = \emptyset$ the manifold X is called <u>q-complete.</u>

<u>Theorem</u> (7.3.2). <u>If</u> X <u>is q-pseudoconvex then for any coherent sheaf</u> $\mathcal{F}$ <u>on</u> X <u>we have</u>

$$\dim_{\mathbb{C}} H^r(X, \mathcal{F}) < \infty \quad \text{if} \quad r > q.$$

<u>Moreover if</u> X <u>is q-complete then for</u> $r > q$ $H^r(X, \mathcal{F}) = 0$.

Analogously we have

Theorem (7.3.3). <u>If X is q-pseudoconcave, for any locally free sheaf $\mathcal{F}$ on X we have</u>

$$\dim_{\mathbb{C}} H^r(X, \mathcal{F}) < \infty \quad \text{if} \quad r < n-q-1$$

Due to the maximum principle there cannot be an analogue of "q-complete" manifolds for the pseudoconcave case.
Theorem (7.3.3) is only true for locally free sheaves. Actually the real analogue of theorem (7.3.2) is that "<u>for a q-pseudoconcave manifold and any coherent sheaf $\mathcal{F}$ on X we have</u>

$$\dim_{\mathbb{C}} H^r_k(X, \mathcal{F}) < \infty \quad \underline{\text{for}} \quad r > q+1 ,$$

<u>where H^r_k denotes cohomology with compact supports.</u>"
Theorem (7.3.3) is obtained from the one quoted above by duality and thus, if stated for any coherent sheaf, the number $n = \dim_{\mathbb{C}} X$ has to be replaced by the depth of the sheaf $\mathcal{F}$.
As reference see [2], [7].

7.4. <u>Applications to projective algebraic manifolds</u> (cf. [9]).

a) We revert to the case $d = 0$ to begin with.

Let D be a complex manifold. Let $C_0^+(D)$ denote the space of positive 0-dimensional cycles i.e. the free abelian monoid generated by the points of D. We have

$$C_0^+(D) = \{\Sigma\, n_i p_i \mid n_i \in \mathbb{N}, \text{ almost all zero;}$$

$$p_i \in D\} = D \cup D^{(1)} \cup D^{(2)} \cup \ldots$$

where $D^{(k)}$ denotes the k-fold symmetric product of D. For each $k \geq 1$ $D^{(k)}$ is the quotient of the cartesian product $D \times \ldots \times D$ (k-times) by the action of the symmetric group on k letters permuting the factors of this cartesian product in all possible ways. As such $D^{(k)}$, and therefore $C_0^+(D)$, carries a natural structure of a complex space.

Moreover we have a natural map

$$\rho_0 : H^0(D, \mathcal{O}) \to H^0(C_0^+(D), \mathcal{O}) \quad \text{given by}$$

$$\rho_0(f)(\Sigma n_i p_i) = \Sigma n_i f(p_i) \quad \text{for} \quad f \in H^0(D, \mathcal{O}).$$

Note that the image of ρ_0 is a set of holomorphic functions on $C_0^+(D)$ which are additive with respect to the monoid structure of $C_0^+(D)$.

<u>Proposition</u> (7.4.1). <u>The space $C_0^+(D)$ is holomorphically convex (holomorphically complete) iff D is holomorphically convex (holomorphically complete).</u>

<u>Proof</u>. If $C_0^+(D)$ is holomorphically convex (complete) then D as a connected component of $C_0^+(D)$ has the same property.

If D is holomorphically convex then also $C_0^+(D)$ is. In fact let $\{c_\nu\} \subset C_0^+(D)$ be a divergent sequence. Let for every $c = \Sigma n_i p_i$ $\deg(c) = \Sigma n_i$. Then deg is a holomorphic function on $C_0^+(D)$ telling us on which symmetric product $D^{(k)}$ the considered cycles stay. If $\deg(c_\nu) \to \infty$ there is nothing to prove. If $\deg(c_\nu) \leq k$, then we can extract a divergent sequence $\{c_{\nu'}\}$ contained in a given product $D^{(l)}$ (with $l \leq k$).
Now $D^l = D \times D \times \ldots \times D$ (l-times) is holomorphically convex. Let $\pi : D^l \to D^{(l)}$ be the natural projection and let Σ_l be the symmetric group acting on D^l. For each $c_{\nu'}$ let $s_{\nu'} \in D^l$ be such that $\pi(s_{\nu'}) = c_{\nu'}$. The sequence $\{s_{\nu'}\}$ must be divergent in D^l. Therefore there exists a holomorphic function $g \in H(D^l)$ such that $\sup_\nu |g(s_{\nu'})| = \infty$. Consider the polynomial in t

$$\prod_{\gamma \in \Sigma_l} (t - g(\gamma x)) = t^{l!} + a_1(x)\, t^{l!-1} + \ldots + a_{l!}(x).$$

The coefficients $a_i(x)$ are holomorphic functions which are Σ_l-invariant thus for each i there exists a holomorphic function $f_i \in H(D^{(l)})$ with $\pi^* f_i = a_i$, $1 \leq i \leq l!$. On the sequence s_ν, at least one of these functions a_i, must have unbounded absolute value. Therefore $\sup |f_i(s_\nu')| = \infty$.

If D is Stein then also $C_0^+(D)$ is Stein. This follows at once from the fact that on a Stein manifold we can prescribe the values of a holomorphic function on any finite set.

b) Suppose now that D is an open subset of a projective algebraic manifold X of $\dim_{\mathbb{C}} X = n$. Given an integer d, $0 \leq d \leq n$, we can consider the space $C_d^+(D)$ of positive compact d-dimensional cycles of D i.e. the free abelian monoid generated by irreducible compact analytic subsets of D of dimension d

$$C_d^+(D) = \{\sum n_i A_i \mid n_i \in \mathbb{N} \text{ almost all zero, } A_i \text{ compact irreducible analytic subset of D of } \dim_{\mathbb{C}} A_i = d\}$$

It is known that $C_d^+(D)$ carries a complex structure of a weakly normal complex space.

<u>Examples</u>.

1. $D = X = \mathbb{P}_n(\mathbb{C})$, $d = n-1$.

$$C_{n-1}^+(\mathbb{P}_n(\mathbb{C})) = \mathbb{P}_n(\mathbb{C}) \cup \mathbb{P}_{\binom{n+2}{2}-1}(\mathbb{C}) \cup \mathbb{P}_{\binom{n+3}{3}-1}(\mathbb{C}) \cup \ldots$$

2. $X = \mathbb{P}_n(\mathbb{C})$, $D = \mathbb{P}_n(\mathbb{C}) - \{\text{point}\}$ then

$$C_{n-1}^+(D) = \mathbb{C}^n \cup \mathbb{C}^{\binom{n+2}{2}-1} \cup \mathbb{C}^{\binom{n+3}{3}-1} \cup \ldots$$

Let us consider the cohomology group $H^d(D, \Omega^d)$. Every cohomology class ξ on it can be represented by a $\bar{\partial}$-closed differential form $\phi^{d,d}$ of type (d, d) modulo $\bar{\partial}$ of forms of type (d, d-1). Given $c \in C_d^+(D)$ we can consider (by virtue of a theorem of Lelong) the integral

$$\xi(c) = \int_c \phi^{d,d}.$$

One verifies that this integral is independent of the choice of the representative ϕ of the class ξ.

As analogue of proposition (7.4.1) we then have the following

Theorem (7.4.2).

a) The function $\xi(c)$ for c a variable on $C_d^+(D)$ is a holomorphic function so that we get a linear map

$$\rho_d : H^d(D, \Omega^d) \to H^0(C_d^+(D), \mathcal{O}).$$

b) If D is strictly d-Levi-convex then for any divergent sequence $\{c_\nu\} \subset C_d^+(D)$ there exists a class $\xi \in H^d(D, \Omega^d)$ such that

$$\sup_\nu |\xi(c_\nu)| = \infty.$$

In particular $C_d^+(D)$ is holomorphically convex.

c) If D in addition is d-complete then given c_1, c_2 in $C_d^+(D)$ with $c_1 \neq c_2$ there exists a class $\xi \in H^d(D, \Omega^d)$ such that $\xi(c_1) \neq \xi(c_2)$.

In particular $C_d^+(D)$ is holomorphically complete.

We limit ourself to a very brief sketch of the proof (cf. [9]).

(α) First one shows that the function $\xi(c)$ is a continuous function of c (this is the most difficult part).

(β) Since $C_d^+(D)$ is weakly normal it is enough to show that $\xi(c)$ is holomorphic at non-singular points of $C_d^+(D)$. This is essentially done via Morera's theorem. Loosely speaking, if Δ is an analytic 1-dimensional disc in $C_d^+(D)$, if $\gamma = \partial\Delta$ and if $F \xrightarrow{\pi} C_d^+(D)$ is the fibered space of d-dimensional cycles over their parameter space $C_d^+(D)$, we have

$$\int_{t\in\gamma} \int_{c(t)} \{\phi\}\, dt = \int_{\pi^{-1}(\gamma)} \phi^{d,d} \wedge dt = \int_{\pi^{-1}(\Delta)} d(\phi \wedge dt) =$$

$$\int_{\pi^{-1}(\Delta)} \bar\partial\phi \wedge dt = 0$$

where t denotes the variable on Δ and $c(t) = \pi^{-1}(t)$. The last equality sign is for reason of degree as $\pi^{-1}(\Delta)$ is a $d+1$ dimensional space.

(γ) If D is strictly d-Levi convex then one constructs the class $\mathcal{E}$ using the local non-extendable cohomology classes as we did in $(\text{Levi-problem})_d$.

(δ) The separation of points on $C_d^+(D)$ when D is d-complete follows by an exact sequence argument from the vanishing of $H^{d+1}(D, \Omega^d)$ and the existence of a $\bar{\partial}$-closed (d,d) form which has non-vanishing integral over every cycle $c \in C_d^+(D)$ (this is for instance the d-th exterior power of the exterior form of a Kahler metric on X).

c) Consider, as an application, the following situation. Let $f : X \to Y$ be a proper holomorphic map between the projective algebraic manifold X onto the projective algebraic variety Y. Let A be a compact irreducible analytic subset of X and B a compact irreducible analytic subset of Y such that $f^{-1}(B) = A$ and $f : X - A \to Y - B$ is an isomorphism.
Set $a = \dim_{\mathbb{C}} A$, $b = \dim_{\mathbb{C}} B$ and assume that $a > b$.
Then there exists a neighborhood W of $C_b^+(A)$ in $C_b^+(X)$ which is holomorphically convex.
In fact f induces a map $f_b : (C_b^+(X) - C_b^+(A)) \to (C_b^+(Y) - C_b^+(B))$.
The cycles $B, 2B, 3B, \ldots$ are isolated points of $C_b^+(Y)$ thus $\bigcup_{n=1}^{\infty} nB$ has a holomorphically convex neighborhood U in $C_b^+(Y)$.
One then verifies that $W = C_b^+(A) \cup f_b^{-1}(U)$ has the required property.
Now the existence of a neighborhood W of that sort is certainly garantied by the existence of a neighborhood $N(A)$ of A in X which is strictly b-convex, by virtue of the previous theorem.

A. Andreotti

Chapter VIII. Duality theorems on complex manifolds.

8.1 Preliminaries.

a) Čech homology with value in a cosheaf.

As in the case of Poincaré duality on topological manifolds, it is better understood as a duality between cohomology and homology, so in the case of complex manifolds duality should be a pairing between cohomology and homology.
For this reason we develop here Čech homology theory as a preparation for duality theorems (cf. [7] which we follow in this exposition).

Precosheaves (cf. [18]). A precosheaf on a topological space X is a covariant functor from the category of open sets $U \subset X$ to abelian groups, i.e. for every open set $U \subset X$ an abelian group $\mathcal{S}(U)$ is given and if $V \subset U$ are open a homomorphism

$$i^V_U : \mathcal{S}(V) \to \mathcal{S}(U)$$

is given, such that if $W \subset V \subset U$ are open subsets of X we have,

$$i^V_U \circ i^W_V = i^W_U .$$

A precosheaf is called a cosheaf if for every open set $\Omega \subset X$ and every open covering $\mathcal{U} = \{U_i\}_{i\in I}$ of Ω, the following sequence is exact.

$$\amalg\, \mathcal{S}(U_i \cap U_j) \xrightarrow{\partial_0} \amalg\, \mathcal{S}(U_i) \xrightarrow{\varepsilon} \mathcal{S}(\Omega) \to 0$$

where ∂_0 is defined by

$$\partial_0\{f_{ij}\} = \{\sum_\gamma (i^{U_\gamma \cap U_i}_{U_i} f_{\gamma i} - i^{U_i \cap U_\gamma}_{U_i} f_{i\gamma})\}$$

and

$$\varepsilon\{(f_i) = \Sigma\, i^{U_i}_{\Omega} f_i .$$

Example. Set $\mathcal{S}(U)$ = continuous functions on U with compact support in U. Define the "extension maps" as the natural inclusions $\mathcal{S}(V) \subset \mathcal{S}(U)$. We obtain in this way a precosheaf and also a cosheaf.

<u>Čech homology</u>. Given a precosheaf $\mathcal{S} = \{\mathcal{S}(U), i^V_U\}$ and an open covering $\mathcal{U} = \{U_i\}_{i\in I}$ of X, one defines the groups

$$C_q(\mathcal{U},\mathcal{S}) = \coprod_{(i_0,\dots,i_q)} \mathcal{S}(U_{i_0\dots i_q})$$

and the homomorphisms

$$\partial_{q-1} : C_q(\mathcal{U},\mathcal{S}) \to C_{q-1}(\mathcal{U},\mathcal{S})$$

by

$$\partial_{q-1}\{g_{i_0\dots i_q}\} = \Sigma(-1)^h \; i^{U_{i_0\dots i_q}}_{U_{i_0\dots\hat{i}_h\dots i_q}} g_{i_0\dots i_q}$$

$$\text{for } g = \{g_{i_0\dots i_q}\} \in C_q(\mathcal{U},\mathcal{S}).$$

We have $\partial_{q-1} \circ \partial_q = 0$ for all $q \geq 1$ and we put $\partial_{-1} = 0$. We thus get a complex with differential operator of degree -1 and an augmentation

$$\varepsilon_X : C_0(\mathcal{U},\mathcal{S}) \to \mathcal{S}(X).$$

We define $H_q(\mathcal{U},\mathcal{S})$ the q-th homology group of this complex,

$$H_q(\mathcal{U},\mathcal{S}) = \frac{\operatorname{Ker}\partial_{q-1}}{\operatorname{Im}\partial_q} .$$

If $\mathcal{V} = \{V_j\}_{j\in J}$ is a refinement of $\mathcal{U}$, to each choice of the refinement function $\tau : J \to I$ there is associated a homomorphism

$$\tau_* : C_q(\mathcal{V},\mathcal{S}) \to C_q(\mathcal{U},\mathcal{S}) \quad \text{for every } q \geq 0.$$

which is compatible with the differential operator. Thus τ_* induces a homomorphism

$$\tau^{\mathcal{V}}_{\mathcal{U}} : H_q(\mathcal{V},\mathcal{S}) \to H_q(\mathcal{U},\mathcal{S})$$

which, as one verifies, is independent from the choice of the refinement function. One can then define

$$H_q(X,\mathcal{S}) = \varprojlim_{\mathcal{U}} H_q(\mathcal{U},\mathcal{S}) .$$

Remark. We may not wish to use all open subsets of X but only the open sets of a particular class $\mathcal{M}$. This can be done provided $\mathcal{M}$ is stable by finite intersections and contains arbitrarily fine coverings of X.

b) A cosheaf $\mathcal{S} = \{\mathcal{S}(U), i^V_U\}$ is called flabby if

$$i^U_X : \mathcal{S}(U) \to \mathcal{S}(X)$$

is injective for all open sets U.

For example let Σ be a soft sheaf on X. On every open set $U \subset X$ consider the group $\Gamma_k(U, \Sigma)$ of sections of Σ, compactly supported in U and set $\mathcal{S}(U) = \Gamma_k(U, \Sigma)$. Define i^V_U in the obvious way. We thus obtain a flabby cosheaf on X, $\mathcal{S} = \{\mathcal{S}(U), i^V_U\}$. For every cosheaf of this sort and for every locally finite covering $\mathcal{U} = \{U_i\}_{i \in I}$ of X, we have

$$H_q(U, \mathcal{S}) = 0 \text{ if } q \geq 1.$$

Proof. Define the sheaf $\Sigma_{U_{i_0 \ldots i_q}}$ by

$$(U_{i_0 \ldots i_q})_x = \begin{cases} \Sigma_x & \text{if } x \in U_{i_0 \ldots i_q} \\ 0 & \text{if } x \notin U_{i_0 \ldots i_q} \end{cases}$$

and set $\Sigma_q = \coprod_{(i_0 \ldots i_q)} \Sigma_{U_{i_0 \ldots i_q}}$ as the sheaf which has for fiber at a point x the direct sum of the fibers of the sheaf $\Sigma_{U_{i_0 \ldots i_q}}$ at the same point.

We have an exact sequence of sheaves

$$(*) \quad \ldots \to \Sigma_q \xrightarrow{\alpha_q} \Sigma_{q-1} \xrightarrow{\alpha_{q-1}} \ldots \xrightarrow{\alpha_1} \Sigma_0 \xrightarrow{\alpha_0} \Sigma \to 0$$

where

$$\alpha_q((g_{i_0 \ldots i_q})_x) = \coprod_h (-1)^h (g_{i_0 \ldots \hat{i}_h \ldots i_q})_x .$$

The exactness follows from the fact that at each point x the homology of that complex is the homology of a simplex with coefficient in Σ_x.

Now the sheaves Σ_q are soft. Taking sections with compact supports in the sequence (*) we get an exact sequence as the functor Γ_k is exact on soft sheaves. We thus get exactness of the sequence

$$\ldots \to C_q(\mathcal{U}, \mathcal{S}) \to C_{q-1}(\mathcal{U}, \mathcal{S}) \to \ldots \to C_0(\mathcal{U}, \mathcal{S}) \to \mathcal{S}(X) \to 0$$

and this proves our contention.

One can prove that any flabby cosheaf is of the sort described above.

We have for cosheaves the corresponding statement to the Leray theorem:

Let $\mathcal{U} = \{U_i\}_{i \in I}$ <u>be a locally finite covering of</u> X <u>and let</u> $\mathcal{S}$ <u>be a cosheaf on</u> X <u>with the following property</u> <u>for every open covering</u> $\mathcal{V} = \{V_j\}_{j \in J}$ we have

$$H_q(U_i \cap \mathcal{V}, \mathcal{S}) = H_q(U_i \cap U_j \cap \mathcal{V}, \mathcal{S}) = \ldots = 0 \text{ for every } q > 0$$

<u>then the natural homomorphism</u>

$$H_q(\mathcal{U}, \mathcal{S}) \tilde{\to} H_q(X, \mathcal{S})$$

<u>is an isomorphism.</u>

A covering $\mathcal{U}$ of this sort will be called a <u>Leray covering</u> for the cosheaf $\mathcal{S}$. As a consequence of the Leray theorem, we mention the following fact:

<u>Let</u> $\mathcal{S}'$, $\mathcal{S}$, $\mathcal{S}''$ <u>be cosheaves on</u> X <u>and let</u> $\mathcal{U} = \{U_i\}_{i \in I}$ <u>be a Leray covering for</u> $\mathcal{S}'$, $\mathcal{S}$ and $\mathcal{S}''$. <u>Suppose that for each open set</u> $U = U_{i_0 \ldots i_q}$ <u>homomorphisms</u> h_U, k_U <u>are given such that the sequence</u>

$$0 \to \mathcal{S}'(U) \overset{h_U}{\to} \mathcal{S}(U) \overset{k_U}{\to} \mathcal{S}''(U) \to 0$$

is exact, and compatible with the extension maps. Then one has an exact homology sequence

$$\ldots \to H_1(X, \mathcal{S}') \xrightarrow{h_*} H_1(X, \mathcal{S}) \xrightarrow{k_*} H_1(X, \mathcal{S}'') \xrightarrow{\partial}$$

$$\to H_0(X, \mathcal{S}') \xrightarrow{h_*} H_0(X, \mathcal{S}) \xrightarrow{k_*} H_0(X, \mathcal{S}'') \to 0$$

Note that exact sequences do not commute in general with inverse limits, thus the Leray theorem is essential to replace here holomogy on the covering $\mathcal{U}$ by that of X.

8.2. **Čech homology on complex manifolds.**

The following lemma is a consequence of the Hahn - Banach theorem (cf. [47]).

Lemma (8.2.1). **Let**

$$A \xrightarrow{u} B \xrightarrow{v} C$$

be a sequence of locally convex topological vector spaces over $\mathbb{C}$ **and continuous linear maps** u, v **with** $v \circ u = 0$. **Let**

$$A' \xleftarrow{{}^t u} B' \xleftarrow{{}^t v} C'$$

be the corresponding sequence of dual spaces and transposed maps. Then ${}^t u \circ {}^t v = 0$ **and we have a natural algebraic homomorphism**

$$\sigma : \frac{\operatorname{Ker} {}^t u}{\operatorname{Im} {}^t v} \to \operatorname{Hom\,cont} \left(\frac{\operatorname{Ker} v}{\operatorname{Im} u}, \mathbb{C}\right).$$

If v is a topological homomorphism then σ is an isomorphism.

Proof. Let $\beta \in \frac{\operatorname{Ker} {}^t u}{\operatorname{Im} {}^t v}$, $\beta = b' + {}^t v(C')$ with $b' \circ u = 0$.

For every $k \in \frac{\operatorname{Ker} v}{\operatorname{Im} u}$, $k = b + u(A)$ with $v(b) = 0$, we define $\beta(k) = b'(b)$. This does not depend on the choice of the representative b' and b and thus it defines a linear map $\sigma(\beta) : \frac{\operatorname{Ker} v}{\operatorname{Im} u} \to \mathbb{C}$. For every $\epsilon > 0$ the set $\{ b \in B \mid |b'(b)| < \varepsilon \}$ is open in B and u(A) - saturated. Therefore $\sigma(\beta)$ is continuous.

Conversely given $\lambda \in \operatorname{Hom\,cont} (\frac{\operatorname{Ker} v}{\operatorname{Im} u}, \mathbb{C})$ it lifts to a continuous linear map $\lambda' : \operatorname{Ker} v \to \mathbb{C}$. By Hahn - Banach we can extend λ' to a continuous linear map $\hat{\lambda}' : B \to \mathbb{C}$. Since $\lambda' \mid \operatorname{Im} u = 0$ then $\hat{\lambda}' \mid \operatorname{Im} u = 0$ therefore $\hat{\lambda}' \in \operatorname{Ker} {}^t u$, and thus it defines an element of $\frac{\mathrm{Ker}\ {}^t u}{\mathrm{Im}\ {}^t v}$. If $\tilde{\lambda}'$ is another extension of λ' to B then $\hat{\lambda}' - \tilde{\lambda}' = 0$ on Ker v. Then $\hat{\lambda}' - \tilde{\lambda}'$ defines a linear map of v(B) into $\mathbb{C}$. If v is topological, then the map of v(B) into $\mathbb{C}$ is continuous and by Hahn - Banach can be extended to a map $\mu : C \to \mathbb{C}$, linear and continuous. This shows that $\hat{\lambda}' - \tilde{\lambda}' = {}^t v(\mu)$. Therefore $\hat{\lambda}'$ and $\tilde{\lambda}'$ define the same element $\tau(\lambda)$ in $\frac{\mathrm{Ker}\ {}^t u}{\mathrm{Im}\ {}^t v}$.

One then verifies that σ and τ are each others inverse.

Let X be a complex manifold. We consider only those open set $U \subset\subset X$ which are open sets of holomorphy. This collection $\mathcal{M}$ of sets if stable by finite intersections and contains arbitrarily fine coverings.

Let $\mathcal{O}$ be the structure sheaf of X and let $\mathcal{F}$ be any coherent sheaf. For every $U \in \mathcal{M}$, we can find a surjection $\mathcal{O}^p \xrightarrow{\alpha} \mathcal{F} \to 0$(1) from which we derive (by theorem B of H. Cartan and J.P. Serre) a surjection $\Gamma(U, \mathcal{O}^p) \xrightarrow{\alpha_*} \Gamma(U, \mathcal{F}) \to 0$. The space $\Gamma(U, \mathcal{O}^p)$ has the structure of a space of Fréchet-Schwartz. One verifies that there is a unique structure of a space of Fréchet-Schwartz on $\Gamma(U, \mathcal{F})$ which makes the map α_* continuous. Moreover this structure on $\Gamma(U, \mathcal{F})$ is independent of the presentation $\mathcal{O}^p \xrightarrow{\alpha} \mathcal{F} \to 0$ we have chosen. (See also [28]). For any $U \in \mathcal{M}$ we set

$$\mathcal{F}_*(U) = \mathrm{Hom\ cont}\ (\Gamma(U, \mathcal{F}), \mathbb{C})$$

As transposed s of the restriction maps $r^U_V : \Gamma(U, \mathcal{F}) \to \Gamma(V, \mathcal{F})$, for $V \subset U$, $V, U \in \mathcal{M}$ (which are continuous) we get extension maps

$$i^V_U = {}^t r^U_V : \qquad i^V_U : \mathcal{F}_*(V) \to \mathcal{F}_*(U)$$

(1) This fact can be proved but it is not obvious (cf. [22]). It is obvious for sufficiently small U's, indeed it is part of the definition of coherence.

and therefore $\{\mathcal{F}_*(U), i^V_U\}$ is a precosheaf.

<u>Proposition</u> (8.2.2). <u>For any coherent sheaf $\mathcal{F}$, $\mathcal{F}_*$ is a cosheaf. Moreover for any $\Omega \in \mathcal{M}$ and any countable locally finite covering $\mathcal{U} = \{U_i\}_{i \in I}$ of Ω, with $\mathcal{U} \subset \mathcal{M}$, we have</u>

$$H_q(\mathcal{U}, \mathcal{F}_*) = 0 \quad \underline{\text{for}} \quad q > 0.$$

<u>Proof</u>. The augmented Čech conplex

$$0 \to \Gamma(\Omega, \mathcal{F}) \to C^0(\mathcal{U}, \mathcal{F}) \to C^1(\mathcal{U}, \mathcal{F}) \to \dots$$

is a complex of Fréchet spaces (as $\mathcal{U}$ is countable) and continuous maps. By theorem B this complex is acyclic, i.e. the sequence is exact. By the lemma (8.2.1) the dual sequence is exact. But that is the sequence of the augmented homology complex.

$$0 \leftarrow \mathcal{F}_*(\Omega) \leftarrow C_0(\mathcal{U}, \mathcal{F}_*) \leftarrow C_1(\mathcal{U}, \mathcal{F}_*) \leftarrow \dots .$$

In particular any countable locally finite covering $\mathcal{U} \subset \mathcal{M}$ is a Leray covering for the dual cosheaves $\mathcal{F}_*$ of coherent sheaves $\mathcal{F}$.

8.3. <u>Duality between cohomology and homology</u>.

This duality results by comparison of the two sequences

$$\text{(I)} \quad C^{q-1}(\mathcal{U}, \mathcal{F}) \xrightarrow{\delta_{q-1}} C^q(\mathcal{U}, \mathcal{F}) \xrightarrow{\delta_q} C^{q+1}(\mathcal{U}, \mathcal{F})$$

$$\text{(II)} \quad C_{q-1}(\mathcal{U}, \mathcal{F}_*) \xleftarrow{\partial_{q-1}} C_q(\mathcal{U}, \mathcal{F}_*) \xleftarrow{\partial_q} C_{q+1}(\mathcal{U}, \mathcal{F}_*)$$

where $\mathcal{F}$ is a coherent sheaf and $\mathcal{U} \subset \mathcal{M}$ is a countable finite covering of X.

In (I) the spaces are spaces of Fréchet-Schwartz and the maps are continuous, in (II) the spaces are strong duals of spaces of Fréchet-Schwartz and the maps as transposeds of the previous ones are continuous. Moreover each sequence is the dual of the other as the spaces of Fréchet-Schwartz and their strong duals are reflexive spaces.

By application of the duality lemma (8.2.1) we obtain

<u>Theorem</u> (8.3.1)

(a) <u>If δ_q is a topological homomorphism then</u>

$$H_q(X, \mathcal{F}_*) = \text{Hom cont } (H^q(X, \mathcal{F}), \mathbb{C}).$$

<u>Moreover</u> $H^{q+1}(X, \mathcal{F})$ <u>and</u> $H_q(X, \mathcal{F}_*)$ <u>are separated.</u>

(b) <u>If ∂_{q-1} is a topological homomorphism then</u>

$$H^q(X, \mathcal{F}) = \text{Hom cont } (H_q(X, \mathcal{F}_*), \mathbb{C}) .$$

<u>Moreover</u> $H_{q-1}(X, \mathcal{F}_*)$ <u>and</u> $H^q(X, \mathcal{F})$ <u>are separated.</u>

Note that the assumption in (a) is equivalent to the separation of $H^{q+1}(X, \mathcal{F})$ and that the assumption in (b) is equivalent to the separation of $H_{q-1}(X, \mathcal{F}_*)$. These conditions are certainly satisfied if $H^{q+1}(X, \mathcal{F})$ or, respectively, $H_{q-1}(X, \mathcal{F}_*)$ are finite-dimensional.

8.4. <u>Čech homology and the functor EXT.</u>

a) Let $\mathcal{F}$, $\mathcal{G}$ be sheaves of $\mathcal{O}$-modules on X. Then the sheaf $\mathcal{H}om_{\mathcal{O}}(\mathcal{F}, \mathcal{G})$ is defined as the sheaf associated to the presheaf

$$U \to \text{Hom}_{\mathcal{O}|U} \; (\mathcal{F}|_U, \mathcal{G}|_U)$$

If $\mathcal{F}$ and $\mathcal{G}$ are coherent so is the sheaf $\mathcal{H}om_{\mathcal{O}}(\mathcal{F}, \mathcal{G})$. If ϕ is a family of supports, we set

$$\text{HOM}_\phi(U, \mathcal{F}, \mathcal{G}) = \Gamma_\phi(U, \mathcal{H}om_{\mathcal{O}}(\mathcal{F}, \mathcal{G})).$$

A sheaf $\mathcal{J}$ of $\mathcal{O}$-modules on X is called <u>injective</u> if $\text{HOM}(X; \cdot, \mathcal{J})$ is an exact functor on sheaves of $\mathcal{O}$-modules i.e. for any exact sequence of sheaves of $\mathcal{O}$-modules

$$0 \to \mathcal{F}' \to \mathcal{F} \to \mathcal{F}'' \to 0$$

$$0 \to \text{HOM}\,(X; \mathcal{F}'', \mathcal{J}) \to \text{HOM}\,(X: \mathcal{F}, \mathcal{J}) \to \text{HOM}\,(X; \mathcal{F}', \mathcal{J}) \to 0$$

is exact.

We have the following facts:

i) an injective sheaf is flabby

ii) if $\mathcal{J}$ is injective, for any sheaf $\mathcal{F}$ of $\mathcal{O}$-modules $\mathcal{H}om_{\mathcal{O}}(\mathcal{F},\mathcal{J})$ is flabby

iii) for any sheaf $\mathcal{G}$ of $\mathcal{O}$-modules one can find an injective resolution $(_*)$ $0 \to \mathcal{G} \to \mathcal{G}_0 \to \mathcal{G}_1 \to \mathcal{G}_2 \to \dots$ i.e. the sequence is exact and every $\mathcal{G}_i$, $i \geq 0$, is an injective sheaf.

These facts follow directly from the definitions and the possibility to imbed every module in an jnjective module.

Applying the functor $\mathcal{H}om_{\mathcal{O}}(\mathcal{F},\ \ )$ to the sequence $(_*)$ we get a complex of sheaves and homomorphisms

$$0 \to \mathcal{H}om_{\mathcal{O}}(\mathcal{F},\mathcal{G}_0) \to \mathcal{H}om_{\mathcal{O}}(\mathcal{F},\mathcal{G}_1) \to \mathcal{H}om_{\mathcal{O}}(\mathcal{F},\mathcal{G}_2) \to \dots$$

The q-th cohomology group of this complex is denoted by

$$\mathcal{E}xt^q_{\mathcal{O}}(\mathcal{F},\mathcal{G}).$$

This is a sheaf of $\mathcal{O}$-modules and one verifies it is independent of the choice of the resolution (*).

Moreover if $\mathcal{F}$ and $\mathcal{G}$ are coherent then $\mathcal{E}xt^q_{\mathcal{O}}(\mathcal{F},\mathcal{G})$ is a coherent sheaf. This can be seen as follows:

Let

$$\dots \to \mathcal{L}_2 \to \mathcal{L}_1 \to \mathcal{L}_0 \to \mathcal{F} \to 0 \qquad (**)$$

be a resolution of $\mathcal{F}$ be locally free sheaves(1). Applying

---

(1)

On any open set of holomorphy $U \subset\subset X$ we have such a resolution, as any coherent sheaf on X is the quotient sheaf on U of a free sheaf by the theorem of Coen (cf. [22]). However since here we need only this resolution locally one can invoque the theorem of syzygies of Hilbert to derive the existence in a sufficiently small neighborhood of a given point $x \in X$ of a finite free resolution:

$$0 \to \mathcal{O}^{p_d} \to \mathcal{O}^{p_{d-1}} \to \dots \to \mathcal{O}^{p_0} \to \mathcal{F} \to 0 \text{ where } d \leq \dim_x X \text{ (cf.[2])}.$$

the functor $\mathcal{H}om_{\mathcal{O}}(\ .,\mathcal{G})$ we get another complex

$$0 \to \mathcal{H}om_{\mathcal{O}}(\mathcal{L}_0,\mathcal{G}) \to \mathcal{H}om_{\mathcal{O}}(\mathcal{L}_1,\mathcal{G}) \to \mathcal{H}om_{\mathcal{O}}(\mathcal{L}_2,\mathcal{G}) \to \dots$$

whose q-th cohomology group is again $\mathcal{E}xt^q_{\mathcal{O}}(\mathcal{F},\mathcal{G})$, as it follows by a standard spectral sequence argument. By this construction $\mathcal{E}xt^q_{\mathcal{O}}(\mathcal{F},\mathcal{G})$ is a coherent sheaf.
In particular if $\mathcal{F}$ is locally free we get $\mathcal{E}xt^q_{\mathcal{O}}(\mathcal{F},\mathcal{G}) = 0$ if $q > 0$ as we can take $\mathcal{L}_0 = \mathcal{F}, \mathcal{L}_1 = \mathcal{L}_2 = \dots = 0$.

Note that in any case $\mathcal{E}st^0_{\mathcal{O}}(\mathcal{F},\mathcal{G}) = \mathcal{H}om_{\mathcal{O}}(\mathcal{F},\mathcal{G})$ because the sequence

$$0 \to \mathcal{H}om_{\mathcal{O}}(\mathcal{F},\mathcal{G}) \to \mathcal{H}om_{\mathcal{O}}(\mathcal{F},\mathcal{G}_0) \to \mathcal{H}om_{\mathcal{O}}(\mathcal{F},\mathcal{G}_1)$$

is exact as the functor $\mathcal{H}om_{\mathcal{O}}(\mathcal{F},.)$ is left exact.

Applying to the resolution (*) the functor $HOM_{\phi}(X;\mathcal{F},\ .)$ we obtain a complex whose q-th cohomology group is denoted by

$$EXT^q_{\phi}(X;\mathcal{F},\mathcal{G}).$$

For $q = 0$ we have analogously to the previous case

$$EXT^0_{\phi}(X;\mathcal{F},\mathcal{G}) = HOM_{\phi}(X;\mathcal{F},\mathcal{G}).$$

The spectral sequence of the double complex $K = \{K^{p,q}, d\}$, where

$$K^{p,q} = \Gamma_{\phi}(X;\mathcal{H}om_{\mathcal{O}}(\mathcal{L}_p,\mathcal{G}_q))$$

and d is induced by the maps of the resolution (*) and (**), leads to a spectral sequence connecting the global with the local extension functors

$$E_2^{p,q} \Rightarrow EXT^n(X;\mathcal{F},\mathcal{G}) \qquad (n = p+q)$$

where

$$E_2^{p,q} = H^p_{\phi}(X,\mathcal{E}xt^q_{\mathcal{O}}(\mathcal{F},\mathcal{G})).$$

b) The connection of homology and the functor EXT is established by the following

Lemma (8.4.1). Let X be a complex manifold of pure dimension n. Let Ω^n be the sheaf of germs of holomorphic n-forms on X.

For any open set of holomorphy $U \subset\subset X$ and for any coherent sheaf $\mathcal{F}$ on X one has

$$\mathcal{F}_*(U) = \mathrm{EXT}^n_k\ (U;\mathcal{F},\Omega^n)\ ;\quad \mathrm{EXT}^q_k\ U;\mathcal{F},\Omega^n) = 0 \text{ if } q \neq n,$$

the suffix k denoting the family of compact supports.

Proof.

(α) Let $\mathcal{F}$ be a locally free sheaf on X and let U be an open set of holomorphy in X. The sheaf $\mathcal{F}$ can be considered as the sheaf of germs of holomorphic sections of a holomorphic vector bundle E, $\mathcal{F} = \quad (E)$. Let E^* denote the dual bundle of E; if E is defined by the transition functions $\{g_{ij}\}$ then E^* is defined by the transition functions $\{{}^t g_{ij}^{-1}\}$.

Let $\mathcal{A}^{r,s}(E)$ denote the sheaf of germs of C^∞ forms of type (r,s) with values in E and let $\mathcal{K}^{u,v}(E^*)$ denote the sheaf of germs of forms with distribution coefficients, of type (u, v), with value in E^*.

Since U is an open set of holomorphy, the sequence

$$0 \to \Gamma(U, \mathcal{O}(E)) \xrightarrow{i} (U,\mathcal{A}^{0,0}(E)) \xrightarrow{\bar\partial} (U,\mathcal{A}^{0,1}(E)) \xrightarrow{\bar\partial} \ldots$$

$$\ldots \xrightarrow{\bar\partial} \Gamma(U,\mathcal{A}^{0,n}(E)) \to 0$$

is an exact sequence of spaces of Fréchet-Schwartz and continuous maps. By the duality lemma (8.2.1) the dual sequence is also exact. But this is the sequence

$$0 \leftarrow (U,\mathcal{O}(E))' \xleftarrow{{}^t i} {}_k(U,\mathcal{K}^{n,n}(E^*)) \xleftarrow{\bar\partial} {}_k(U,\mathcal{K}^{n,n-1}(E^*)) \xleftarrow{\bar\partial} \ldots$$

$$\ldots \xleftarrow{\bar\partial} \Gamma_k(U,\mathcal{K}^{n,0}(E^*)) \leftarrow 0.$$

Now for any vector bundle E^*, denoting by $\Omega^n(E^*)$ the sheaf of germs of holomorphic n-forms with values in E^*, we have in the exact sequence of sheaves

$$0 \to \Omega^n(E^*) \to \mathcal{K}^{n,0}(E^*) \to \mathcal{K}^{n,1}(E^*) \to \ldots$$

A soft resolution of $\Omega^n(E^*)$. Therefore, since E^* can be any holomorphic vector bundle and hence $\Omega^n(E^*)$ any locally free sheaf, we get :

for any locally free sheaf $\mathcal{G} = \Omega^n(E^*) = \mathcal{H}om_{\mathcal{O}}(\mathcal{F}, \Omega^n)$

$$H^r_k(U,\mathcal{G}) = \begin{cases} 0 \text{ if } r \neq n \\ \text{Hom cont } (\Gamma(U,\mathcal{O}(E)), \mathbb{E}) \text{ if } r = n. \end{cases}$$

(β) Let $\mathcal{F}$ and $\mathcal{G}$ be locally free then we have

$$\mathrm{EXT}^p_k(U;\mathcal{F},\mathcal{G}) = 0 \text{ if } p \neq n$$

$$\mathrm{EXT}^n_k(U;\mathcal{F},\mathcal{G}) = H^n_k(U, \mathcal{H}om_{\mathcal{O}}(\mathcal{F},\mathcal{G})).$$

In fact the spectral sequence converging to $\mathrm{EXT}^s_k(U;\mathcal{F},\mathcal{G})$ has the term $E_2^{p,q} = H^p_k(U; \mathcal{E}xt^q(\mathcal{F},\mathcal{G}))$ $(p+q=s)$. Thus $E_2^{p,q} = 0$ if $q \neq 0$ as $\mathcal{F}$ is locally free. Moreover $\mathcal{E}xt^0_{\mathcal{O}}(\mathcal{F},\mathcal{G}) = \mathcal{H}om_{\mathcal{O}}(\mathcal{F},\mathcal{G})$ is also locally free, hence by (α) we get $E_2^{p,0} = 0$ if $p \neq n$ and

$E_2^{n,0} = H^n_k(U;\mathcal{H}om_{\mathcal{O}}(\mathcal{F},\mathcal{G}))$. But the spectral sequence is degenerated so that we have $\mathrm{EXT}^p_k(U;\mathcal{F},\mathcal{G}) = E_2^{p,0}$. This proves our contention.

(γ) Suppose now that $\mathcal{F}$ is coherent and $\mathcal{G}$ locally free. The we have

$$\mathrm{EXT}^p_k(U;\mathcal{F},\mathcal{G}) = 0 \text{ if } p < n$$

and an exact sequence

$$0 \to \mathrm{EXT}^n_k(U;\mathcal{F},\mathcal{G}) \to H^n_k(U;\mathcal{H}om_{\mathcal{O}}(\mathcal{L}_0,\mathcal{G}) \to H^n_k(U;\mathcal{H}om_{\mathcal{O}}(\mathcal{L}_1,\mathcal{G})).$$

In fact consider the double complex $K^{p,q} = \Gamma_k(U,\mathcal{H}om_{\mathcal{O}}(\mathcal{L}_p,\mathcal{G}_q))$ has for cohomology groups the groups $\mathrm{EXT}^*{}_k(U;\mathcal{F},\mathcal{G})$. For any p the sequence of sheaves

$$0 \to \mathcal{H}om_{\mathcal{O}}(\mathcal{L}_p,\mathcal{G}) \to \mathcal{H}om_{\mathcal{O}}(\mathcal{L}_p,\mathcal{G}_0) \to \mathcal{H}om_{\mathcal{O}}(\mathcal{L}_p,\mathcal{G}_1) \to \cdots$$

is exact as $\mathcal{L}_p$ is locally free and provides an injective resolution of $\mathcal{H}om_{\mathcal{O}}(\mathcal{L}_p,\mathcal{G})$. Taking cohomology with respect to the differential coming from the resolution (*) we get

$$E_1^{p,q} = H_k^q(U, \mathcal{H}om_{\mathcal{O}}(\mathcal{L}_p, \mathcal{G})) = 0 \quad \text{if} \quad q \neq n$$

as $\mathcal{H}om_{\mathcal{O}}(\mathcal{L}_p, \mathcal{G})$ is locally free.

It follows that $EXT_k^p(U; \mathcal{F}, \mathcal{G}) = 0$ if $p < n$ and $EXT_k^{n+1}(U; \mathcal{F}, \mathcal{G})$ is the 1-th cohomology group of the complex:

$$(0) \quad H_k^n(U; \mathcal{H}om_{\mathcal{O}}(\mathcal{L}_0, \mathcal{G})) \to H_k^n(U; \mathcal{H}om_{\mathcal{O}}(\mathcal{L}_1, \mathcal{G}) \to \\ \to H_k^n(U; \mathcal{H}om_{\mathcal{O}}(\mathcal{L}_2, \mathcal{G}) \to \cdots$$

In particular we get the exact sequence

$$(1) \quad 0 \to EXT_k^n(U; \mathcal{F}, \mathcal{G}) \to H_k^n(U; \mathcal{H}om_{\mathcal{O}}(\mathcal{L}_0, \mathcal{G}) \to H_k^n(U; \mathcal{H}om_{\mathcal{O}}(\mathcal{L}_1, \mathcal{G}))$$

(δ) Consider the exact sequence of spaces of Fréchet-Schwartz obtained by applying the functor Γ to $(_{**})$

$$\cdots \to \Gamma(U, \mathcal{L}_1) \to \Gamma(U, \mathcal{L}_0) \to \Gamma(U, \mathcal{F}) \to 0 .$$

Exactness follows from the assumption that U is an open set of holomorphy. By the duality lemma and (α) we get an exact sequence

$$(2) \quad 0 \to \mathcal{F}_*(U) \to H_k^n(U, \mathcal{H}om_{\mathcal{O}}(\mathcal{L}_0, \Omega^n)) \to H_k^n(U, \mathcal{H}om_{\mathcal{O}}(\mathcal{L}_1, \Omega^n) \ldots$$

Comparing (0), (1) and (2) we obtain

$$\mathcal{F}_*(U) = EXT_k^n(U; \mathcal{F}, \Omega^n) \quad \text{and} \quad EXT_k^q(U; \mathcal{F}, \Omega^n) = 0 \text{ if } q \neq n.$$

Given $\mathcal{F}$ and $\mathcal{G}$ coherent sheaves, define a precosheaf by $U \to EXT_k^q(U; \mathcal{F}, \mathcal{G})$ for $U \in \mathcal{M}$. We denote this precosheaf as $\mathcal{E}xt_k^q(\mathcal{F}, \mathcal{G})$. We do not need to verify it is a cosheaf,

Lemma (8. 4.2). <u>There exists a spectral sequence</u>

$$E_2^{-p,q} \Rightarrow EXT_k^{-p+q}(X; \mathcal{F}, \mathcal{G})$$

<u>with</u> $E_2^{-p,q} = H_p(\mathcal{U}, \mathcal{E}xt_k^q(\mathcal{F}, \mathcal{G}))$,
<u>where</u> $\mathcal{U} \subset \mathcal{M}$ <u>is a locally finite covering of</u> X.

Proof. Consider the double complex

$K^{-p,q} = C_p(\mathcal{U}, \mathcal{H}om_k(\mathcal{F}, \mathcal{G}_q))$. If we take cohomology with respect to the differential coming from (*) we get $'E_1^{-p,q} = C_p(\mathcal{U}, \mathcal{E}xt_k^q(\mathcal{F}, \mathcal{G}))$ and then taking homology with respect to the Čech differential we get

$$'E_2^{-p,q} = H_p(\mathcal{U}, \mathcal{E}xt_k^q(\mathcal{F}, \mathcal{G})).$$

Now we remark that $\mathcal{H}om_k(\mathcal{F}, \mathcal{G}_q)\ (U) = \Gamma_k(U; \mathcal{H}om\ (\mathcal{F}, \mathcal{G}_q))$. Thus we have in $\mathcal{H}om_k(\mathcal{F}, \mathcal{G}_q)$ a flabby cosheaf as $\mathcal{H}om\ (\mathcal{F}, \mathcal{G}_q)$ is soft, $\mathcal{G}_q$ being injective. Therefore taking first homology with respect to the Čech differential we get

$$''E_1^{-p,q} = 0 \quad \text{if} \quad p \neq 0$$

$$''E_1^{0,q} = H_0(\mathcal{U}, \mathcal{H}om_k(\mathcal{F}, \mathcal{G}_q)) = HOM_k(X; \mathcal{F}, \mathcal{G}_q).$$

Taking now the cohomology with respect to the differential coming from (*) we get that the spectral sequence degenerates having as total cohomology the groups

$$EXT_k^q(X; \mathcal{F}, \mathcal{G}).$$

We apply this lemma to $\mathcal{G} = \Omega^n$. Then $\mathcal{E}xt_k^q(\mathcal{F}, \Omega^n) = 0$ if $q \neq n$ and $\mathcal{E}xt_k^n(\mathcal{F}, \Omega^n) = \mathcal{F}_*$ as it follows from lemma (8.4.1). Therefore we get the following

Theorem (8.4.3). Let X be a complex manifold of pure dimension n and let $\mathcal{F}$ be any coherent sheaf on X. Then we have

$$H_p(X, \mathcal{F}_*) \simeq EXT_k^{n-p}(X; \mathcal{F}, \Omega^n).$$

The combination of this theorem (8.4.3) with theorem (8,3.1) is what is usually called the "duality theorem"; in this form it is due to Malgrange and Serre (cf. [39], [47], [44]).

Remarks.

1. If $\mathcal{F}$ is locally free i.e. the sheaf of germs of holomorphic sections of a holomorphic vector bundle E, $\mathcal{F} = \mathcal{O}(E)$, then

$$\mathcal{E}xt^q(\mathcal{F}, \Omega^n) = 0 \text{ if } q \neq 0 \text{ and } \mathcal{E}xt^0_{\mathcal{O}}(\mathcal{F}, \Omega^n) = \mathcal{H}om_{\mathcal{O}}(\mathcal{F}, \Omega^n) = \Omega^n(E^*),$$

E^* being the dual bundle of E. Therefore

$$EXT_k^{n-p}(X; \mathcal{O}(E), \Omega^n) = H_k^{n-p}(X, \Omega^n(E^*))$$

so that theorem (8.4.3) gives

$$H_p(X, \mathcal{O}(E)_*) \simeq H_k^{n-p}(X, \Omega^n(E^*)).$$

In particular if E is a line bundle, $E^* = E^{-1}$ and we get in this case

$$H_p(X, \mathcal{O}(E)_*) \simeq H_k^{n-p}(X, \Omega^n(E^{-1})).$$

2. If $\mathcal{F} = \mathcal{O}(E)$ is locally free and if X is a compact manifold then the theorem (8.3.1) can be applied without restrictions and we get

$$H^p(X, \mathcal{O}(E)) = \mathrm{Hom}\,(H^{n-p}(X, \Omega^n(E^*)), \mathbb{C})$$

as the cohomology groups being finite-dimensional any linear map into $\mathbb{C}$ is automatically continuous.

8.5. Divisors and Riemann-Roch theorem.

a) Let X be a connected complex manifold of complex dimension n. The sheaf $\mathcal{O}^*$ of germs of never vanishing holomorphic functions as a sheaf of multiplicative groups, can be considered as a subsheaf of the following sheaves:

the sheaf $\mathcal{M}^*$ of germs of non identically zero meromorphic functions,

the sheaf $\mathcal{O}_o^*$ of germs of non identically zero homolorphic functions.

While the sheaf $\mathcal{M}^*$ is a sheaf of multiplicative groups, the sheaf $\mathcal{O}_o^*$ is a sheaf of multiplicative monoids.
We thus get two exact sequences of sheaves

(1) $0 \to \mathcal{O}^* \to \mathcal{M}^* \to \mathcal{D} \to 0$

(2) $0 \to \mathcal{O}^* \to \mathcal{O}_o^* \to \mathcal{D}_+ \to 0$

where the sheaves $\mathcal{D}$ and $\mathcal{D}_+$ are defined by the exactness of the sequences. The sheaf $\mathcal{D}$ is called the <u>sheaf of germs of (meromorphic) divisors.</u> The sheaf $\mathcal{D}_+$ is called the <u>sheaf of germs of holomorphic or positive divisors</u>.
The elements of $H^0(X, \mathcal{D})$ are called <u>meromorphic divisors</u> on X and the elements of $H^0(X, \mathcal{D}_+)$ are called <u>holomorphic or positive divisors on</u> X.
An element $D \in H^0(X, \mathcal{D})$ (resp. $D \in H^0(X, \mathcal{D}_+)$) is given on a sufficiently fine open covering $\mathcal{U} = \{U_i\}_{i \in I}$ of X by a collection $\{f_i\}$ of meromorphic (resp. holomorphic) functions, not identically zero, on each U_i, and such that

$$\frac{f_i}{f_j} = g_{ij} : U_i \cap U_j \to \mathbb{C}^* \quad i, j \in I$$

is holomorphic and never zero.
The functions $\{g_{ij}\}$ are transition functions of a holomorphic line bundle $\{D\}$ and represent the element $\delta(D) \in H^1(X, \mathcal{O}^*)$, where δ is the connecting homomorphism of the exact cohomology sequences associated to (1) or (2).

In every meromorphic divisor D we can distinguish the <u>0-part</u> D_0, and the <u>polar part</u> D_∞, these are holomorphic divisors and we usually write

$$D = D_0 - D_\infty .$$

Correspondingly we have

$$\{D\} = \{D_0\} \cdot \{D_\infty\}^{-1} .$$

The study of general (i.e. meromorphic) divisors can thus be reduced to the study of holomorphic divisors.

Consider the space $H^0(X,\mathcal{O}(D))$ where $\mathcal{O}(D)$ is the sheaf of germs of holomorphic sections of the bundle $\{D\}$. To every s $H^0(X,\mathcal{O}(D))$ there corresponds a holomorphic divisor, this in its turn determines s up to multiplication by a global never vanishing holomorphic function.

If X is compact (or pseudoconcave) then the divisor of a section $s \in H^0(X,\mathcal{O}(D))$ determines the section up to multiplication by a non-zero constant. The set of positive divisors corresponding to elements of $H^0(X,\mathcal{O}(D))$ is called the linear system of D and is denoted by $|D|$. Its elements correspond one-to-one to the point of the projective space

$$(H^0(X,\mathcal{O}(D)) - \{0\}) / \mathbb{C}^* .$$

We attribute to $|D|$ the dimension of this space Thus

$$\dim |D| = \dim_{\mathbb{C}} H^0(X,\mathcal{O}(D)) - 1 .$$

The problems we want to consider is the following: given on X a divisor D compute $\dim |D|$. Note that if we are able to establish that $\dim |D| \geq 0$ then we have proved that the holomorphic line bundle $\{D\}$ admits a holomorphic section different from the 0-section.

b) Let us suppose now that X is compact and $\dim_{\mathbb{C}} X = 1$. Then every divisor D is given by a finite sum $D = \sum n_i p_i$ with $n_i \in \mathbb{Z}$ and p_i being the divisors associated to points of X. A divisor is positive iff all n_i are ≥ 0. The integer $\sum n_i$ is called the degree of the divisor D.
Assume first that $D = \sum n_i p_i$ is a positive divisor. In this case the bundle $\{D\}$ has certainly a holomorphic section $s \neq 0$, the section corresponding to the divisor itself. Given s we have a sheaf homomorphism $\mathcal{O} \to \mathcal{O}(D)$ given by multiplication by s. The homomorphism is certainly injective and the quotient sheaf is concentrated in the points p_i and given at each point p_i by the vector space $V(n_i, p_i)$ of dimension n_i representing the Taylor expansion at p_i of holomorphic functions, truncated at the order n_i-1.

In other words we have an exact sequence of sheaves:

$$0 \to \mathcal{O} \xrightarrow{s} \mathcal{O}(D) \to \coprod V(n_i, p_i) \to 0 .$$

This gives the exact cohomology sequence

$$0 \to H^0(X, \mathcal{O}) \to H^0(X, \mathcal{O}(D)) \to \mathbb{C}^{\Sigma n_i}$$

$$\to H^1(X, \mathcal{O}) \to H^1(X, \mathcal{O}(D)) \to 0$$

as $H^1(X, \coprod V(n_i, p_i)) = 0$ the sheaf having 0-dimensional support. Therefore we get, in particular:

$$\dim_{\mathbb{C}} H^0(X, \mathcal{O}(D)) - \dim_{\mathbb{C}} H^1(X, \mathcal{O}(D)) = \deg(D) + \dim_{\mathbb{C}} H^0(X, \mathcal{O}) - \dim_{\mathbb{C}} H^1(X, \mathcal{O}).$$

Now:

$H^0(X, \mathcal{O}) = \mathbb{C}$ as X is compact thus $\dim_{\mathbb{C}} H^0(X, \mathcal{O}) = 1$.

$H^1(X, \mathcal{O})$ is finite-dimensional and, by duality, $\simeq H^0(X, \Omega^1)$, the space of holomorphic differentials on X. Its dimension is called the <u>genus</u> $g(X)$ of X.

$H^1(X, \mathcal{O}(D))$ if finite-dimensional and, by duality $\simeq H^0(X, \Omega^1(-D)) = H^0(X, \mathcal{O}(K-D))$ where $\{K\}$ is the line bundle associated to the sheaf Ω^1. The bundle $\{K\}$ is called the <u>canonical bundle</u>, and $\dim_{\mathbb{C}} H^0(X, \mathcal{O}(K-D)) = i(D)$ is called the <u>speciality index</u> of the divisor D.

We thus have the following formula (Riemann - Roch theorem)

(1) $\qquad \dim |D| = \deg (D) - g(x) + i(D)$

This formula can be extended to any divisor $D = D_0 - D_\infty$ even if not positive maintaining for $i(D)$ the meaning of $\dim_{\mathbb{C}} H^0(X, \mathcal{O}(K-D))$.

Indeed let s denote the section, corresponding to D_∞ of $H^0(X, \mathcal{O}(D_\infty))$. We have an exact sequence of sheaves

$$0 \to \mathcal{O}(D) \to \mathcal{O}(D_0) \to \coprod V(p_i, n_i) \to 0$$

where $D_\infty = \Sigma\, n_i p_i$.

Therefore

$$\dim_{\mathbb{C}} H^0(X, \mathcal{O}(D_0)) - \dim_{\mathbb{C}} H(X, \mathcal{O}(D_0)) = \dim_{\mathbb{C}} H^0(X, \mathcal{O}(D)) -$$

$$- \dim_{\mathbb{C}} H^1(X, \mathcal{O}(D)) + \deg(D_\infty)$$

Thus using for D_0 the formula already established we get

$$\deg (D_0) - g(X) = \dim |D| - i(D) + \deg (D_\infty).$$

From this the assertion follows as we have $\deg(D) = \deg (D_0) - \deg (D_\infty)$.

c) we add a few remarks to the Riemann-Roch theorem in dimension one.

<u>The genus</u> $g(X)$ <u>of</u> X <u>equals</u> $\frac{1}{2}$ <u>of the first Betti number</u> $b_1(X)$ <u>of</u> X :

(1) $2g(X) = b_1(X)$.

First we show that $b_1(X) \geq 2g(X)$. From the exact sequence

$$0 \to \mathbb{C} \to \mathcal{O} \xrightarrow{d} \Omega^1 \to 0$$

Since $H^0(X, \mathbb{C}) \simeq \mathbb{C}$, $H^0(X, \mathcal{O}) \simeq \mathbb{C}$, we get an injective map

$$0 \to H^0(X, \Omega^1) \to H^1(X, \mathbb{C}) .$$

This map associates to every holomorphic 1-form its cohomology class as a closed form.

Now $H^0(X, \Omega^1)$ and $H^0(X, \bar{\Omega}^1)$ can be considered as subspaces of $H^1(X, \mathbb{C})$. Their intersection is reduced to $\{0\}$. In fact if $\alpha \in H^0(, \Omega^1)$, $\beta \in H^0(X, \bar{\Omega}^1)$ and if $\alpha = \beta + dg$ where g is a C^∞ function on X, by reason of bidegree we get $\alpha = \partial g$. Thus $\bar{\partial}\partial g = 0$ i.e. g is (pluri-)harmonic. By the maximum principle g must be constant hence $\alpha = 0$.

Therefore $\dim_{\mathbb{C}} H^1(X, \mathbb{C}) \geq 2 \dim_{\mathbb{C}} H^0(X, \Omega^1)$ i.e. $b_1(X) \geq 2g(X)$.

We show now that $b_1(X) \leq 2g(X)$. From the exact sequence

$$0 \to \mathbb{C} \to \mathcal{O} \oplus \bar{\mathcal{O}} \to \mathcal{H} \to 0$$

we get an injection $0 \to H^1(X, \mathbb{C}) \to H^1(X, \mathcal{O}) \oplus H^1(X, \bar{\mathcal{O}})$.

In fact $H^0(X, \mathbb{C}) \simeq \mathbb{C}$, $H^0(X, \mathcal{O}) \simeq \mathbb{C}$, $H^0(X, \mathcal{H}) \simeq \mathbb{C}$ (by the maximum principle). Therefore

$$\dim_{\mathbb{C}} H^1(X, \mathbb{C}) \leq 2 \dim_{\mathbb{C}} H^1(X, \mathcal{O}) \quad \text{i.e.} \quad b_1(X) \leq 2g(X).$$

If we apply the Riemann-Roch theorem to $D = 0$ we get

$$(2) \quad \dim |K| = g(X) - 1$$

If we apply the same theorem to $D = K$ we get

$$(3) \quad \deg (K) = 2g(X) - 2.$$

To do this we have to know that $\{K\}$ comes from a divisor. Now if $g(X) \geq 1$ this follows from (2). If $g(X) = 0$ then for any positive divisor D we have $\dim |D| = \deg (D) + i(D)$. But then necessarily $i(D) = 0$.
There exists on X a rational function with a single pole of first order. This function extablishes an isomorphism of X onto the Riemann sphere and on this manifold one verifies immediately that $K = -2p$, p being a point of X. In particular <u>for a divisor</u> D <u>with</u> $\deg(D) > 2g-2$ <u>we must have</u>

$$i(D) = 0.$$

As an exercise one can show now that <u>every compact manifold</u> X <u>of complex dimension one admits a porjective imbedding</u> i.e. <u>is projective algebraic</u>. Indeed if $s_0, \ldots, s_t$ is a basis of $H^0(X, \mathcal{O}(D))$ the map $X \to \mathbb{P}_t(\mathbb{C})$ defined by $x \mapsto (s_0(x), \ldots, s_t(x))$ is holomorphic everywhere. If $(D) > 2g$ then one verifies by means of the Riemann-Roch theorem that the map is one-to-one and biholomorphic.

d) Let X be a connected compact manifold of $\dim_{\mathbb{C}} X = 2$. Let D be a holomorphic divisor on X which will be supposed of "multiplicity" one at each point and non-singular.
We have now an exact sequence

$$0 \to \mathcal{O} \xrightarrow{s} \mathcal{O}(D) \to \mathcal{O}(D)\big|_D \to 0$$

where s is a section of $\{D\}$ corresponding to the divisor D.

We get an exact cohomology sequence:

$$0 \to H^0(X,\mathcal{O}) \to H^0(X, \mathcal{O}(D)) \to H^0(D, \mathcal{O}(D)|_D) \to$$
$$\to H^1(X,\mathcal{O}) \to H^1(X, \mathcal{O}(D)) \to H^1(D, \mathcal{O}(D)|_D) \to$$
$$\to H^2(X,\mathcal{O}) \to H^2(X, \mathcal{O}(D)) \to 0.$$

We have

$$\dim_{(\mathbb{C})} H^0(D, \mathcal{O}(D)/_D) - \dim_{\mathbb{C}} H^1(D, \mathcal{O}(D)/_D) = \deg \{D\}|_D - \text{genus of } (D) + 1$$

$$\dim_{\mathbb{C}} H^2(X,\mathcal{O}) = \dim_{\mathbb{C}} H^0(X, \Omega^2) = p_g(X) = \text{geometric genus of } X$$

$$\dim_{\mathbb{C}} H^1(X,\mathcal{O}) = h^{0,1}$$

$\dim_{\mathbb{C}} H^2(X,\mathcal{O}(D)) = \dim_{\mathbb{C}} H^0(X, \mathcal{O}(K-D))$ where $\{K\}$ denotes the bundle corresponding to the sheaf of holomorphic 2-forms $\Omega^2 = \mathcal{O}(K)$. This dimension is denoted by $i(D)$ and called the <u>speciality index</u> of D.

By the same argument used for dimension one we get now

$$\dim |D| \geq \deg \{D\}|_D - \text{genus } (D) + (p_g(X) - h^{0,1}) - i(D) + 1.$$

This is Castelnuovo's theorem. The difference of the left and right side is $\dim_{\mathbb{C}} H^1(X,\mathcal{O}(D))$ which is called the "superabundance".

If X is Kahler then $q = h^{0,1} = h^{1,0} = \dim_{\mathbb{C}} H^0(X,\Omega^1)$ = number of linearly independent holomorphic 1-forms, and $p_g(X) - q = p_a(X)$ is called the arithmetic genus.

The inequality can be extablished without the restrictive assumptions we have made on D. In particular for a multiple lD of D we get the inequality

$$\dim |lD| \geq l^2 \deg \{D\}|_D - \left(\frac{l(l-1)}{2} \deg \{D\}|_D + l \text{ genus } (D) - l + 1\right) + (p_g(X) - h^{0,1}) - i(lD) + 1.$$

If D is positive and l large enough and positive, then $i(lD) = 0$.

Therefore if $\deg\{D\}|_D > 0$, $\dim |lD|$ grows like l^2. Therefore

<u>If</u> X <u>contains a divisor</u> D <u>with deg</u> $\{D\}|_D > 0$ <u>then trans. degree</u> $\mathcal{K}(X) = 2$. One could show that X is in this case a projective algebraic variety as it is always the case for a complex surface X with $\mathcal{K}(X)$ of transcendence degree 2. (cf. [21]).

A. Andreotti

Chapter IX. The H. Lewy problem.

9.1. Preliminaries.

To simplify the exposition we restrict ourselves to the space $\mathbb{C}^n$ although the results will hold on any complex manifold with only formal changes of notation.

Let U be an open set in $\mathbb{C}^n$ and let $\rho : U \to \mathbb{R}$ be a C^∞ function. We define

$$U^+ = \{z \in U \mid \rho(z) \geq 0\}$$

$$U^- = \{z \in U \mid \rho(z) \leq 0\}$$

$$S = \{z \in U \mid \rho(z) = 0\}$$

and we will assume $d\rho \neq 0$ on S, so that S is a smooth hypersurface. On U we consider the Dolbeault complex

$$C^*(U) = \{C^{0,0}(U) \xrightarrow{\bar\partial} C^{0,1}(U) \xrightarrow{\bar\partial} C^{0,2}(U) \to \dots\}$$

where $C^{0,s}(U)$ denotes the space of C^∞ forms on U of type (0,s) and where $\bar\partial$ is the exterior differentiation with respect to antiholomorphic coordinates.

Analogously we define the spaces $C^{0,s}(U^+)$, resp. $C^{0,s}(U^-)$, as the spaces of those forms of type (0,s) on $\bar{U}^+$, resp $\bar{U}^-$, having C^∞ coefficients with all partial deriviatives continuous up to the boundary S. In this way we obtain two similar complexes, $C^*(U^+)$ and $C^*(U^-)$.

Define

$$\mathcal{J}^{0,s}(U) = \{\phi \in C^{0,s}(U) \mid \phi = \rho\alpha + \bar\partial\rho \wedge \beta, \alpha \in C^{0,s}(U), \beta \in C^{0,s-1}(U)\}.$$

We have $\bar\partial\mathcal{J}^{0s}(U) \subset \mathcal{J}^{0,s+1}(U)$, therefore $\mathcal{J}^*(U) = \coprod_s \mathcal{J}^{0,s}(U)$ is a subcomplex of $C^*(U)$ and indeed a differnetial ideal.

in a similar way one defines the subcomplexes $\mathcal{J}^*(U^\pm)$ of $C^*(U^\pm)$.

Finally one defines the quotient complex

$$Q^*(S) = \{Q^{0,0}(S) \xrightarrow{\bar\partial_S} Q^{0,1}(S) \xrightarrow{\bar\partial_S} Q^{0,2}(S) \xrightarrow{\bar\partial_S} \cdots$$

by the exact sequence

$$0 \to \mathcal{J}^*(U) \to C^*(U) \to Q^*(S) \to 0.$$

The quotient complex is denoted by $Q^*(S)$ as each one of its spaces is concentrated on S. The operator $\bar\partial_S$ is by definition induced by the operator $\bar\partial$ on $C^*(U)$ and $\mathcal{J}^*(U)$.

One could define the quotient complex also for the inclusions $\mathcal{J}^*(U^\pm) \subset C^*(U^\pm)$ but one obtains in this way the same complex $Q^*(S)$ as only the values on S of the coefficients of the forms considered are of importance.

We thus can consider four types of cohomology groups

$H^*(U)$ the cohomology of $C^*(U)$

$H^*(U^\pm)$ the cohomology of $C^*(U^\pm)$

$H^*(S)$ the cohomology of $Q^*(S)$.

Note that while the cohomology $H^*(U)$ is the cohomology of U with values in the sheaf $\mathcal{O}$ the same is not true for $H^*(U^\pm)$.

<u>Remark.</u>

We have $\mathcal{J}^{0,0}(U) = \{\phi^{0,0} : U \to \mathbb{C} \mid \phi^{0,0}|_S = 0\}$ thus $Q^{0,0}(S)$ represents the space of C^∞ functions on S. For $u \in Q^{0,0}(S)$ the condition $\bar\partial_S u = 0$ is a necessary condition for u to be the trace on S of a function $\tilde u \in C^{0,0}(U^+)$ (or $\tilde u \in C^{0,0}(U^-)$) which is holomorphic in U^+ (or U^-). Indeed if $\hat u$ is a C^∞ extension of $\tilde u$ to U^- (or U^+) we have $\bar\partial \hat u = 0$ on U^+ (or U^-) and therefore $\bar\partial \hat u = \rho\alpha$ for some $\alpha \in C^{0,1}(U)$, i.e. $\bar\partial_S u = 0$.

In general calling the image of a form of $C^{0,s}(U)$ on $Q^{0,s}(S)$ the trace of that form on S, we can state that for $u \in Q^{0,s}$ the condition $\bar\partial_S u = 0$ is a necessary condition for

u to be the trace on S of a form $\hat{u} \in C^{0,s}(U^+)$ with $\bar{\partial}\hat{u} = 0$
Indeed let $\tilde{u}$ be any form of $C^{0,s}(U^+)$ having u as trace on S.
Then

$$\tilde{u} = \hat{u} + \rho\alpha + \bar{\partial}\rho \wedge \beta \qquad \text{for some } \alpha \text{ and } \beta \text{ in } C^*(U^+).$$

Thus

$$\bar{\partial}\tilde{u} = \rho\bar{\alpha} + \bar{\partial}\rho(\alpha + \bar{\partial}\beta) \in \mathcal{J}^{0,s+1}(U^+)$$

i.e.

$\bar{\partial}_S u = 0.$ Same argument for U^-.

<u>Example.</u>

Take $U = \mathbb{C}^2$, $z_1 = x_1 + ix_2$, $z_2 = x_3 + ix_4$ as co ordinates on $\mathbb{C}^2$. Let

$$\rho \equiv x_4 - (x^2 + x^2) = \frac{1}{2i}(z_2 - \bar{z}_2) - |z_1|^2 .$$

Then S is the product of the paraboloid $x_4 = x_1^2 + x_2^2$ in $\mathbb{R}^3$ where x_1, x_2, x_4 are coordinates, by the x_3-axis.

At each point $d\bar{z}_1$ and $\bar{\partial}\rho = -\frac{1}{2i} d\bar{z}_2 - z_1 d\bar{z}_1$ form a basis for the (0,1)-forms. Thus we have

$$Q^{0,0}(S) = C^\infty(S) = C^\infty \text{ functions on } S$$

$$Q^{0,1}(S) \simeq C^\infty(S) \wedge d\bar{z}_1$$

$$Q^{0,2}(S) = 0.$$

Thus

$$Q^*(S) = \equiv \{C^\infty(S) \xrightarrow{\bar{\partial}_S} C(S) \wedge d\bar{z}_1 \to 0\}.$$

To compute $\bar{\partial}_S$, by its definition, we have to do the following

- given $u \in C^\infty(S)$ extend (in any way) to a C^∞ function $\tilde{u}$ on $\mathbb{C}^2$. Due to the shape of S we may assume $\tilde{u}$ independent of the variable x_4;
- compute $\bar{\partial}\tilde{u}$;

- compute $\bar\partial\tilde{u}$;

$$\bar\partial\tilde{u} = \frac{\partial\tilde{u}}{\bar{z}_1} d\bar{z}_1 + \frac{\partial\tilde{u}}{\bar{z}_2} d\bar{z}_2$$

$$\equiv \frac{\partial\tilde{u}}{\partial\bar{z}_1} d\bar{z}_1 - \frac{\partial\tilde{u}}{\partial\bar{z}_2} (2i\bar\partial\rho + 2iz_1 d\bar{z}_1)$$

$$= (\frac{\partial\tilde{u}}{\bar{z}_1} - 2iz_1 \frac{\partial\tilde{u}}{\partial\bar{z}_2}) d\bar{z}_1 - 2i \frac{\partial\tilde{u}}{\partial\bar{z}_2} \bar\partial\rho;$$

- "restrict" the form thus obtained to S i.e. compute modulo $\mathcal{J}^{0,1}(\mathbb{C}^2))$ to get

$$\bar\partial_S u = (\frac{\partial u}{\partial\bar{z}_1} - iz_1 \frac{\partial u}{\partial x_3}) d\bar{z}_1 ,$$

as $\tilde{u}$ is independent of x_4 and x_1, x_2, x_3 can be taken can be taken as coordinates on S.

In conclusion the complex $Q^*(S)$ is isomorphic to the complex on $\mathbb{R}^3$ where $z_1 = x_1 + ix_2$, and x_3 are coordinates

$$C^\infty(\mathbb{R}^3) \xrightarrow{L} C(\mathbb{R}^3) \to 0$$

where

$$L = \frac{\partial}{\partial\bar{z}_1} - iz_1 \frac{\partial}{\partial x_3} .$$

The operator L is the operator of H. Lewy, of [37].

9.2. <u>Mayer-Vietrois sequence</u>.

a) A C^∞ function on U is called <u>flat on</u> S if it vanishes on S with all of its partial derivatives. Set

$\mathcal{F}^{0,s}(U) = \{\phi \in C^{0,s}(U) \mid$ all coefficients of ϕ are flat on $S\}$.

We have $\bar\partial \mathcal{F}^{0,s}(U) \subset \mathcal{F}^{0,s+1}(U)$ therefore $\mathcal{F}^*(U) = \oplus_s \mathcal{F}^{0,s}(U)$ is another subcomplex of $C^*(U)$ and in fact a subcomplex of $\mathcal{G}^*(U)$. The quotient complex $C^*(S) = C^*(U)/\mathcal{F}^*(U)$ is concentrated on S and is obtained by restricting to S the coefficients of the C^∞ forms on U.

We have

$$C^{0,s}(S) = \{ \sum_{\alpha_1<\dots<\alpha_s} a_{\alpha_1\dots\alpha_s}(x)\, d\bar z_{\alpha_1} \dots d\bar z_{\alpha_s} \mid a_{\alpha_1\dots\alpha_s} \in C^{0,0}(S) \text{ for all } (\alpha_1\dots\alpha_s) \}.$$

$0 \to C^*(U) \to C^*(U^+) \oplus C^*(U^-) \to C^*(S) \to 0$ is an exact sequence from which we get a cohomology sequence

$$(1)\quad \begin{aligned} &0 \to H^{0,0}(U) \to H^{0,0}(U^+) \oplus H^{0,0}(U^-) \to H^{0,0}(C^*(S)) \to \\ &\to H^{0,1}(U) \to H^{0,1}(U^+) \oplus H^{0,1}(U^-) \to H^{0,1}(C^*(S)) \to \dots \end{aligned}$$

To connect the cohomology groups $H^{0,s}(C^*(S))$ with the groups $H^{0,s}(S) = H^{0,s}(Q^*(S))$ we may use the following exact sequence

$$0 \to \mathcal{G}^*(U)/\mathcal{F}^*(U) \to C^*(U)/\mathcal{F}^*(U) \to Q^*(S) \to 0$$

i.e.

$$(2)\quad 0 \to \mathcal{G}^*(U)/\mathcal{F}^*(U) \to C^*(S) \to Q^*(S) \to 0.$$

Lemma (9.2.1). For any choice of U and S the sequence

$$0 \to \frac{\mathcal{G}^{0,0}(U)}{\mathcal{F}^{0,0}(U)} \xrightarrow{\bar\partial} \frac{\mathcal{G}^{0,1}(U)}{\mathcal{F}^{0,1}(U)} \xrightarrow{\bar\partial} \frac{\mathcal{G}^{0,2}(U)}{\mathcal{F}^{0,2}(U)} \xrightarrow{\bar\partial} \dots \text{ is exact.}$$

Proof.

(α) Let $u \in \mathcal{G}^{0,0}(U)$ thus $u = \rho a_1$ for some $\alpha_1 \in C^{0,0}(U)$. Assume that the coefficients of $\bar\partial u$ are flat on S. Then $\bar\partial \rho a_1|_S = 0$.

This means that $\alpha_1 = \rho\alpha_2$ for some $\alpha_2 \in C^{0,0}(U)$, thus $u = \rho^2\alpha_2$. But then

$$\bar{\partial}\rho\, \alpha_2|_S = 0.$$

thus $\alpha_2 = \rho\alpha_3$ for some $\alpha_3 \in C^{0,0}(U)$, thus $u = \rho^3\alpha_3$. Continuing in this way we see that u must be flat on S i.e. we have exactness at $\mathcal{J}^{0,0}/\mathcal{F}^{0,0}$.

(β) To treat the general case we will make use of the following fact

<u>Given on</u> S <u>a sequence</u> $f_0, f_1, f_2, \ldots$ <u>of</u> C^∞ <u>functions, there exists a</u> C^∞ <u>function</u> F <u>on</u> U <u>such that</u>

$$\frac{\partial^k F}{\partial \rho^k}\Big|_S = f_k \qquad \text{for } k = 0, 1, 2, \ldots .$$

This can be derived as a particular case from the Whitney extension theorem. Also direct proofs are available.[(1)]

Let $f \in \mathcal{J}^{0,s}(U)$, $s \geq 1$. Then $f = \rho\alpha + \bar{\partial}\rho \wedge \beta$ for some and β. Using the above remark, we can find a form $\beta_1 \in C^{0,s-1}(U)$ such that

$$\beta|_{1\,S} = \beta|_S$$

$$(*) \quad \frac{\partial^k \beta_1}{\partial \rho^k}\Big|_S = 0 \qquad \text{for } k = 1, 2, 3, \ldots$$

Thus we can write $f = (f - \bar{\partial}\rho \wedge \beta_1) + \bar{\partial}\rho \wedge \beta_1 = \rho\alpha_1 + \bar{\partial}\rho \wedge \beta_1$ as the coefficient of $f - \bar{\partial}\rho \wedge \beta_1$ vanish on S while β_1 satisfies the conditions (*). Let now $u \in \mathcal{J}^{0,s}(U)$ and assume that the coefficients of $\bar{\partial}u$ are flat on S. Write $u = \rho\alpha_1 + \bar{\partial}\rho \wedge \beta_1$ with β_1 satisfying (*).

---

(1) In particular it follows from this lemma that the space $C^{0,0}(S)$ can be identified with the space $\mathcal{E}(S)\{\rho\}$ of formal power series in ρ with C^∞ coefficients on S.

Then

$$u - \bar\partial(\rho\beta_1) = \rho(\alpha_1 - \bar\partial\beta_1).$$

Set $\gamma_1 = \alpha_1 - \bar\partial\beta_1$. By the assumption we get $\bar\partial\rho \wedge \gamma_1|_S = 0$ thus

$\gamma_1 = \rho\alpha_2 + \bar\partial\rho \wedge \beta_2$ with β_2 satisfying (*). Then

$$u - \bar\partial(\rho\beta_1 + \tfrac{1}{2}\rho^2\beta_2) = \rho^2(\alpha_2 - \tfrac{1}{2}\bar\partial\beta_2).$$

Set $\gamma_2 = \alpha_2 - \frac{1}{2}\bar\partial\beta_2$. By the assumption $\bar\partial\rho \wedge \gamma_2|_S = 0$ thus

$\gamma_2 = \rho\alpha_3 + \bar\partial\rho \wedge \beta_3$ with β_3 satisfying (*). Then

$$u - \bar\partial(\rho\beta_1 + \tfrac{1}{2}\rho^2\beta_2 + \tfrac{1}{3}\rho^3\beta_3) = \rho^3(\alpha_3 - \tfrac{1}{3}\bar\partial\beta_3).$$

Proceeding in this way we construct a formal power series in

$$\phi = \sum_1^\infty \frac{\rho^m}{m}\beta_m$$

with β_m satisfying (*) and with the property that

$$u - \bar\partial\left(\sum_1^{m+1} \frac{\rho^k}{k}\beta_k\right) = \rho^{m+1}\gamma_{m+1\ m+1} \in C^{0,s}(U).$$

Using the remark made of the beginning, we can find $f \in C^{0,s-1}(U)$ such that

$$\frac{\partial^k}{\partial\rho^k} f|_S = \frac{\partial^k \phi}{\partial\rho^k} = 0 \quad \text{for } k = 0, 1, 2, \ldots .$$

Set $v = \rho f$ then $\rho f \in \mathcal{G}^{0,s-1}(U)$ and we have that $u - \bar\partial v$ has flat coefficients on S. This proves exactness at $\mathcal{G}^{0,s}(U)/\mathcal{F}^{0,s}(U)$.

This lemma tells us that the cohomology of $\mathcal{G}^*(U)/\mathcal{F}^*(U)$ is zero in any dimension and therefore the cohomology sequence of (2) gives a set of isomorphism

$$H^{0,s}(C^*(S)) \simeq H^{0,s}(Q^*(S)) = H^{0,s}(S).$$

Introducing this result in the sequence (1) we obtain an exact sequence

$$0 \to H^{0,0}(U) \to H^{0,0}(U^+) \oplus H^{0,0}(U^-) \to H^{0,0}(S) \to$$

$$\to H^{0,1}(U) \to H^{0,1}(U^+) \oplus H^{0,1}(U^-) \to H^{0,1}(S) \to \dots$$

which is called the <u>Mayer-Vietoris</u> sequence for U and S.

b) The previous considerations can be repeated replacing the space of C^∞ functions on U, $U^\pm$ with the space of those C functions with compact support on U, $U^\pm$ respectively. This leads to an exact sequence

$$0 \to H^{0,0}_k(U) \to H^{0,0}_k(U^+) \oplus H^{0,0}_k(U^-) \to H^{0,0}_k(S) \to$$

$$\to H^{0,1}_k(U) \to H^{0,1}_k(U^+) \oplus H^{0,1}_k(U^-) \to H^{0,1}_k(S) \to \dots$$

(Mayer-Vietoris sequence with compact supports).

9.3. <u>Bochner theorem.</u>

Let X be any $(n-2)$-complete connected manifold of complex dimension $n \geq 2$, for instance a Stein manifold. Let S be any connected closed C^∞ hypersurface in X such that $X-S = \mathring{X}^- \cup \mathring{X}^+$ is the union of two connected open sets $\mathring{X}^-$ and $\mathring{X}^+$, of which say $\mathring{X}^-$ is relatively compact.

<u>Theorem</u> (9.3.1). <u>Under the above assumptions, any</u> C^∞ <u>function</u> f <u>on</u> S <u>satisfying the compatibility condition</u> $\bar\partial_S f = 0$ <u>is the trace on</u> S <u>of a</u> C^∞ <u>function on</u> X^- <u>which is holomorphic in</u> $\mathring{X}^-$.

<u>Proof.</u> The Mayer-Vietoris sequence with compact supports gives

$$0 \to H^{0,0}_k(X) \to H^{0,0}_k(X^-) \oplus H^{0,0}_k(X^+) \to H^{0,0}_k(S) \to$$

$$\to H^{0,0}_k(X) \to \dots$$

Now $H^{0,0}_k(X) = H^{0,0}_k(X^+) = 0$ as X and X^+ are connected and and not compact.

Moreover $H^{0,0}_k(X^-) = H^{0,0}_k(X^-)$ as X^- is compact. By the duality theorem we get

$$H^{0,1}_k(X) \simeq \text{Hom cont } (H^{n-1}(X, \ ^n), \mathbb{C}) = 0$$

by the assumption that $n \geq 2$ and X is $(n-2)$-complete.

Since $H^{n-1}(X, \Omega^n)$ and $H^n(X, \Omega^n)$ are finite-dimensional, in

$$C^{n-2}(\mathcal{U}, \Omega^n) \quad C^{n-1}(\mathcal{U}, \Omega^n) \quad C^n(\mathcal{U}, \Omega^n)$$

the maps δ are topological homomorphisms.

Hence $H_{n-1}(X, \Omega^n_*) \simeq \text{Hom cont } (H^{n-1}(X, \Omega^n), \mathbb{C})$

$$\simeq \text{Hom } (H^{n-1}(X, \Omega^n), \mathbb{C}).$$

And $H_{n-1}(X, \Omega^n_*) = \text{Ext}^1_k \ (X; \Omega^n, \Omega^n) = H^1_k(X, \Omega^n(E^*)) = H^1_k(X \mathcal{O})$.

Here E^* is the dual bundle of the canonical bundle E.

Hence the restriction map

$$H^{0,0}(X^-) \to H^{0,0}(S) \quad \text{is surjective.}$$

For $X = \mathbb{C}^n$ this theorem is due to Bochner, Fichera, Martinelli (cf. [16,] [24,] [41]).

Note that no assumption is made on the shape or convexity of S.

9.4. Riemann-Hilbert and Cauchy problem.

These problems are generalizations of a problem concerning holomorphic functions in several variables first considered and solved by Hans Lewy (cf. [37], [38]).

Assume that U is an open set of holomorphy in $\mathbb{C}^n$. Then the Mayer-Vietoris sequence (with closed supports) splits in the short exact sequences

$$0 \to H^{0,0}(U) \to H^{0,0}(U^+) \oplus H^{0,0}(U^-) \to H^{0,0}(S) \to 0$$

$$0 \to H^{0,s}(U^+) \oplus H^{0,s}(U^-) \to H^{0,s}(S) \to 0, \quad \text{for } s > 0.$$

This shows that

(a) Every cohomology class on S can be written as a jump of a cohomology class on U^+ and a cohomology class on U^- i.e. the so-called Riemann-Hilbert problem is always solvable for an open set of holomorphy.
Moreover if $s > 0$ the solution is unique while for $s = 0$ it is determined up to addition to the functions on U^+ and U^- of a global holomorphic function on U.

(b) Let us agree to say that the Cauchy-problem is solvable from the side of U^+ if $H^{0,s}(U^+) \to H^{0,s}(S)$ is surjective. Then we realize that if U is an open set of holomorphy the solvability of Cauchy-problem for $s > 0$ is equivalent to the vanishing theorem

$$H^{0,s}(U^-) = 0$$

and in this case the solution is unique.
If $s = 0$ the Cauchy-problem is solvable if and only if

$$H^{0,0}(U) \to H^{0,0}(U^-)$$

is surjective i.e., loosely speaking, iff U is contained in the envelope of holomorphy of U^-. The solution will be unique if for instance, U^+ is connected.

Example.

Let us consider the situation of the example given at the end of section 9.1. Here $U^+ = \{x_4 - x_1^2 + x_2^2\}$ is an elementary convex set while U^- is the closure of the complement in $\mathbb{C}^2$. The boundary S of U^+ is strongly Levi-convex. Let $a \in S$ and let Ω be any neighborhood of a which is also a domain of holomorphy. Writing for $\Omega, \Omega^+, \Omega^-$ and $S_\Omega = \Omega \cap S$ the Mayer-Vietoris sequence, we get the exact sequence

$$0 \to H^{0,0}(\Omega) \to H^{0,0}(\Omega^+) \oplus H^{0,0}(\Omega^-) \to H^{0,0}(S_\Omega)$$

$$\to H^{0,1}(\Omega) \to H^{0,1}(\Omega^+) \oplus H^{0,1}(\Omega^-) \to H^{0,1}(S_\Omega)$$

$$\to H^{0,2}(\Omega) \to H^{0,2}(\Omega^+) \oplus H^{0,2}(\Omega^-) \to 0$$

We have

i) $H^{0,1}(\Omega) = 0$, $H^{0,2}(\Omega) = 0$, as Ω is a domain of holomorphy.

ii) $H^{0,1}(\Omega^+) = 0$.
This fact is a consequence of the regularity theorem of Kohn and Nirenberg (cf. next section). It would be desirable to obtain a direct proof in this special case.

iii) Since U^- has a pseudoconcave boundary one realizes that for every point $a \in S$ we can find a fundamental sequence of neighborhoods Ω of a, which are domains of holomorphy and such that

$$H^{0,0}(\Omega) \to H^{0,0}(\Omega^-)$$

is surjective (and an isomorphism).

Making use of this information the sequence of Mayer-Vietoris gives us the following isomorphism

$$H^{0,0}(\Omega^+) \xrightarrow{\sim} H^{0,0}(S_\Omega)$$

$$H^{0,1}(\Omega^-) \xrightarrow{\sim} H^{0,1}(S_\Omega)$$

The first tells us that <u>given any</u> C^∞ <u>functions on</u> S_Ω <u>satisfying the compatibility condition</u> $\bar\partial_S u = 0$ <u>there exists a</u> C^∞ <u>function</u> $\hat u$ <u>in</u> Ω^+ <u>which is holomorphic in</u> Ω^+ <u>and such that</u>

$$\hat u|_S = u .$$

(use is made of the assumption iii)).

In connection with the second isomorphism (which is valid even if the assumption iii) is not satisfied) we remark that

iv) $H^{0,1}(\Omega^-) \neq 0$ and in fact $\dim_{\mathbb{C}} H^{0,1}(\Omega^-) = \infty$, provided Ω is sufficiently small. In fact by a local change of holomorphic coordinates at a we may assume that S_Ω is strongly elementary convex (1.4, exercise 2). Then the statement is a straightforward consequence of lemma (7.2.2).

An immediate consequence of this fact and the isomorphism established above is the following theorem first proved by H. Lewy [37].

<u>Given on $\mathbb{R}^3$ the equation</u>

$$Lu \equiv \frac{\partial u}{\partial \bar{z}_1} - i z_1 \frac{\partial u}{\partial x_3} = f ;$$

<u>for any point</u> $a \in \mathbb{R}^3$ <u>we can find a fundamental sequence of neighborhoods</u> ω_ν <u>such that for infinitely many</u> $f \in C^\infty(\omega_\nu)$ <u>the equation does not admit any solution</u> $u \in C^\infty(\omega_\nu)$.

9.5. <u>Cauchy-problem as a vanishing theorem for cohomology.</u>

Let us now consider the Levi form restricted to the analytic tangent plane of the hypersurface S.
Using the methods of proof of the vanishing theorem (7.1.1) and the regularization theorem of Kohn and Nirenberg, (see [34], [32] and [5]) we obtain the following result

<u>Theorem</u> (9.5.1). <u>For any point</u> $z_0 \in S$ <u>at which the Levi form has</u> p <u>positive and</u> q <u>negative eigenvalues on the analytic tangent plane to</u> S at z_0, <u>we can find a fundamental sequence of neighborhoods</u> $\{U_\nu\}_{\nu\in\mathbb{N}}$ <u>of</u> z_0, <u>all domains of holomorphy, such that</u>

$$H^{0,s}(U^+_\nu) = 0 \quad \text{if} \quad \begin{cases} s > n-q-1 \\ \text{or} \\ 0 < s < p \end{cases} \quad \nu \in \mathbb{N}$$

Analogously one can find a similar fundamental sequence of neighborhoods $\{U_\nu\}_{\nu \in \mathbb{N}}$ *such that*

$$H^{0,s}(U_\nu^-) = 0 \quad \text{\underline{if}} \quad \begin{cases} s > n-p-1 \\ \text{\underline{or}} \\ 0 < s < q \end{cases} \qquad \nu \in \mathbb{N}.$$

Moreover, if $p > 0$, *we can select the sequence* $\{U_\nu\}$ *in such a way that the restriction*

$$H^{0,0}(U_\nu) \to H^{0,0}(U_\nu^+)$$

is surjective, i.e. U_ν *is in the "envelope of holomorphy" of* U_ν^+. *Analogously, if* $q > 0$, *we can select the sequence* $\{U_\nu\}$ *in such a way that the restriction*

$$H^{0,0}(U_\nu) \to H^{0,0}(U_\nu^-)$$

is surjective, i.e. U_ν *is in the "envelope of holomorphy" of* U_ν^-.

According to the remarks made in the previous sections this theorem tells us when locally the Cauchy problem for cohomology classes is solvable. Two special cases will serve as an illustration

Case 1. Assume that the Levi-form is non degenerate with

$$0 < p < q = n-1-p$$

Marking only the cohomology groups which are (possible) different from zero, the situation is illustrated by the following picture

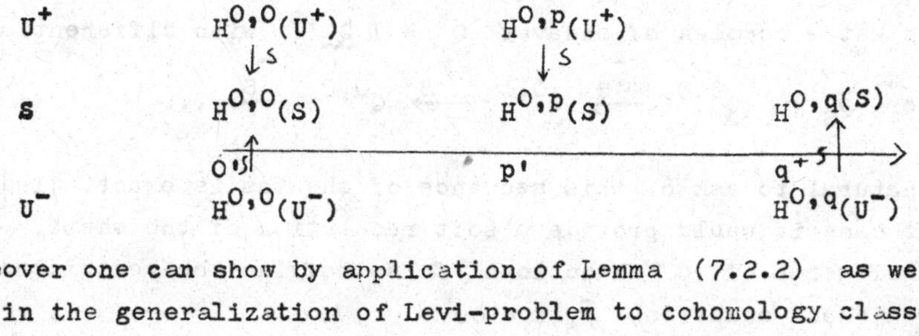

Moreover one can show by application of Lemma (7.2.2) as we did in the generalization of Levi-problem to cohomology classes

(point 8) in the proof) that the cohomology groups we have marked are all infinite-dimensional.

Case 2. Assume that the Levi-form is non-degenerate, n is odd and

$$0 < p = q = \frac{n-1}{2},$$

then with the same conventions the situation is illustrated by the following picture

$$\begin{array}{llll}
U^+ & & H^{0,0}(U^+) & H^{0,p}(U^+) \\
 & & \downarrow s & \downarrow \\
S & & H^{0,0}(S) & H^{0,p}(S) \simeq H^{0,p}(U^+) \oplus H^{0,p}(U^-) \\
 & 0 & \uparrow s & p\uparrow \\
U^- & & H^{0,0}(U^-) & H^{0,p}(U^-)
\end{array}$$

Again the groups marked in the picture are all infinite-dimensional In this case in dimension p the Cauchy problem is not solvable from either side, only the Riemann-Hilbert problem is solvable in that dimension.

Remark. In both cases in dimensions p and q we are in the presence of systems of first order partial differential equations $Lu = f$ on S which for infinitely many C^∞ functions f satisfying the integrability conditions have no solution u of class C^∞.

9.6. Non-validity of Poincaré lemma for the complex $\{\underset{\sim}{Q}^*(S), \bar\partial_S\}$.

On S we can consider for any s the sheaf $Q^{0,s}$ defined by the presheaf

$$\Omega \to Q^{0,s}(\Omega) .$$

We thus get a complex of sheaves $\underset{\sim}{Q}^* = \amalg \underset{\sim}{Q}^{0,s}$ with differential operator $\bar\partial_S$: $\underset{\sim}{Q}^{0,0} \xrightarrow{\bar\partial_S} \underset{\sim}{Q}^{0,1} \xrightarrow{\bar\partial_S} \underset{\sim}{Q}^{0,2} \xrightarrow{\bar\partial_S} \dots$

It is natural to ask of this sequence of sheaves is exact. Indeed in that case it would provide a soft resolution of the sheaf $\mathcal{O}(S)$ of germs of C^∞ functions f satisfying the restricted Cauchy-Riemann equations $\bar\partial_S f = 0$.

The answer to this question is in general negative as does show the following

Theorem (9.6.1). Let S be a locally closed hypersurface in $\mathbb{C}^n$ and let $z_o \in S$ be a point at which the Levi form on the analytic tangent space of S at z_o has p positive and q negative eigenvalues and is non-degenerate (so that $p+q = n-1$). Then in the complex

$$\underset{\sim}{Q}^{0,0} \xrightarrow{\bar\partial_S} \underset{\sim}{Q}^{0,1} \xrightarrow{\bar\partial_S} \underset{\sim}{Q}^{0,2} \xrightarrow{\bar\partial_S} \cdots$$

the Poincaré lemma is not valid in dimensions p and q but holds in any other dimension.

Proof. From theorem (9.5.1) and the Mayer-Vietoris sequence we deduce that there exists a fundamental sequence of neighborhoods ω_ν of z_o in S, $\nu = 1, 2, 3, \ldots,$ such that $H^{0,s}(\omega_\nu) = 0$ if $s \neq p,q$. Thus the Poincaré lemma for $\bar\partial_S$ is valid in dimensions different from p and q.

We have to show that this is no longer true in dimensions p and q. Let ω be any neighborhood of z_o in S. First we remark that

$$Q^{0,p}(\omega) \simeq C^\infty(\omega)^{\alpha(p)}$$

where $(p) = \binom{n-1}{p}$. Indeed we can select a basis for the space of $(0,1)$ forms in a neighborhood Ω of z_o in $\mathbb{C}^n$, $\Omega \cap S = \omega$, of the form $\eta_1,\ldots,\eta_{n-1}, \bar\partial\rho$.

Then

$$Q^{0,p}(\omega) \simeq \Big\{ \sum_{\alpha_1<\ldots<\alpha_p} a_{\alpha_1\ldots\alpha_p}\eta_{\alpha_1}\wedge\ldots\wedge\eta_{\alpha_p} \Bigm| a_{\alpha_1\ldots\alpha_p} \in C^\infty(\omega) \Big\} .$$

As such $Q^{0,p}(\omega)$ has a natural structure of a Fréchet space.

Given ω we can select a fundamental sequence of neighborhoods $\omega \subset \varpi, \nu = 1,2,3,\ldots,$ of z_o in S such that $H^p(\omega_\nu) \neq 0$ as we have remarked in section 9.5.
Set for $\omega = \omega$ or ω_ν

$$Z^p(\omega) = \mathrm{Ker}\{\bar\partial_S : Q^{0,p}(\omega) \to Q^{0,p+1}(\omega)\}$$

$$B^p(\omega) = \mathrm{Im}\{\bar\partial_S : Q^{0,p-1}(\omega) \to Q^{0,p}(\omega)\}$$

Since $\bar\partial_S$ is a differential operator, it is continuous for the topology of $Q^*(\omega)$. Therefore $Z^p(\omega)$ is a closed subspace of $Q^{0,p}(\omega)$ and therefore a Fréchet space. On its turn $B^p(\omega) = Q^{0,p-1}(\omega) / Z^{p-1}(\omega)$ inherits a quotient structure of a Fréchet space.
Consider for every ν the following set of continuous maps

$$\begin{array}{ccc} Z^p(\omega) & \xrightarrow{r_\nu} & Z^p(\omega_\nu) \\ & & \uparrow i_\nu \\ & & B^p(\omega_\nu) \end{array}$$

where i_ν is the natural injection and r_ν the restriction map.
Set

$$E_\nu = \{(\alpha,\beta) \in Z^p(\omega) \times B^p(\omega_\nu) \mid r_\nu(\alpha) = i_\nu(\beta)\}.$$

Then E_ν as a closed subspace of $Z^p(\omega) \times B^p(\omega_\nu)$ has the structure of a Fréchet space and we can complete the previous maps with the following commutative diagram

$$\begin{array}{ccc} Z^p(\omega) & \xrightarrow{r_\nu} & Z^p(\omega_\nu) \\ \uparrow j & & \uparrow i \\ E_\nu & \xrightarrow[\sigma]{} & B^p(\omega_\nu) \end{array}$$

with j and σ continuous. We have $j(E_\nu) = r_\nu^{-1}\, i_\nu(B^p(\omega_\nu))$. Thus this space is a continuous image by a linear map of a Fréchet space. By the Banach open mapping theorem we must have one of these two properties

i) either $r_\nu^{-1}\, i_\nu(B^p(\omega_\nu)) = Z^p(\omega)$

ii) or else $r_\nu^{-1}\, i_\nu(B^p(\omega_\nu)$ is of first category in $Z^p(\omega)$.

Now as we have remarked in section 9.5. we can construct for each ν an element $\phi_\nu \in Z^p(\omega)$ such that $r_\nu(\phi_\nu) \notin i_\nu(B^p(\omega_\nu))$. This rules out possibility i) for every ν. Hence

$$\bigcup_{\nu=1}^{\infty} r_\nu^{-1}\, i_\nu(B^p(\omega_\nu)) \quad \text{is of first category.}$$

Therefore there exists an element $g \in Z^p(\omega)$ such that for every ν

$$g \notin r_\nu^{-1}\, i_\nu(B^p(\omega_\nu))$$

i.e. such that the equation

$$\bar\partial_S u_\nu = g$$

cannot be solved in ω_ν although the integrability condition $\bar\partial_S\, g = 0$ is satisfied in the whole of ω.

This shows that Poincaré lemma cannot hold in dimension p. For dimension q the argument is the same.

<u>Example</u>. In the particular case of the Lewy operator $Lu = f$ in $\mathbb{R}^3$ it follows that given $a \in \mathbb{R}^3$ there exists a neighborhood ω of a and $f \in C^\infty(\omega)$ wuch that for any $\omega_\nu \subset \omega$ the equation $Lu = f$ cannot be solved.

9.7. <u>Global results</u>.

We illustrate the type of global results one can obtain by the following example. For the proofs we refer to [5].

Let X be a compact connected manifold of complex dimension n. Let $\rho : X \to \mathbb{R}$ be a C^∞ function and assume that

$$S = \{x \in X \mid \rho(x) = 0\}$$

is a smooth hypersurface ($d\rho \neq 0$ on S) dividing X into the two regions

$$X^- = \{x \in X \mid \rho(x) \leq 0\}, \quad X^+ = \{x \in X \mid \rho(x) \geq 0\}.$$

We assume that the Levi form of ρ restricted to the analytic tangent plane to S is nowhere degenerate and has p positive and q negative eigenvalues ($p+q = n-1$).
Then one can prove that

$$\dim_{\mathbb{C}} H^{0,s}(X^-) < \infty \quad \text{if} \quad s \neq q$$

$$\dim_{\mathbb{C}} H^{0,q}(X^-) = \infty$$

and similarly

$$\dim_{\mathbb{C}} H^{0,s}(X^+) < \infty \quad \text{if} \quad s \neq p$$

$$\dim_{\mathbb{C}} H^{0,p}(X^+) = \infty .$$

Let us agree to say that the <u>global Riemann-Hilbert problem is almost always solvable in dimension</u> r iff

$$H^{0,r}(X^+) \oplus H^{0,r}(X^-) \to H^{0,r}(S)$$

has finite-dimensional kernel and cokernel.
Similarly we agree to say that the <u>Cauchy-problem is almost always solvable in dimension</u> r <u>from the side</u> X^+ iff

$$H^{0,r}(X^+) \to H^{0,r}(S)$$

has finite-dimensional kernel and cokernel.

A straightforward application of the Mayer-Vietoris sequence in connection with the description of the groups $H^{0,r}(X^{\pm})$ given above leads to the following conclusion:

If $p \neq q$ the Cauchy problem is of interest[(1)] only in dimension p from the side X^+ and in dimension q from the side X^- and it is almost always solvable. If $p = q = \frac{n-1}{2}$ (n must be odd) then the Cauchy problem is of interest in dimension p but not almost always solvable on either side, while the Riemann-Hilbert problem is of interest in dimension p and almost always solvable.

It is worth noticing that these considerations can be extended to more general complexes of partial differential operators.

The operator $\bar{\partial}_S$ was first introduced in a different form in [35]. In this exposition we have followed [4] and [5].

---

(1)

In the sense that it leads to maps between infinite-dimensional spaces.

A. Andreotti

BIBLIOGRAPHY.

[1] Andreotti, A.: Théorèmes de dependance algébrique sur les espaces complexes pseudo-concaves. Bull. Soc. Math. France 91 (1963), 1-38.

[2] Andreotti, A. and Grauert, H.: Théorèmes de finitude pour la cohomologie des espaces complexes. Bull. Soc. Math. France 90 (1962), 193-259.

[3] Andreotti, A. and Grauert, H.: Algebraische Körper von automorphen Funktionen. Nachr. Ak. Wiss. Göttingen (1961), 39-48.

[4] Andreotti, A. and Hill, C.D.: E.E.Levi convexity and the Hans Lewy problem. Part I: Reduction to vanishing theorems. Ann. Sc. Norm. Sup. Pisa 26 (1972), 325-363.

[5] Andreotti, A. and Hill, C.D.: E.E. Levi convexity and the Hans Lewy problem. Part II: Vanishing theorems. Ann. Sc. Norm. Sup. Pisa 26 (1972), 747-806.

[6] Andreotti, A. and Huckleberry, A.: Pseudoconcave Lie groups. Compositio Mathematica 25 (1972), 109-115.

[7] Andreotti, A. and Kas, A.: Duality theorems for complex spaces. Ann. Sc. Norm. Sup. Pisa, to appear.

[8] Andreotti, A. and Norguet, F.: Problème de Levi et convexité holomorphe pour les classes de cohomologie. Ann. Sc. Norm. Sup. Pisa, s. 3, 20 (1966), 197-241.

[9] Andreotti, A. and Norguet, F.: La convexité holomorphe dans l'espace analytique des cycles d'une variété algébrique. Ann. Sc. Norm. Sup. Pisa, s. 3, 21 (1967), 31-82.

[10] Andreotti, A. and Yum-Tong Siu: Projective embedding of pseudoconcave spaces. Ann. Sc. Norm. Sup. Pisa, s. 3, 24 (1970), 231-278.

[11] Andreotti, A. and Stoll, W.: Analytic and algebraic dependence of meromorphic functions. Lecture notes in Mathematics 234. Springer, Berlin, 1971.

[12] Andreotti, A. and Tomassini, G.: Some remarks on pseudoconcave manifolds. Essays in topology and related topics. Mémoires dédiés à G. de Rham, 1970, 85-104.

[13] Andreotti, A. and Vesentini, E.: Sopra un teorema di Kodaira. Ann. Sc. Norm. Sup. Pisa, s. 3, 15 (1961), 283-309.

[14] Andreotti, A. and Vesentini, E.: Carleman estimates for the Laplace-Beltrami equation on complex manifolds. Publications Mathématiques I.H.E.S. 25 (1965), 81-130.

[15] Barth, W.: Der Abstand von einer algebraischen Mannigfaltigkeit im komplex-projectiven Raum. Math. Ann. 187 (1970), 150-162.

[16] Bochner, S.: Analytic and meromorphic continuation by means of Green's formula. Ann. of Math. 44 (1943), 652-673.

[17] Borel, A.: Pseudo-concavité et groupes arithmétiques. Essays in topology and related topics. Mémoires dédiés a G. de Rham, 1970, 70-84.

[18] Bredon, G.: Sheaf theory. Mc. Graw-Hill series in higher mathematics, 1967.

[19] Cartan, H.: Quotient d'un espace analytique par un groupe d'automorphismes. Algebraic geometry and topology. A symposium in honor of S. Lefschetz. Princeton U.P. 1957, 90-102.

[20] Cartan, H.: Quotients of complex analytic spaces, Contributions to function theory. Tata Institute of Fundamental Research, Bombay, 1960, 1-15.

[21] Chow, W.L. and Kodaira, K.: On analytic surfaces with two independent meromorphic functions. Proc. Nat. Acad. Sci. U.S.A. 38 (1952), 319-325.

[22] Coen, S.: Sul rango dei fasci coerenti, Boll. U.M.I. 22 (1967), 373-382.

[23] Docquier, F. and Grauert, H.: Levisches Problem und Rungescher Satz für Teilgebiete Steinscher Mannigfaltigkeiten. Math. Ann. 140 (1960), 94-123.

[24] Fichera, G.: Caratterizzazione della traccia, sulla frontiera di un campo, di una funzione analitica di più variabili complesse. Atti Accad. Naz. Lincei. Rend. Cl. Sci. Fis. Mat. Nat. 22 (1957), 706-715.

[25] Godement, R.: Topologie algébrique et théorie des faisceaux. Hermann, Paris, 1958.

[26] Grauert, H.: On Levi's problem and the imbedding of real-analytic manifolds. Ann. of Math 68 (1858) 460-472.

[27] Grauert, H.: Über Modifikationen und exzeptionelle analytische Mengen. Math. Ann. 146 (1962). 331-368.

[28] Gunning, R.: and Rossi, H.: Analytic Functions of Several Complex Variables. Prentice-Hall Inc. Englewood Cliffs, N.J., 1965.

[29] Hervé, M.: Several complex variables. Local theory. Tata Institute of Fundamental Research, Bombay, Oxford U.P., 1963.

[30] Hirzebruch, F.: Topological methods in algebraic geometry. Springer, Berlin, 1966.

[31] Hörmander, L.: An introduction to complex analysis in several variables. Van Nostrand, Princeton, N.J., 1966.

[32] Hörmander, L.: L^2 estimates and existence theorems for the $\bar{\partial}$ operator. Acta Math. 113 (1965), 89-152.

[33] Kodaira, K.: On Kähler varieties of restricted type (an intrinsic characterization of algebraic varieties). Ann. of Math. 60 (1954), 28-48.

[34] Kohn, J.J.: and Nirenberg, L.: Non-coercive boundary value problems. Comm. Pure Appl. Math. 18 (1965), 443-492.

[35] Kohn, J.J. and Rossi, H.: On the extension of holomorphic functions from the boundary of a complex manifold. Ann. of Math. 81 (1965), 451-472.

[36] Levi, E.E.: Studii sui punti singolari essenziali delle funzioni analitiche di due o più variabili. Opere, Cremonese, Roma, 1958, 187-213.

[37] Lewy, H.: An example of a smooth linear partial differential equation without solution. Ann. of Math. 66(1957), 155-158.

[38] Lewy, H.: On hulls of holomorphy. Comm. Pure Appl. Math. 13 (1960), 587-591.

[39] Malgrange, B.: Systèmes différentiels à coefficients constants. Sém. Bourbaki 1962, 265.

[40] Kulhmann, N.: Über holomorphe Abbildungen komplexer Räume. Archiv Math. 15 (1964), 81-90.

[41] Martinelli, E.: Sopra un teorema di F. Severi nella teoria delle funzioni analitiche di più variabili complesse. Rend. Mat. e Appl. 20 (1961), 81-96.

[42] Narasimhan, R.: Introduction to the theory of analytic spaces. Lecture notes in Mathematics 25. Springer, Berlin, 1966.

[43] Narasimhan, R.: Several complex variables. U. of Chicago Press, 1971.

[44] Ramis, J.P. and Ruget, G.: Complexe dualisant et théorèmes de dualité en géométrie analytique complexe. Publications Mathématiques I.H.E.S. 38 (1970), 77-91.

[45] Remmert, R.: Projektionen analytischer Mengen. Math. Ann. 130 (1956), 410-441.

[46] Serre, J.P.: Fonctions automorphes, quelques majorations dans le cas où X/G est compact. Sém. H. Cartan, 1953-54. Benjamin 1957.

[47] Serre, J.P.: Un théorème de dualité. Comm. Math. Helvetici 29 (1955), 9-26.

[48] Siegel, C.L.: Analytic functions of several complex variables. Lectures delivered at the Institute for advanced study, 1948-49. Notes by P. Bateman.

[49] Siegel, C.L.: Einführung in die Theorie der Modulformen n' ten Grades. Math. Ann. 116 (1939), 617-657.

[50] Siegel, C.L.: Meromorphe Funktionen auf kompakten analytischen Mannigfaltigkeiten. Nachr. Ak. Wiss. Gottingen, 1955, 71-77.

[51] Weil, A.: Variétés Kahleriennes. Hermann, Paris, 1958.

[52] Zariski, O.: Some results in the arithmetic theory of algebraic varieties. Am. J. Math. 61 (1939), 249-294.

[53] Zariski, O.: Sur la normalité analytique des variétés normales. Ann. Inst. Fourier 2 (1950), 161-164.

CENTRO INTERNAZIONALE MATEMATICO ESTIVO
(C. I. M. E.)

PROPAGATION OF SINGULARITIES FOR THE CAUCHY-RIEMANN EQUATIONS

J. J. KOHN

Corso tenuto a Bressanone dal 3 al 12 giugno 1973

J. J. Kohn

Introduction.

These lectures are intended as an introduction to the study of several complex variables from the point of view of partial differential equations. More specifically here we take the approach of the calculus of variations known as the $\bar{\partial}$-Neumann problem. Most of the material covered here is contained in Folland and Kohn, [4], Hörmander [11] and in the more recent work of the author (see [16], [17] and [20]). We consistently use the Laplace operator as in Kohn [14], since we believe that this method is particularly suitable for the study of regularity and for the study of the induced Cauchy-Riemann equations. Our main emphasis is in finding regular solutions of the inhomogenious Cauchy-Riemann equations. We wish to call attention to the extensive research on this problem by different methods from the ones mentioned above (see Ramirez [29], Grauert and Lieb [8], Kerzman [13], Øvrelid [28], Henkin [9], Folland and Stein [5]). It would take us too far afield to present these matters here. Another closely related subject which we cannot take up here is the theory of approximations by holomorphic functions (see R. Nirenberg and O. Wells [27], R. Nirenberg [26]. Hörmander and Wermer [12], etc.).

J. J. Kohn

Lecture 1. The $\bar\partial$ - problem and Hartog's theorem

The purpose of these lectures is to serve as an introduction to the use of the methods of partial differential equations in the theory of several complex variables. Let $z_1,\ldots,z_n$ be the coordinate functions in $\mathbb{C}^n$ and let

$$x_j = \mathrm{Re}(z_j), \quad y_j = \mathrm{Im}(z_j),$$

for a function u we define the derivatives u_{z_j} and $u_{\bar z_j}$ by:

$$(1.1) \qquad u_{z_j} = \tfrac{1}{2}(u_{x_j} - iu_{y_j}) \quad \text{and} \quad u_{\bar z_j} = \tfrac{1}{2}(u_{x_j} + iu_{y_j}).$$

The Cauchy-Riemann equations are then the equations

$$(1.2) \qquad u_{\bar z_j} = 0 \quad \text{for} \quad j = 1,\ldots,n.$$

A function satisfying the system (2) is called holomorphic and the theory of several complex variables consists of the study of these functions. A classical theorem states that a function is holomorphic if and only if it can be represented locally as a power series in the coordinates $z_1,\ldots z_n$ (see, for example, Hörmander [11] theorem 2.2.6). Here we will prove the existence of globally defined holomorphic functions, these functions will not be constructed by piecing together local solutions of (2) but rather by studying the inhomogeneous Cauchy-Riemann equations. More precisely, given functions $\alpha_1,\ldots\alpha_n$ we wish to solve the equations

$$(1.3) \qquad u_{z_j} = \alpha_j \qquad j = 1,\ldots,n.$$

Furthermore, we will want to investigate the dependence of

Furthermore, we will want to investigate the dependence of the solution u on the α_j. For example, in the discussion of Hartog's theorem given below, the crucial step is to find a compactly supported u when the α_j are compactly supported. Since the operators in (1) have constant coefficients we have $u_{\bar{z}_j\bar{z}_k} = u_{\bar{z}_k\bar{z}_j}$ and hence the following compatibility conditions are necessary for the existence of a solution u of (3):

$$(1.4) \qquad \alpha_{j\bar{z}_k} = \alpha_{k\bar{z}_j}, \quad \text{for all } k, \quad j = 1,2,\ldots,n.$$

The equations (3) and (4) are best expressed in terms of differential forms setting

$$\alpha = \sum \alpha_j dz_j, \qquad \bar{\partial} u = \sum u_{\bar{z}_j} dz_j$$

and

$$\bar{\partial}\alpha = \sum_{k<j} (\alpha_{j\bar{z}_k} - \alpha_{k\bar{z}_j}) d\bar{z}_k \wedge d\bar{z}_j$$

we have

$$(1.3') \qquad \bar{\partial} u = \alpha$$

$$(1.4') \qquad \bar{\partial}\alpha = 0$$

The problem of finding u satisfying (3') with given that satisfies (4') is called the $\bar{\partial}$ - problem.

The following classical theorem due to Hartog indicates the profound difference between the theory of several complex variables and one complex variable. The proof is an example of how the $\bar{\partial}$-problem can be used to prove existence theorems of holomorphic functions.

Theorem (Hartog). If $\delta > 0$ and

$$U = \left\{ (z_1, \ldots, z_n) \in \mathbb{C}^n \,\middle|\, 1-\delta < \sum |z_j|^2 < 1 \right\}$$

and if $n \geq 2$, then every holomorphic function defined on U has a unique holomorphic extension to the ball

$$B = \left\{ (z_1, \ldots, z_n) \in \mathbb{C}^n \,\middle|\, \sum |z_j|^2 < 1 \right\}.$$

That is, if h is a holomorphic function on U then there exists a holomorphic function $\tilde{h}$ on B such that $\tilde{h} = h$ on U.

In case $n = 1$ the function $h(z) = z^{-1}$ shows that, in the above theorem, the restriction $n \geq 2$ is necessary Ehrenpreis in [3] showed that the required extension property can be obtainedfrom the existence of a compactly supported solution u whenever α has compact support. The argument is as follows: let ρ be a C^∞ function which equals 1 when $\sum |z_j|^2 \geq 1-\delta/4$ and equals 0 if $\sum |z_j|^2 \leq 1-\delta/2$ we set

$$\alpha_j = \begin{cases} \rho_{z_j} h & \text{for } 1-\delta < \sum |z_j|^2 \leq 1-\delta/2 \\ 0 & \text{elsewhere} \end{cases} \tag{1.5}$$

Clearly α satisfies (3) and has compact support. If there exists u with compact support satisfying (3) then, first of all by (3) u is holomorphic in the domain

$$\sum |z_j|^2 > 1 - \delta/2,$$

and hence by analytic continuation u is zero in this domain.

In B we define $\tilde{h}$ by

(1.6) $$\tilde{h} = \rho h - u.$$

Then $\tilde{h}$ is the required solution since, by (3) and (5) it is holomorphic and since $\tilde{h} = h$ in an non-empty open subset of U and hence throughout U .

It remains to be shown that when $n \geq 2$ then there exist $u \in C_o^\infty(\mathbb{C}^n)$ whenever the $\alpha_j \in C_o^\infty(\mathbb{C}^n)$. Following Hörmander (see [11] 2.3), we deduce this from the following classical solution of (3) in one variable. Namely, if $\Omega \subset \mathbb{C}^1$ and $b\Omega$, the boundary of Ω is smooth and if $u \in C^\infty(\bar{\Omega})$ then

(1.7) $$u(z) = \frac{1}{2\pi i}\int_{b\Omega} \frac{u(\tau)}{\tau - z}\, d\tau + \frac{1}{2\pi i}\iint_{\Omega} \frac{u_{\bar{\tau}}}{\tau - z}\, d\tau \wedge d\bar{\tau}.$$

This formula is easily derived by using Stoke's theorem on the domain obtained by removing a disc of radius ε and center z from Ω and then letting $\varepsilon \rightarrow 0$. It then follows that if $\alpha \in C_o^\infty(\Omega)$ and if we define u by

(1.8) $$\begin{aligned} u(z) &= \frac{1}{2\pi i}\iint_{\Omega} \frac{\alpha(\tau)}{\tau - z}\, d\tau \wedge d\bar{\tau} \\ &= \frac{1}{2\pi i}\iint_{\mathbb{C}} \frac{\alpha(\tau)}{\tau - z}\, d\tau \wedge d\bar{\tau} \\ &= \frac{1}{2\pi i}\iint_{\mathbb{C}} \frac{\alpha_{\bar{\tau}}(\tau + z)}{\tau}\, d\tau \wedge d\bar{\tau} \end{aligned}$$

differentiating and changing variables again, we obtain

$$(1.9)\qquad u_{\bar z}(z) = \frac{1}{2\pi i}\iint_{\mathbb{C}} \frac{\alpha_{\bar\tau}(\tau + z)}{\tau}\, d\tau \wedge d\bar\tau = \frac{1}{2\pi i}\iint_{\Omega} \frac{\alpha_{\bar\tau}}{\tau - z}\, d\tau \wedge d\bar\tau$$

Since $\alpha = 0$ on $b\Omega$, applying (1.7 to α we obtain

$$(1.10)\qquad u_{\bar z} = \alpha .$$

It is clear that if a solution of (10) has compact support then the integral of α is zero. In the case of $n \geq 2$, however, we obtain a solution u of (3) with compact support whenever the α_j have compact support and satisfy (4). The desired u is then defined by:

$$(1.11)\qquad u(z_1, \ldots z_n) = \frac{1}{2\pi i}\iint_{\mathbb{C}} \frac{\alpha_1(\tau, z_2, \ldots, z_n)}{\tau - z_1}\, d\tau \wedge d\bar\tau = \frac{1}{2\pi i}\iint_{\mathbb{C}} \frac{\alpha_1(\tau + z_1, z_2, \ldots z_n)}{\tau}\, d\tau \wedge d\bar\tau$$

To see that u satisfies (3) we first note that by (10) we have $u_{\bar z_1} = \alpha_1$ further, using (4) and (7) we obtain

$$(1.12)\qquad u_{\bar z_j} = \frac{1}{2\pi i}\iint_{\mathbb{C}} \frac{\alpha_{1\bar z_j}(\tau + z_1, z_2, \ldots, z_n)}{\tau}\, d\tau \wedge d\bar\tau = \frac{1}{2\pi i}\iint_{\mathbb{C}} \frac{\alpha_{j z_1}(\tau + z_1, z_2, \ldots, z_n)}{\tau}\, d\tau \wedge d\bar\tau = \frac{1}{2\pi i}\iint_{\mathbb{C}} \frac{\alpha_{j\bar z}(\tau, z_2, \ldots, z_n)}{\tau - z_1}\, d\tau \wedge d\bar\tau = \alpha_j .$$

In particular we see that u is holomorphic outside of $K = \bigcup \text{supp}(\alpha_j)$. From (11) we see that u is 0 if $|z_2|$ is sufficiently large and hence, by analytic continuation, u = 0 in the unbounded component of the compliment of K; therefore u has compact support.

The above argument can be used to prove the following generalization of Hartog's theorem, due to Bochner (see [2]).

<u>Theorem.</u> If $n \geq 2$ and if $\Omega \subset \mathbb{C}^n$ is a bounded connected open set then every holomorphic function on Ω has a unique holomorphic extension to $\widetilde{\Omega}$; where, $\widetilde{\Omega}$ is the union of Ω and the bounded components of the compliment of Ω.

If we suppose that $b\Omega$, the boundary of Ω, is smooth and that a function h is defined on $b\Omega$ it is natural to ask, when does there exist a smooth function $\tilde{h}$ defined on Ω such that $\tilde{h} = h$ on $b\Omega$ and h is holomorphic in Ω. The obvious necessary condition on h is that it satisfies the so-called "tangential Cauchy-Riemann equations", i.e. those linear combinations of the $\frac{\partial}{\partial \bar{z}_j}$ which are tangential to $b\Omega$. More precisely, if r is a real-valued smooth function defined in a neighborhood of $b\Omega$ such that $dr \neq 0$ and if $a_1, \ldots, a_n$ are functions such that on $b\Omega$

(1.13) $$\sum a_j r_{\bar{z}_j} = 0,$$

and if Uf is any smooth extension of h, then,

(1.14) $$\sum a_j f_{\bar{z}_j} = 0 \quad \text{on } b\Omega.$$

An equivalent way of writing this is:

(1.14')
$$\bar{\partial} r \wedge \bar{\partial} f = 0 \text{ on } \Omega .$$

<u>Theorem.</u> If $\Omega \subset \mathbb{C}^n$, $n \geq 2$, is a domain with a smooth boundary $b\Omega$, if the complement of Ω is connected and if Ω is bounded. Then a smooth function h on $b\Omega$ can be extended to a smooth function $\tilde{h}$ on $\bar{\Omega}$ such that h is holomorphic in Ω, if and only if h satisfies (14) for all $a_1, \ldots, a_n$ that satisfy (13).

Following Hörmander [11] (theorem 2.3.2') we can prove the above by first constructing a smooth extension H such that

(1.15)
$$H_{\bar{z}_j} = O(r^2) \qquad j = 1, \ldots, n.$$

Since the $\bar{\partial}$-problem can be solved with a compactly supported u for $\alpha_j \in C_0^1(\mathbb{C}^n)$, we set $\alpha = \bar{\partial}(\rho H)$ where $\rho = 1$ in a small neighborhood of $b\Omega$ and vanishes outside a slightly larger neighborhood. Using $\alpha = \bar{\partial} u$ with a compactly supported u we obtain the desired extension $\tilde{h} = \rho H - u$ as in the previous theorem. The function H can be constructed by starting with any smooth extension f of h and noting that (14) implies that:

(1.16)
$$\bar{\partial} f = f_0 \bar{\partial} r + r \psi \qquad \text{near } b\Omega$$

rewriting this we have

(1.17)
$$\bar{\partial}(f - f_0 r) = r(\psi - \bar{\partial} f_0)$$

which implies that

$$\bar\partial r \wedge \bar\partial r \wedge (\psi - \bar\partial f_0) = 0$$

and hence

(1.18) $$\psi - \bar\partial f_o = f_1 \bar\partial r + r\theta .$$

Setting

(1.19) $$H = f - f_o r - f_1 r^2/2$$

we obtain

(1.20) $$\bar\partial H = r^2 (\theta - \tfrac{1}{2}\bar\partial f_1)$$

as required.

It should be mentioned that the tangential Cauchy-Riemann equations (14) have been studied extensively (see Bochner [2] , Lewy [22] , Kohn and Rossi [21] , Kohn [15] , Andreotti and Hill [1]. etc.).

In fact, the famous example of Lewy of an equation without solutions is one of these. We will return to this equation later.

Our main concern will be to study (3) on a domain without any restriction on the support of α. Suppose, for example, that the $\alpha_j \in L_2(\Omega)$ then we wish to find $u \in L_2(\Omega)$ satisfying (3), by this we mean that we want to find a sequence of smooth u_ν defined on Ω such that in $L_2(\Omega)$ we have $u = \lim u_\nu$ and $\alpha_j = \lim u_{\nu \bar z_j}$.

Suppose that there exists a point $P \in b\Omega$ and a holomorphic function f defined in a neighborhood U of P such that $f(P) = 0$ and $f \neq 0$ in $U \cap \bar\Omega - \{P\}$. Then we claim that if a solution of (3) exists in $L_2(\Omega)$ then there exists a holomorphic function h defined on Ω, such that h cannot be continued analytically over the point P. That is, if V is an open set containing P then there exists no holomorphic function g defined on V such that $g = h$ on $V \cap \Omega$. To see this let $\rho \in C_0^\infty(U)$ such that $\rho = 1$ in a neighborhood of P. Now we define F by:

$$(1.21) \qquad F = \begin{cases} \rho f^{-N} & \text{in } U \cap \Omega \\ 0 & \text{in } \Omega - U \end{cases}$$

and we chose N so large that $F \notin L_2(\Omega)$. We set $\alpha = \bar\partial F$ and note that $\alpha_j \in L_2(\Omega)$ (in fact, $\alpha_j = 0$ in a neighborhood of P). So if there exists $u \in L_2(\Omega)$ such that $\bar\partial u = \alpha$ then we define the desired h by:

$$(1.22) \qquad h = F - u$$

Now $h \notin L_2(\Omega)$ but for any neighborhood W of P $h \in L_2(\Omega - W)$ and thus h cannot be continued past P.

The problem of finding a holomorphic function h on which cannot be continued past a point $P \in b\Omega$ is called the <u>Levi problem</u> for Ω at P. A holomorphic function f defined in a neighborhood of $P \in b\Omega$ which has the properties hypothesized above is called a <u>local holomorphic separating function</u> at P. We have thus proven the following:

<u>Theorem</u>. If the $\bar\partial$ - problem has a solution $u \in L_2(\Omega)$ whenever the $\alpha_j \in L_2(\Omega)$ and satisfies (4) and if there exists a holomorphic separating function at $P \in b\Omega$ then the Levi problem has a solution at P.

It is easy to see that a holomorphic separating function exists at P if the domain Ω is convex in a neighborhood of P (in that case there is a linear holomorphic separating function). Let Ω_1, be the domain given by the inequalities

$$(1.23) \qquad \tfrac{1}{4} < \sum |z_j|^2 < 1$$

and Ω_2 is the ball of radius $\frac{1}{2}$ and center $(\frac{1}{2},0,\ldots,0)$. Let $\Omega = \Omega_1 \cup \Omega_2$ then the origin is in $b\Omega$ and Ω is convex in a neighborhood of the origin. Thus we see that the equation (3) cannot be solved in $L_2(\Omega)$ for if it were possible to solve it we would obtain also a solution of Levi's problem at the origin and this would contradict Hartog's theorem. A modification of the above argument shows that for the above Ω the range of the operator $\bar\partial$ $u \longmapsto (u_{\bar z}, u_{\bar z_2}, \ldots, u_{\bar z_n})$ is not closed in any of the natural topologies (i.e. L_p, distribution) and thus the $\bar\partial$ - problem cannot be solved in any satisfactory sense.

J. J. Kohn

Lecture 2. Pseudo-convexity

We will now study the properties of $\Omega \subset \mathbb{C}^n$ which insure that there is a local holomorphic separating function at $P \in b\Omega$. As was pointed out in the previous lecture it is enough that Ω be convex in a neighborhood of P. However, the notion of holomorphic separating function is invariant under holomorphic transformations whereas the notion of convexity is not. We will introduce the property of "pseudo-convexity" by, roughly speaking, isolating that feature of convexity which is invariant under holomorphic transformations.

We assume that $b\Omega$ is smooth; that is, that there exists a real-valued function r defined in a neighborhood of $b\Omega$ such that $dr \neq 0$ and $r = 0$ on $b\Omega$. We will fix the sign of r so that:

(2.1) $\qquad r > 0$ outside of $\bar{\Omega}$ and $r < 0$ in Ω.

The domain Ω is convex if the Hessian of r (i.e. the matrix of second partial derivatives) is non-negative acting as a quadratic form on the tangent vectors to $b\Omega$; if it is positive - definite then Ω is strictly convex. In $\mathbb{R}^2$ this reduces to the familiar fact that the graph of a function is convex if the second derivative is non - positive, the general case may also be deduced from this by taking intersections with planes. In terms of the complex coordinate in $\mathbb{C}^n$ this condition is expressed by

(2.2) $$\operatorname{Re}\left(\sum r_{z_i z_j}(P)\, a_i\, a_j\right) + \sum r_{z_i \bar{z}_j}(P) a_i a_j \geq 0$$

wherever

(2.3) $\quad \mathrm{Re}\ (\sum r_{z_i}(P)a_i) = 0, \quad \text{where} \quad P \in b\Omega.$

In other words we are restricting the $2n \times 2n$ quadratic form

$$(a_1 \ldots a_n\ b_1 \ldots b_n) \begin{pmatrix} r_{z_i z_j}(P) & r_{z_i \bar{z}_j}(P) \\ r_{\bar{z}_i z_j}(P) & r_{z_i z_j}(P) \end{pmatrix} \begin{pmatrix} \bar{a}_1 \\ \vdots \\ \bar{a}_n \\ \bar{b}_1 \\ \vdots \\ \bar{b}_n \end{pmatrix}$$

To those vectors which are real (i.e. $b_j = \bar{a}_j$) and tangential to $b\Omega$ at P. It is clear that this condition is not invariant under holomorphic transformations. However, observe that the condition implies that

(2.4) $$\sum r_{z_i \bar{z}_j}(P)a_i\bar{a}_j \geq 0,$$

whenever

(2.5) $$\sum r_{z_i}(P)a_i = 0 \quad \text{for} \quad P \in b\Omega.$$

Note that the above is invariant under holomorphic transformations and, in fact, we will express this last condition in invariant form. Let $T_P^{1,0}(b\Omega)$ be the subspace of the complex tangent vectors at $P \in b\Omega$ which are of type (1,0); that is, they annihilate the anti-holomorphic functions. In terms of coordinates $T_P^{1,0}(b\Omega)$ if and only if it can be expressed as

(2.6) $$L = \sum a_j \frac{\partial}{\partial z_j}, \quad \text{where} \quad \sum r_{z_i}(P)a_i = 0.$$

For each $P \in b\Omega$ we have the hermitian form

(2.7) $$\lambda_P : T_P^{1,0}(b\Omega) \times T_P^{1,0}(b\Omega) \longrightarrow \mathbb{C}$$

by

(2.8) $$\lambda_P(L, L') = \langle (\partial\bar{\partial} r)_P, L \wedge \bar{L}' \rangle,$$

where $\langle \, , \, \rangle$ denotes the contraction between contravariant and covariant tensors. The form λ_P is called the <u>Levi form</u> and the condition given by (4) and (5) is equivalent to

(2.9) $$\lambda_P(L, L) \geq 0 \quad \text{for all } L \in {}_P^{1,0}(b\Omega).$$

This condition is clearly invariant and we say that Ω is <u>pseudo-convex</u> if (9) is satisfied for all $P \in b\Omega$ and that it is <u>strongly pseudo-convex</u> if the Levi form is positive definite at all $P \in b\Omega$.

Conversely, if Ω is strongly pseudo-convex and if $P \in b\Omega$ then there exists a holomorphic coordinate system on a neighborhood U of P such that $\Omega \cap U$ is strictly convex. To prove this we first note that strong pseudo-convexity is independent of the choice of r; provided, of course, that r satisfy (1). Then we note that for any coordinate system r can be chosen so that $(r_{z_i\bar{z}_j}(P))$ is a positive definite hermitian form on all n - tuples $(a_1, \ldots, a_n)$, i.e. without the restriction (5). To achieve this we set

(2.10) $$r = e^{\tau R} - 1,$$

where R is any fixed function satisfying (1) and τ is a sufficiently large number. Then we have

$$(2.11)\qquad r_{z_i\bar{z}_j} = (R_{z_i\bar{z}_j} + \tau^2 R_{z_i} R_{\bar{z}_j}) e^{\tau R} .$$

We decompose an arbitrary n-tuples as follows:

$$(2.12)\qquad (b_1,\ldots,b_n) = (a_1,\ldots,a_n) + \mu(R_{\bar{z}_1}(P),\ldots,R_{\bar{z}_n}(P)),$$

where

$$\mu = \frac{\sum R_{z_i}(P)b_i}{|R_{z_i}(P)|^2} .$$

so that $(a_1,\ldots,a_n)$ satisfies (5). Thus we have

$$(2.13)\qquad e^{-\tau R}\sum r_{z_i\bar{z}_j}(P)b_i\bar{b}_j = \tau\sum R_{z_i\bar{z}_j}(P)a_i a_j + \tau^2 \quad {}^2$$

$$+ O(\tau|\mu|(\sum|a_i|^2)^{\frac{1}{2}}.$$

The error term is bounded by:

$$(2.14)\qquad (\text{large const. } |\mu|^2 + \text{small const. } \sum|a_i|^2).$$

Choosing τ sufficiently large this is smaller than the first two terms on the right of (13) - since, by hypothesis, the first term is larger than τ const. $\sum|a_i|^2$. Therefore, we have

$$(2.15)\qquad \sum r_{z_i\bar{z}_j}(P)b_i\bar{b}_j \geq \text{const.}\sum|b_i|^2$$

as desired.

We are now ready to prove the following classical result.

<u>Theorem.</u> If Ω is a strongly pseudo-convex domain and

$P \in b\Omega$ then there exists a neighborhood U of P and a holomorphic coordinate system whose domain contains such that $U \cap \Omega$ is strictly convex with respect to the linear structure given by these coordinates.

<u>Proof</u>: It suffices to find a coordinate system such that the form (2) is positive-definite even without restriction (3). Let $u_1, \ldots, u_n$ be any holomorphic coordinate system with origin at P and let r be a function satisfying (1) with $(r_{u_i \bar{u}_j}(0))$ positive definite. Expanding r in a Taylor series, we have

(2.16)
$$r = 2\,\mathrm{Re}(\sum r_{u_i}(0)u_i + r_{u_i u_j}(0)u_i u_j) + \sum r_{u_i \bar{u}_j}(0)u_i \bar{u}_j + ((\sum |u_i|)^3).$$

Setting

(2.17)
$$z_i = u_i \quad \text{for } i = 1,\ldots,n-1, \text{ and}$$
$$z_n = 2\sum r_{u_i}(P)u_i + 2\sum r_{u_i u_j}(P)u_i u_j$$

we have

(2.18)
$$r = \mathrm{Re}(z_n) + \sum r_{z_i \bar{z}_j}(0)z_i \bar{z}_j + ((\sum |z_i|)^3).$$

Since $r_{u_i \bar{u}_j}(0)$ is positive definite we also have $r_{z_i \bar{z}_j}(0)$ positive definite and thus by (18) the Hessian is also positive definite.

The following classical theorem shows that in case the Levi form is identically zero the domain is also locally convex. The proof of this theorem is less elementary than the one above.

Theorem. If r is real, analytic and

If in a neighborhood U of $P \in b\Omega$ the Levi-form is identically zero (i.e. on $U \cap b\Omega$) then there exists a coordinate neighborhood V of P with holomorphic coordinates $z_1, \ldots, z_n$ on V, such that the set $V \cap b\Omega$ consists of the points for which $\mathrm{Re}(z_n) = 0$.

The two theorems above make it seem plausible that whenever a domain is pseudo-convex then in a neighborhood of each boundary point these exist coordinates with respect to which it is convex. However, this is not true as is shown in the following example (see Kohn and Nirenberg [14]) Let $\Omega \subset \mathbb{C}^2$ so that near the origin the function r is given by:

$$(2.19) \qquad r = \mathrm{Re}(w) + |z|^2 |w|^2 + |z|^8 + \frac{15}{7} |z|^2 \mathrm{Re}(z^6).$$

Since complex dimension 2 the space $T_P^{1,0}(b\Omega)$ is one-dimensional the Levi form 1×1 matrix. In the above example it can be shown that the Levi form is larger than const. $(|w|^2 + |z|^6)$ near the origin. Thus, by suitably extending r we have Ω pseudo-convex and strongly pseudo-convex everywhere except at $(0,0)$. Now if there were holomorphic coordinates on a neighborhood U of the origin relative to which $U \cap \Omega$ is convex, then we could find a linear function h such that $\mathrm{Re}\, h(0,0) = 0$ and the zeros of $\mathrm{Re}\, h$ which are in U remain outside of Ω. That this is not possible in this example is shown by the following result.

<u>Theorem</u>. If h is a holomorphic function which is defined on a neighborhood U of the origin and if $h(0,0) = 0$ then there exist points (z_1,w_1), $(z_2,w_2) \in U$ such that $h(z_1,w_1) = h(z_2,w_2) = 0$, $r(z_1,w_1) > 0$ and $r(z_2,w_2) < 0$; where r is defined by (19).

The proof of this theorem depends on the following lemma.

<u>Lemma</u>. For each $a \in \mathbb{C}$ the function f, defined by

$$f_a(z) = |z|^8 + \frac{15}{7}|z|^2 \operatorname{Re}(z^6) + \operatorname{Re}(az^8), \tag{2.20}$$

changes sign in every neighborhood of the origin.

<u>Proof</u>: since for $t \geq 0$, $f_a(tz) = t^8 f_a(z)$ it suffices to that the function

$$g(\theta) = f_a(e^{i\theta}) = 1 + \frac{15}{7}\operatorname{Re}(e^{6i\theta}) + \operatorname{Re}(ae^{8i}) \tag{2.21}$$

changes sign. Then we obtain

$$\frac{15}{7}\pi = \int_0^{2\pi} g(\theta)e^{-6i\theta}\,d\theta \leq \int_0^{2\pi}|g(\theta)|\,d\theta\,. \tag{2.22}$$

Since $\int_0^{2\pi} g(\theta)d\theta = 2\pi$, we see that if g does not change sign then $\int_0^{2\pi}|g(\theta)|\,d = 2\pi$ which contradicts the above inequality.

<u>Proof of the theorem</u>: Suppose h is a holomorphic function defined in a neighborhood of $(0,0)$ with $h(0,0) = 0$. If $h(0,w) \equiv 0$ then $r(0,w) = \operatorname{Re}(w)$ which changes sign as required. If $h(z,0) \equiv 0$ then $h(z,0) = f_0(z)$ which changes sign by the above lemma. Otherwise some of the zeroes of h can be paranretrized as follows:

(2.23) $z = \bar{q}$

$$w = \sum_{j=p}^{\infty} a_j \zeta^j \quad \text{with} \quad a_p \neq 0.$$

That is, for every ζ we have $h(z,w) = 0$ when z and w are defined by (23). Evaluating r on these zeroes, we obtain

(2.24) $$r = \operatorname{Re}(a_p \zeta^p) + |\zeta|^{8q} + \frac{15}{7}|\zeta|^{2q}\operatorname{Re}(\zeta^{6q}) + (|\zeta|^{p+1}).$$

Now, if $p < 8q$ then the first term on the right controls the sign of r for $|\zeta|$ small and hence r changes sign. If $p = 8q$, then the first three terms control the sign and so r changes sign by the lemma, with $a = a_p$. If $p > 8q$, then the second and third terms control the sign and the lemma can be applied with $a = 0$.

At this point we wish to discuss invariants of pseudo-convex domains which are not strongly pseudo-convex. These invariants play an important role in the study of boundary regularity of the $\bar{\partial}$ - problem (see Kohn [16]). Let L be a C^{∞} vector field defined in a neighborhood U of $P \in b\Omega$. We suppose that L is of degree (1,0) and that for each $Q \in U \cap b\Omega$, $L_Q \in \Gamma^{1,0}_Q(b\Omega)$, i.e.

(2.25) $$L = \sum a_j \frac{\partial}{\partial z_j} \quad \text{and}$$

$$L(r) = \sum a_j r_{z_j} = 0 \quad \text{on } b\Omega.$$

Denote by $\bar{L}$ the conjuge of L (defined by

$\bar{L}(u) = \overline{L(u)}$, in terms of local coordinates

$$L = \sum a_j \frac{\partial}{\partial z_j} . \tag{2.26}$$

For each non-negative integer k we define a space of vector fields $\mathcal{L}^k(L)$ on U as follows

$$\mathcal{L}^0(L) = \{fL + g\bar{L}\} \tag{2.27}$$

i.e. the space of all $fL + g\bar{L}$ with $f, g \in C^\infty(V)$;

$$\mathcal{L}^k(L) = \mathcal{L}^{k-1}(L) + \left[\mathcal{L}^0(L), \mathcal{L}^{k-1}(L)\right], \tag{2.28}$$

that is the space of all vector fields of the form

$$V = V + [T, V'] \tag{2.29}$$

where $V, V' \in \mathcal{L}^{k-1}(L)$, $T \in \mathcal{L}^0(L)$ and

$$[T, V'] = TV' - V'T.$$

We say that L is of <u>finite type at P</u> if for some k we have

$$\mathcal{L}^k_P(L) \not\subset T_P^{1,0}(b\Omega) + T_P^{0,1}(b\Omega), \tag{2.30}$$

where $\mathcal{L}^k_P(L)$ denotes the vector space obtained by evaluating all the vector fields in $\mathcal{L}^k(L)$ at P. The lowest integer k for which (30) is true is called the <u>order of L at P</u> and if (30) is not ture for any k then we say that L is of <u>infinite order at P</u>. The following hold:

(A) If Ω is pseudo-convex, then it is strongly pseudo-convex if and only if for each $P \in b\Omega$ and each non-zero L defined in a neighborhood of P, L is of order 1 at P.

(B) If U is open and all vector fields on U are of infinite order at $P \in U \cap b\Omega$, then the Levi form is identically zero on $\Gamma_P^{1,0}(b\Omega)$ for all $P \in U \cap b\Omega$.

The basis for the above properties is the following expression for the Levi form in terms of vector fields L, L' with values on $\Gamma^{1,0}(b\Omega)$.

$$\langle \partial\bar{\partial} r , L \wedge \bar{L} \rangle = \langle \partial r , [L, \bar{L}] \rangle \tag{2.31}$$

This formula follows from the fact that $\partial\bar{\partial} r = - d\partial r$ and the classical expression for the exterior derivative:

$$\langle \partial\bar{\partial} r, L \wedge \bar{L} \rangle = - \langle d\partial r, L \wedge \bar{L} \rangle$$

$$= -L(\langle \partial r , \bar{L} \rangle) + \bar{L}(\langle \partial r , L \rangle) + \langle \partial r, [L, \bar{L}] \rangle$$

and we have $\langle \partial r, \bar{L} \rangle = 0$ since ∂r is of degree $(1,0)$ and $\bar{L}$ of degree $(0,1)$, further

$$\langle \partial r, L \rangle = \langle dr , L \rangle = L(r) = 0. \tag{2.32}$$

<u>Lemma:</u> If $V \in \mathbb{C}\Gamma(b\Omega)$, i.e. $V(r) = 0$ on $b\Omega$, then $V \in \Gamma_P^{1,0}(b\Omega) + \Gamma_P^{0,1}(b\Omega)$ if and only if

$$\langle (\partial r)_P, V \rangle = 0 \tag{2.33}$$

<u>Proof:</u> By (32) and by considerations of degree it is clear that all $V \in \Gamma_P^{1,0}(b\Omega) + \Gamma_P^{0,1}(b\Omega)$ satisfy (33). On the other hand

$$\dim \left(\Gamma_P^{1,0}(b\Omega) + \Gamma_P^{0,1}(b\Omega) \right) = 2n - 2 \tag{2.34}$$

$$\dim_{\mathbb{C}} \mathbb{C}\,\Gamma_P(b\Omega) = 2n - 1$$

and since $(\partial r)_p$ is considered as a linear functional on $\mathbb{C}T_P(b\Omega)$ and is non-trivial (if it were trivial it would be a multiple of dr) we see that the subspace of $\mathbb{C}T_P(b\Omega)$ of those V satisfying (33) is of dimension $2n-2$ which concludes the proof.

The above properties are then easy consequences of (31), the lemma and the definition of the Levi form (8).

We conclude the present discussion of order with the following lemma.

Lemma. If every non-zero local vector field on $b\Omega$, with values in $T_P^{1,0}(b\Omega)$, is of finite order then there exists no non-trivial connected analytic vari[illegible] contained in $b\Omega$.

Proof: Suppose V is such an analytic variety and that $P \in V$ is a regular point of V. Then there is a non-zero vector field L of degree (1,0) defined in a neighborhood U of P such that

$$L_Q \in T_P^{1,0}(b\Omega) \text{ for } Q \in U \cap b\Omega$$ and such that $L_Q \in T_Q^{1,0}(V)$ for $Q \in U \cap V$. Then, clearly

$$\mathcal{L}_Q^k(L) \subset \mathbb{C}T_Q(V) \text{ for all } Q \in U \cap V$$ and since

$$\mathbb{C}T_Q(V) = T_Q^{1,0}(V) + T_Q^{0,1}(V)$$ we have

$$\mathbb{C}T_Q(V) \subset T_Q^{1,0}(b\Omega) + T_Q^{0,1}(b\Omega)$$ and

therefore, L is of infinite order for all $Q \in U \cap V$.

To conclude this lecture we wish to mention some classical properties of pseudo-convexity. We refer to Hormander [10] section 2.6 for a systematic treatment of these.

Definition. A real-valued function $u \in C^\infty(\Omega)$ is called plurisubharmonic if the quadratic form

$$(35) \quad \sum u_{z_i \bar{z}_j} a_i a_j$$

is positive definite.

The following theorem (see Hormander 11 theorems 2.6.7, 2.6.11 and 2.612) shows how the notion of plurisubharmonicity leads to generalizing pseudo-convexity to domains whose boundary is not smooth.

Theorem. If Ω is an open set in $\mathbb{C}^n$ with a smooth boundary $b\Omega$ then the following are equivalent:

(a) There exists a pluri-subharmonic function u on Ω such that for every $c \in \mathbb{R}$

$$\Omega_c = \left\{ z \in \Omega \,\middle|\, u(z) < c \right\}$$

is compact.

(b) Ω is pseudo-convex, i.e. (4), (5) are satisfied.

As is standard we extend the definition of pseudo-convex to domains satisfying condition (a). This condition says essentially that Ω can be exhausted by strongly pseudo-convex domains. This need not be ture for pseudo-convex domains in complex manifolds. From condition (a) and from the solution of the Levi problem for strongly pseudo-convex domains, it can be deduced that a domain Ω in $\mathbb{C}^n$ is a domain of holomorphy (i.e. there exists a holomorphic function on Ω which cannot be continued to a larger domain) if and only if Ω is pseudo-convex.

Finally we wish to call attention to the following result (see Hormander [11] theorem 2.6.13) which shows that if a domain in $\mathbb{C}^n$ with a smooth boundary is not pseudo-convex then it is not a domain of holomorphy.

<u>**Theorem**</u>. Suppose that $\Omega \subset \mathbb{C}^n$ has a boundary of class C^4, i.e. the function r is of class C^4. Suppose that for some $P \in b\Omega$ and some $(a_1, \ldots, a_n)$ with

$$\sum r_{z_j}(P) a_j = 0$$

we have

$$\sum r_{z_i \bar{z}_j}(P) a_i \bar{a}_j < 0 ,$$

then there exists a neighborhood V U of P such that if $u \in C^4(U)$ and satisfies the tangential Cauchy-Riemann equations ((1.) on $U \cap b\Omega$ then there exists a function $v \in C^1(V)$ such that $v = u$ on $V \cap b\Omega$ and $\bar{\partial} v = 0$ on $V^+ = \{ Q \in V \mid r(Q) \geq 0 \}$. In particular, any holomorphic function on $U^- = \{ Q \; U \mid r(Q) \leq 0 \}$ can be extended to V^+.

J. J. Kohn

Lecture 3. Formulation of the $\bar{\partial}$-Neumann problem.

We return now to the study of the operator $\bar{\partial}$ from the point of view of L_2. Denoting by $\mathcal{Q}^{p,q}(\Omega)$ the forms of type (p,q) which are C^∞ in $\bar{\Omega}$, that is restrictions of C^∞ forms on a neighborhood of $\bar{\Omega}$, we have

(3.1) $$\mathcal{Q}^{0,0}(\Omega) \xrightarrow{\bar{\partial}} \mathcal{Q}^{0,1}(\Omega) \xrightarrow{\bar{\partial}} \mathcal{Q}^{0,2}(\Omega)$$

with $\bar{\partial}^2 = 0$. On these spaces we define L_2 inner products as follows:

(3.2) $$(u,v) = \int_\Omega u\bar{v}dV \quad \text{for} \quad u,v \in \mathcal{Q}^{0,0}(\Omega)$$

$$(\varphi,\psi) = \sum \int_\Omega \varphi_i \bar{\psi}_i dV \quad \text{for} \quad \varphi,\psi \in \mathcal{Q}^{0,1}(\Omega)$$

$$(\theta,\sigma) = \sum \int_\Omega \theta_{ij} \bar{\sigma}_{ij} dV \quad \text{for} \quad \theta,\sigma \in \mathcal{Q}^{0,2}(\Omega).$$

The corresponding norms are denoted by $\| \ \|$. We denote by $L_2^{p,q}(\Omega)$ the completion of $\mathcal{Q}^{p,q}(\Omega)$ under the L_2-norm. Let T and S denote the L_2-closures of the operators $\bar{\partial}$ in (3.1) and T*, S* the adjoints of T, S then we have

(3.3) $$L_2^{0,0}(\Omega) \underset{T^*}{\overset{T}{\rightleftarrows}} L_2^{0,1}(\Omega) \underset{S^*}{\overset{S}{\rightleftarrows}} L_2^{0,2}(\Omega).$$

These operators are defined on their domains (not on the whole space at the beginning of the arrow). The domain of T, denoted by Dom (T), is defined as follows:

(3.4) $$\text{Dom}(T) = \{u \in L_2^{0,0}(\Omega) \mid \exists \{u_j\} \text{ with } u_j \in \mathcal{Q}^{0,0}(\Omega),$$
$$u = \lim u_j \text{ and } \{\bar{\partial} u_j\} \text{ Cauchy in } L_2^{0,1}(\Omega)\}.$$

By taking $\psi \in \mathcal{A}^{0,1}(\Omega)$ with compact support and $w \in C^\infty(\bar{\Omega})$ we have

(3.5) $$(\bar{\partial} w, \psi) = (w, -\sum \psi_{i z_i}).$$

Therefore, if $u \in \text{Dom}(T)$ and $\{u_j\}$, $\{v_j\}$ are two sequences with limit u and such that both $\{\bar{\partial} u_j\}$ and $\{\bar{\partial} v_j\}$ are Cauchy; thus, setting $w = u_j - v_j$ in (3.5), we obtain

(3.6) $$(\bar{\partial} u_j - \bar{\partial} v_j, \psi) \leq \text{const.} \|u_j - v_j\|$$

so that for $u \in \text{Dom}(T)$ we can define Tu by

(3.7) $$Tu = \lim \bar{\partial} u_j$$

and it is independent of the sequence.

Similarly, we define Dom (S) and $S\varphi$ for all $\varphi \in \text{Dom}(S)$.

The domain of T^*, denoted by Dom (T^*), is defined by:

(3.8) $$\text{Dom}(T^*) = \{\varphi \in L_2^{0,1}(\Omega) \mid \exists\, C > 0 \text{ with } |(Tu, \varphi)| \leq C\|u\| \text{ for all } u \in \text{Dom}(T)\}$$

to define T^* on Dom (T^*) we see from (3.8) that the map $u \to (Tu, \varphi)$ is a bounded linear functional on Dom (T) and thus has a unique extension to $L_2^{0,0}(\Omega)$, since Dom (T) is dense. This extension has a unique representative which is by definition T^*, thus we have

(3.9) $$(u, T^*\varphi) = (Tu, \varphi),$$

for all $u \in \text{Dom}(T)$, $\varphi \in \text{Dom}(T^*)$.

If $\varphi \in \mathrm{Dom}\,(T^*)$ is differentiable then we can apply (3.5) with $\psi = \varphi$ and w with compact support and we obtain

$$T^*\varphi = -\Sigma\varphi_{i z_i}. \tag{3.10}$$

This can be thought of as a complex divergence corresponding to the complex gradient and complex curl which are given by $\bar\partial$ on functions and on (0,1)-forms respectively,

Since $\bar\partial^2 = 0$ we have

$$ST = 0 \quad \text{and} \quad T^*S^* = 0. \tag{3.11}$$

This means that if $u \in \mathrm{Dom}\,(T)$ then $Tu \in \mathrm{Dom}\,(S)$ and $STu = 0$ and similarly for S^* and T^*.

The following is a classical result in Hilbert space theory (see Hörmander [11] theorem 1.1.1).

<u>Theorem</u>. If A and B are Hilbert spaces and $T : A \to B$ is a closed densely defined operator, then the following are equivalent:

(a) $\mathcal{R}(T)$, the range of T, is closed

(b) There is a constant C such that

$$\|u\|_A \leq C\|Tu\|_B, \quad u \in \mathrm{Dom}\,(T) \cap [\mathcal{R}(T^*)], \tag{3.12}$$

when $[\ ]$ denotes closure.

(c) $\mathcal{R}(T^*)$ is closed

(d) There is a constant C such that

(3.12') $\|\varphi\|_B \leq C\|T^*\varphi\|_A$, $\varphi \in \mathrm{Dom}\ (T^*) \cap [\mathcal{R}(T)]$.

<u>Proof</u>. Since (T*) is the orthogonal complement of

$$\mathcal{N}(T) = \{h \in \mathrm{Dom}\ (T) \mid Th = 0\}$$

we have

$$T : \mathrm{Dom}\ (T) \cap \mathcal{R}(T^*) \longrightarrow \mathcal{R}(T)$$

is aclosed one-to-one map hence by theclosed graph theorem (a) $\Longrightarrow$ (b) and similarly (since T = T**), (b) $\Longrightarrow$ (d). We wish to prove (b) $\Longrightarrow$ (d). From (b) we obtain:

$$|(\varphi, Tu)_B| = |(T^*\varphi, u)_A| \leq \|T^*\varphi\|_A \|u\|_A \leq C\|T^*\varphi\|_A \|Tu\|_B ,$$

$$\varphi \in \mathrm{Dom}\ (T^*) \text{ and } u \in \mathrm{Dom}\ (T) \cap [\mathcal{R}(T^*)]$$

hence

$$|(\varphi,\psi)_B| \leq C\|T^*\varphi\|_A \|\psi\|_B , \quad \varphi \in \mathrm{Dom}\ (T^*) \text{ and } \psi \in \mathcal{R}(T)$$

which implies (3.12') since

$$\|\varphi\|_B = \inf \frac{|(\varphi,\psi)_B|}{\|\psi\|_B} \quad \text{for } h \in \mathrm{Dom}\ (T^*) \cap [\mathcal{R}(T)] .$$

The proof is then complete since (b) $\Longrightarrow$ (a) and (d) $\Longrightarrow$ (c) are clear.

Now let C be a third Hilbert space and we assume that we have closed operators densely defined T : A $\longrightarrow$ B and S : B $\longrightarrow$ C such that

(3.13) $$ST = 0$$

We define $\mathcal{H} \subset B$ by

(3.14) $$\mathcal{H} = \mathcal{N}(T^*) \cap \mathcal{N}(S)$$

then we have:

<u>Theorem</u>. A necessary and sufficient condition that $\mathcal{R}(T)$ and $\mathcal{R}(S)$ be both closed is that

(3.15) $$\|\varphi\|_B^2 \leq \text{const.} \left(\|T^*\varphi\|_A^2 + \|S\varphi\|_C^2\right)$$

for all $\varphi \in \text{Dom}\,(T^*) \cap \text{Dom}\,(S)$ with $\varphi \perp \mathcal{H}$.

<u>Proof</u>. First we have the weak orthogonal decomposition formula

(3.16) $$B = [\mathcal{R}(T)] \oplus [\mathcal{R}(S^*)] \oplus \mathcal{H}$$

which is an immediate consequence of (3.13). From (3.13) we also conclude that $[\mathcal{R}(T)] \subset \mathcal{H}(S)$ thus if

$$g \in \text{Dom}\,(T^*) \cap [\mathcal{R}(T)] \text{ then } Sg = 0$$

and (3.15) reduces to (3.12'). Similarly, since

$$T^*S^* = 0$$

we obtain (3.12) as a consequence of (3.15). For an arbitrary

$$\varphi \in B \text{ with } \varphi \perp \mathcal{H}$$

we use the orthogonal decomposition (3.16) and obtain

$$\varphi = \varphi_1 + \varphi_2 .$$

To prove (3.15) it suffices to show

$$\|\varphi_1\|_B^2 + \|\varphi_2\|_B^2 \leq \text{const.} \left(\|T^*\varphi_1\|_A^2 + \|S\varphi_2\|^2 \right),$$

which is the sum of (3.12) and (3.12').

The inequality (3.15) is often obtained as a consequence of the following:

<u>Theorem</u>. Suppose that whenever $\{\varphi_k\}$ is a sequence

$$\varphi_k \in \text{Dom}\,(T^*) \cap \text{Dom}\,(S)$$

such that $\|\varphi_k\|_B$ is bounded, if

$$\lim T^*\varphi_k = 0 \text{ in } A \quad \text{and if} \quad \lim S\varphi_k = 0 \text{ in } C$$

there exists a subsequence which converges in B. Then (3.15) holds and $\mathcal{H}$ is finite dimensional.

<u>Proof</u>: The hypothesis implies that the unit sphere in $\mathcal{H}$ is compact, hence $\mathcal{H}$ is finite dimensional. If (3.15) did not hold then we could find a sequence

$$\theta_k \in \text{Dom}\,(T^*) \cap \text{Dom}\,(S)$$

with $\theta_k \perp \mathcal{H}$ such that

$$\|\theta_k\|_B^2 > k(\|T^*\theta_k\|_A^2 + \|S\,\theta_k\|_C^2).$$

Setting

$$\frac{\theta_k}{\|\theta_k\|_B}$$

we obtain

$$\|T^*\varphi_k\|_A^2 + \|S\varphi_k\|_C^2 < \frac{1}{k}$$

hence thereexists a subsequence of $\{\varphi_k\}$ which converges to an element $\varphi \perp \mathcal{H}$ but since $\varphi \in \mathcal{H}$ and $\|\varphi\| = 1$ we obtain a contradiction.

<u>Corollary</u>. If the norm $\|\varphi\|_B + \|T^*\varphi\|_A + \|S\varphi\|_B$ defined on Dom (T*) $\cap$ Dom (S) is compact then (3.15) holds and is finite dimensional.

Once we know that (3.15) is satisfied then we can remove the brackets in (3.16) and thus obtain

$$B = \mathcal{R}(T) \oplus \mathcal{R}(S^*) \oplus \mathcal{H}, \tag{3.17}$$

which is called the orthogonal decomposition. We shall make this more explicit by introducing the laplacian

(3.18) $L\varphi = TT^*\varphi + S^*S\varphi$ for $\varphi \in \mathrm{Dom}\,(T^*) \cap \mathrm{Dom}\,(S)$

with $T^*\varphi \in \mathrm{Dom}\,(T)$ and $S\varphi \in \mathrm{Dom}\,(S^*)$. Then

$$(3.19) \qquad \mathcal{H} = \mathcal{N}(L),$$

since clearly $\mathcal{H} \subset \mathcal{N}(L)$ and if $L\varphi = 0$ then

$$(3.20) \qquad (L\varphi, \varphi)_B = \|T^*\varphi\|_A^2 + \|S\varphi\|_C^2 = 0$$

so that $\varphi \in \mathcal{H}$.

The following theorem is essentially due to Gaffney [6].

__Theorem.__ The operator L, defined by (3.18) where S and T satisfy (3.13) is self-adjoint.

__Proof:__ By a theorem of von Neumann,

$$(I + TT^*)^{-1} \quad \text{and} \quad (I + S^*S)^{-1}$$

are bounded and self-adjoint we set

$$R = (I + TT^*)^{-1} + (I + S^*S)^{-1} - I$$

which is bounded and self-adjoint, we shall prove the theorem by showing that $R = (L + I)^{-1} =$. First,

$$(I + TT^*)^{-1} - I = (I - (I + TT^*))\,(I + TT^*)^{-1}$$
$$= -TT^*(I + TT^*)^{-1}$$

which shows that

$$\mathcal{R}(I + TT^*)^{-1} \subset \mathrm{Dom}\,(TT^*),$$

similarly
$$\mathcal{R}(I + S^*S)^{-1} \subset \mathrm{Dom}\,(S^*S)$$
and we have

$$R = (I + S^*S)^{-1} - TT^*(I + TT^*)^{-1}$$

so since $ST = 0$ we have $\mathcal{R}(R) \subset \mathrm{Dom}\,(S^*S)$ and

$$S^*SR = S^*S(I + S^*S)^{-1}.$$

Similarly
$$\mathcal{R}(R) \subset \mathrm{Dom}\,(TT^*)$$
and

$$TT^*R = TT^*(I + TT^*)^{-1}$$

so that
$$\mathcal{R}(R) \subset \mathrm{Dom}\,(L)$$
and

$$(L + I)R = TT^*(I + TT^*)^{-1} + S^*S(I + S^*S)^{-1} + R = I$$

Finally $L + I$ is injective, since $((L + I)\varphi, \varphi) \geq \|\varphi\|^2$, and therefore, $R = (L + I)^{-1}$.

We define the spaces $\mathcal{D}$ and $\mathcal{F}$ by:

$$\mathcal{D} = \mathrm{Dom}\,(T^*) \cap \mathrm{Dom}\,(S) \tag{3.21}$$

$$\mathcal{F} = \{\varphi \in \mathcal{D} \mid \varphi \perp \mathcal{H}\} \tag{3.22}$$

and the hermitian form $q : \mathcal{F} \times \mathcal{F} \to \mathbb{C}$ by

$$q(\varphi, \psi) = (T^*\varphi, T^*\psi) + (S\varphi, S\psi). \tag{3.22'}$$

Then (3.15) can be written as

$$\|\varphi\|^2 \leq \text{const. } q(\varphi, \varphi) \quad \text{for } \varphi \in \mathcal{F}. \tag{3.23}$$

Now observe that if $\mathcal{D}$ is dense in B then, setting

$$\mathcal{G} = \{\varphi \in B \mid \varphi \perp \mathcal{H}\},$$

we have $\mathcal{F}$ dense in $\mathcal{G}$. Since if $\gamma \in \mathcal{G}$ there exists $\gamma' \in \mathcal{D}$ which is close to γ and $\gamma' = \gamma'' + \theta$ where $\theta \in \mathcal{H}$ and $\gamma'' \perp \mathcal{H}$ hence $\gamma'' \in \mathcal{F}$ and henceis close to γ.

The following result is often called the Friedrich representation theorem.

__Theorem__. If $\mathcal{F} \subset \mathcal{G}$ is dense and $\mathcal{F}$ is a Hilbert space with the norm q, a Hilbert space with the norm $\| \;\|$ and if (3.23) is satisfied then there exists a densely defined $F : \mathcal{G} \to \mathcal{G}$ such that $\quad \text{Dom}(F) \subset \mathcal{F}\quad$ and

$$q(\varphi,\psi) = (F\varphi,\psi) \quad \text{for all } \varphi \in \text{Dom}(F) \tag{3.24}$$

and all $\psi \in \mathcal{F}$. Furthermore, F is self-adjoint, onto and

$$\|F\varphi\| \geq \text{const.}\|\varphi\|, \quad \varphi \in \text{Dom}(F). \tag{3.25}$$

__Proof__: Given $\alpha \in \mathcal{G}$ then consider the function $T_\alpha : \mathcal{F} \to \mathbb{C}$ given by $T_\alpha(\psi) = (\alpha,\psi)$ from (3.23) we have

$$|T_\alpha(\psi)| = |(\alpha,\psi)| \leq \|\alpha\|\,\|\psi\| \leq \text{const.}\sqrt{q(\psi,\psi)}$$

so that T is a bounded functional and hence there exists a unique $K(\alpha) \in \mathcal{F}$ which represents T_α, that is:

$$q(K\alpha,\psi) = (\alpha,\psi) \tag{3.26}$$

for all $\psi \in \mathcal{F}$. Using (3.23) and (3.26) we have

$$\|K\alpha\|^2 \leq \text{const. } q(K\alpha, K\alpha) \leq \text{const. } |(\alpha, K\alpha)|$$

$$\leq \text{const. } \|\alpha\| \, \|K\alpha\|$$

hence

(3.27) $$\|K\alpha\| \leq \text{const. } \|\alpha\|.$$

Furthermore, for $\alpha, \beta \in \mathcal{G}$ we have

$$q(K\alpha, K\beta) = (\alpha, K\beta) = \overline{q(K\beta, K\alpha)} = \overline{(\beta, K\alpha)} = (K\alpha, \beta).$$

Therefore, $K : \mathcal{G} \to \mathcal{F}$ is one-one, bounded and self-adjoint. Setting Dom $(F) = \mathcal{R}(K)$ we define $F = K^{-1}$ and it has the desired properties.

Defining F_1 by setting

(3.28) $$\text{Dom } (F_1) = \text{Dom } (F) + \mathcal{H},$$

$$F_1 = F \text{ on Dom } (F) \text{ and } F_1 = 0 \text{ on } \mathcal{H}$$

we obtain that F_1 is a self-adjoint densely defined operator We claim that Dom (F_1) = Dom (L) and that $F_1 = L$. It is clear that Dom (L) $\subset$ Dom (F_1) and that $F_1 = L$ on Dom (L) therefore Dom $(L^*) \supset$ Dom $(F_1{}^*)$ and since both L and F_1 are self-adjoint, we have Dom (F_1) = Dom (L). Thus we conclude that $\mathcal{R}(L) = \mathcal{G}$ and from (3.17) we can conclude that

$$\mathcal{R}(S^*) = \mathcal{R}(S^*S) \text{ and } \mathcal{R}(T) = \mathcal{R}(TT^*).$$

Further we define the operator $N : B \to B$ such that

$$N = \begin{cases} 0 \text{ on } \mathcal{H} \\ K \text{ on } \mathcal{G}, \end{cases} \tag{3.29}$$

Denoting by $H : B \to \mathcal{H}$ the orthogonal projection into $\mathcal{H}$ we obtain the orthogonal decomposition of

$$\alpha = TT^*N\alpha + S^*SN\alpha + H\alpha \tag{3.30}$$

Now if $S\alpha = 0$ we have from (3.30)

$$SS^*SN\alpha = 0 \tag{3.31}$$

and thus

$$(SS^*SN\alpha, SN\alpha) = \|S^*SN\alpha\|^2 = 0$$

and hence $\quad S^*SN\alpha = 0 \quad$ and

$$\alpha = TT^*N\alpha + H\alpha$$

Thus we see that the necessary and sufficient condition to solve the equation $\alpha = Tu$ is that $S\alpha = 0$ and $H\alpha = 0$. Furthermore, the solution $u = T^*N\alpha$ is the unique solution orthogonal to $\mathcal{N}(T)$.

These facts are summarized in the following theorem.

<u>Theorem.</u> Given that A, B and C are Hilbert spaces and $T : A \to B$, $S : B \to C$ are densely defined closed operators such that $ST = 0$ and setting $\mathcal{D} = \text{Dom}(T^*) \cap \text{Dom}(S)$,

$$\mathcal{H} = \mathcal{N}(T^*) \cap \mathcal{N}(S) \text{ and } \mathcal{F} = \{\varphi \in \mathcal{D} \mid \varphi \perp \mathcal{H}\}$$

we assume that $\mathcal{D}$ is dense in B and that

$$\|\varphi\|_B^2 \le \text{const.}\ (\|T^*\varphi\|_A^2 + \|S\varphi\|_C^2)$$

for all $\varphi \in \mathcal{F}$. Then the space B splits into an orthogonal decomposition

$$B = \mathcal{R}(T) \oplus \mathcal{R}(S) \oplus \mathcal{H}.$$

Furthermore, the operator $L = TT^* + S^*S$ whose domain is

$$\text{Dom}\ (L) = \{\varphi \in \mathcal{D} \mid T^*\varphi \in \text{Dom}\ (T),\ S\varphi \in \text{Dom}\ (S^*)\}$$

is self-adjoint and has a closed range and there exists a unique bounded self-adjoint operator $N : B \rightarrow \text{Dom}\ (L)$ such that

(3.32) $\qquad LN = I - H$ and $HN = NH = 0$.

where $H : B \rightarrow \mathcal{H}$ is the orthogonal projection onto $\mathcal{H}$. It then follows that each $\alpha \in B$ has the orthogonal decomposition (3.30) and that the necessary and sufficient condition for the existance of a solution u satisfying $Tu = \alpha$ is that $S = 0$ and $\alpha \perp \mathcal{H}$, then $u = T^*N\alpha$. It also follows that if $P : A \rightarrow \mathcal{N}(T)$ is the orthogonal projection onto $\mathcal{N}(T)$ then

(3.33) $\qquad P = I - T^*NT.$

<u>Proof</u>: All that remains to be proven is (3.33), that is, we must show that the operator P defined by (3.33) is the orthogonal projection of A onto $\mathcal{N}(T)$. It suffices to show

$$TP = 0 \quad \text{and} \quad \mathcal{R}(P - I) \perp \mathcal{N}(T),$$

now

$$TP = T - TT^*NT = T - LNT = T - T - HT = 0$$

since $HT = 0$. Finally $P - I = -T^*NT$ and hence

$$\mathcal{R}(P - I) \subset \mathcal{R}(T^*) \perp \mathcal{N}(T).$$

The above Hilbert space set-up can be applied whenever we complete the spaces $\mathcal{A}^{p,q}$ with a Hilbert norm, either like the ones introduced in (3.2) or with weights. The same set up works for domains in complex manifolds. In that case we put a hermitian metric on the manifold this induces an inner product on the forms at each point, the integral of this inner product is then the L_2- inner product. We shall also introduce weights here.

<u>Definition</u>. If $\Omega \subset M$, M a complex hermitian manifold, then suppose that λ is a non-negative function, we form the inner product

$$(3.34) \qquad (\varphi,\psi)_{(\lambda)} = \int_\Omega \langle \varphi,\psi\rangle e^{-\lambda}\, dV$$

where $\varphi,\psi \in \mathcal{A}^{p,q}$, $\langle\varphi,\psi\rangle$ denotes the inner product defined at each point and dV denotes the volume element. Denoting by $L^{p,q}(\Omega,\lambda)$ the completion of $\mathcal{A}^{p,q}$ under the norm associated with (3.34) and by T and S the closure of the operators $\bar\partial : \mathcal{A}^{p,q-1} \to \mathcal{A}^{p,q}$ and $\bar\partial : \mathcal{A}^{p,q} \to \mathcal{A}^{p,q+1}$. We then define the operator L as above and we say that the <u>$\bar\partial$-Neumann problem for forms of type (p,q) on Ω and weight λ</u> is to prove the existence of an operator N as above.

As shown in the theorem, the existence of N then implies that the range of $\bar{\partial}$ is closed in L_2 so that N does not exist in general. We shall solve the $\bar{\partial}$ - Neumann problem on certain types of domains by means of establishing estimate (3.16) and this estimate in turn will follow from stronger estimates. Further, we will investigate various kinds of regularity properties for N which will automatically yield regularity results for the $\bar{\partial}$ - problem and for the projection on holomorphic functions (3.33).

J. J. Kohn

Lecture 4. The basic a priori estimates.

We will proceed to solve the $\bar{\partial}$-Neumann problem as set up in the definition of Lecture 3 in the spaces given by (3.4). We will assume that Ω has a smooth boundary defined by a function r as in Lecture 2. First we wish to find the smooth elements in $\mathcal{D} = \text{Dom}(T^*) \cap \text{Dom}(S)$. Denote by $\dot{\mathcal{D}}$ the space of $\varphi \in \mathcal{Q}$ which are C^∞ in $\bar{\Omega}$, i.e. $\dot{\mathcal{D}} = \mathcal{D} \cap \dot{\mathcal{Q}}^{p,q}$. Then $\varphi \in \dot{\mathcal{D}}$ and $\psi \in \mathcal{Q}^{p,q}$ we obtain, by integration by parts:

$$(4.1) \qquad (\varphi, \bar{\partial}\psi)_{(\lambda)} = \int_\Omega \langle e^{-\lambda}\varphi, \bar{\partial}\psi\rangle \, dV$$

$$= \int_\Omega \langle \vartheta(e^{-\lambda}\varphi), \psi\rangle \, dV - \int_{b\Omega} e^{-\lambda} \langle \sigma(\vartheta, dr)\varphi, \psi\rangle \, dV$$

$$= (e^{\lambda}\vartheta(e^{-\lambda}\varphi), \psi)_{(\lambda)} - \int_{b\Omega} e^{-\lambda} \langle \sigma(\vartheta, dr)\varphi, \psi\rangle \, dV,$$

where $\lambda \in C^\infty(\bar{\Omega})$, $\lambda \geq 0$, ϑ denotes the formal adjoint of $\bar{\partial}$ and $\sigma(\vartheta, dr)$ denotes the symbol of ϑ evaluated on dr. Here we assume that r has been normalized so that

$$(4.2) \qquad \langle dr, dr\rangle = 1 \quad \text{on } b\Omega.$$

Before proceeding we will illustrate the formula (4.1) in the case $p = 0$, $q = 1$, $\Omega \subset \mathbb{C}^n$ and $\lambda = 0$; it reduces

$$(4.3) \qquad (\varphi, \bar{\partial}u) = (-\sum_j \varphi_{jz_j}, u) + \sum_j \int_{b\Omega} r_{z_j} \varphi_j \bar{u} \, dS.$$

Since the boundary term in (4.1) vanishes when ψ has compact support in Ω we have

(4.4) $$(\varphi, \bar{\partial}\psi)_{(\lambda)} = (e^{\lambda}\vartheta(e^{-\lambda}\varphi), \varphi)_{(\lambda)}$$

for all $\psi \in \mathcal{Q}^{p,q-1}$ with compact support and since the set of these is dense, we conclude that

(4.5) $$T^*\varphi = e^{\lambda}\vartheta(e^{-\lambda}\varphi) \quad \text{for} \quad \varphi \in \dot{\mathcal{D}}.$$

Then, since $(\varphi, \bar{\partial}\psi)_{(\lambda)} = (T^*\varphi, \psi)_{(\lambda)}$ for $\varphi \in \dot{\mathcal{D}}$ and $\psi \in \mathcal{Q}^{p,q-1}$ we conclude that the boundary term in (4.1) must vanish for all $\psi \in \mathcal{Q}^{p,q-1}$ and hence we obtain

(4.6) $$\sigma(\vartheta, dr)\varphi = 0 \quad \text{on } b\Omega \quad \text{if} \quad \varphi \in \dot{\mathcal{D}}.$$

Conversely it is clear that $\varphi \in \mathrm{Dom}\,(T^*)$ whenever $\varphi \in \mathcal{Q}^{p,q}$ and satisfies (4.6). Therefore, since $\mathrm{Dom}\,(S) \supset \mathcal{Q}^{p,q}$, we conclude that $\dot{\mathcal{D}}$ consists of all $\varphi \in \mathcal{Q}^{p,q}$ for which (4.6) holds.

It will be useful to express the operators $\bar{\partial}$ and ϑ in terms of a special basis of vector fields in neighborhoods of boundary points. Let U be a small neighborhood of $P \in b\Omega$ and on U choose an orthonormal set of vector-fields of type (1.0), $L_1, \ldots, L_2$ such that

(4.7) $$L_j(r) = 0 \quad \text{if} \quad j < n \quad \text{and} \quad L_n(r) = 1$$

Let $\omega^1, \ldots, \omega^n$ be the dual basis of forms of type (1,0) on U. Then if $\varphi \in \mathcal{Q}^{p,q}$ on $U \cap \bar{\Omega}$, φ can be expressed by

(4.8) $$\varphi = \sum \varphi_{IJ}\, \omega^{IJ} \quad \text{with} \quad \varphi_{IJ} \in C^{\infty}(U \cap \bar{\Omega}),$$

where

$$I = (i_1, \ldots, i_p) \quad \text{with} \quad 1 \leq i_1 < i_2 < \ldots < i_p \leq n$$

$$J = (j_1 \ldots, j_q) \quad \text{with} \quad 1 \leq j_1 < \ldots < j_q \leq n$$

and

$$\omega^{IJ} = \omega^{i_1} \wedge \cdots \wedge \omega^{i_p} \wedge \bar{\omega}^{j_1} \wedge \cdots \wedge \bar{\omega}^{j_q}.$$

Then on $U \cap \bar{\Omega}$ we have

(4.9) $$\bar{\partial} u = \sum \bar{L}_j(u)\bar{\omega}^j, \quad u \in C^{\infty}(U \cap \bar{\Omega})$$

$$\bar{\partial}\varphi = \sum \bar{L}_j(\varphi_{IJ})\bar{\omega}^j \wedge \omega^{IJ} + \ldots$$

and

(4.10) $$\vartheta\varphi = \sum_{j \notin H} \pm L_j(\varphi_{I\langle jH\rangle}) \cdot^{IH} + \ldots$$

where the H run over ordered $(q-1)$-tuples, $\langle jH\rangle$ represent ordered q-tuple whose elements are combinations of the φ_{IJ} and the elements of H. The dots represent linear

From (4.10) it is immediate that the condition (4.6) on $U \cap b\Omega$ is equivalent to

(4.11) $$\varphi_{IJ} = 0 \quad \text{on} \quad U \cap b\Omega \quad \text{whenever} \quad n \in J.$$

The fact that the operators T^* and S are of the form given above and that the boundary conditions are of the form (4.11) show that the argument of Lax and Phillips (see [31]) can be applied as in Hörmander [10] (proposition 1.2.4) to prove the following:

<u>Proposition</u>. $\dot{\mathcal{D}}$ is dense in $\mathcal{D}$ under the norm

$$\|T^*\varphi\|^2_{(\lambda)} + \|S\varphi\|^2_{(\lambda)} + \|\varphi\|^2_{(\lambda)}.$$

This proposition enables us to prove estimates for elements in $\dot{\mathcal{D}}$ from which we can then deduce the crucial inequality (3.15) of Lecture 3 for elements in $\dot{\mathcal{D}}$. Let (a^k_{ij}) be the $n \times n$ matrices defined by

(4.12) $$[L_i, \bar{L}_j] = \sum_{k=1}^{n} a^k_{ij} L_k + \sum_{k=1}^{n} b^k_{ij} \bar{L}_k$$

then $$\bar{a}^k_{ij} = -b^k_{ij}.$$

Let (c_{ij}) be the $(n-1) \times (n-1)$ matrix defined by

(4.13) $$c_{ij} = a^n_{ij} \quad \text{for } 1 \leq i,j \leq n-1.$$

It then follows from the formula for the Levi-form ((2.31) of Lecture 2) that c_{ij} is the Levi-form in terms of our basis.

The following formula is at the root of all the estimates which we will derive here. If $\varphi \in \dot{\mathcal{D}}$ and the support of φ lies in $\Omega \cap U$ then

$$(4.14)\qquad \|T^*\varphi\|^2_{(\lambda)} + \|S\varphi\|^2_{(\lambda)} = \|\varphi\|^2_{(\lambda)\bar{z}} + \sum \int_{\Omega} \lambda_{[jk]} \varphi_{I\langle jK\rangle} \bar{\varphi}_{I\langle kK\rangle} e^{-\lambda} dV + \sum \int_{b\Omega} c_{jk} \varphi_{I\langle jk\rangle} \bar{\varphi}_{I\langle kK\rangle} e^{-\lambda} dS + O(\|\varphi\|_{(\lambda)\bar{z}} \|\varphi\|_{(\lambda)}),$$

where

$$(4.15)\qquad \|\varphi\|^2_{(\lambda)\bar{z}} = \sum \|\bar{L}_j \varphi_{IJ}\|^2_{(\lambda)} + \|\varphi\|^2_{(\lambda)}$$

and the $\lambda_{[jk]}$ are defined by:

$$(4.16)\qquad \lambda_{[jk]} = \bar{L}_j L_k(\lambda) + \sum a^i_{jk} L_i(\lambda).$$

Observe that the norm defined by (4.15) is equivalent to

$$(4.17)\qquad \sum \|\varphi_{IJ\bar{z}_j}\|^2_{(\lambda)} + \|\varphi\|^2_{(\lambda)}.$$

We will derive formula (4.14) only in the case $p = 0$, $q = 1$ and $\lambda = 0$, to obtain the general case one proceeds in exactly the same manner, the calculations are then somewhat more complicated.

From (4.9) we obtain

(4.18) $\|\bar\partial\varphi\|^2 = \sum_{j<k} \|L_j \varphi_k - L_k \varphi_j\|^2 + O(\|\varphi\|_z \|\varphi\|)$,

where $\|\varphi\|_{\bar z} = \|\varphi\|_{(0)\bar z}$.

Then

(4.19) $\sum_{j<k} \|\bar L_j \varphi_k - \bar L_k \varphi_j\|^2 = \sum_{j,k} \|\bar L_j \varphi_k\|^2 - \sum_{j,k} (\bar L_j \varphi_k, \bar L_k \varphi_j)$

and

(4.20) $(\bar L_j \varphi_k, \bar L_k \varphi_j) = -(L_k \bar L_j \varphi_k, \varphi_j) + O(\|\varphi\|_{\bar z} \|\varphi\|)$.

To justify (4.20) we observe that if $u, v\ C_0^\infty(\bar\Omega \cap U)$ then, by integration by parts

(4.21) $$(u, \bar L_k v) = -(L_k u, v) + \int_{b\Omega} L_k(r) u \bar v dS$$

$$+ O(\|u\| \|v\|),$$

the boundary term thus appears only when $k = n$ since $L_k(r) = \delta_{kn}$. The boundary does not appear in (4.20) since $\varphi \in \dot{\mathcal{D}}$ implies that $(\bar L_j \varphi_n)$ is zero on $b\Omega$; because then $\varphi_n = 0$ on $b\Omega$ so the term vanishes when $j = n$ and if $j < n$ then $\bar L_j \varphi_n = 0$ on $b\Omega$ since L_j is tangential.

Further, we have

$$(4.22)\quad (L_k \bar{L}_j \varphi_k, \varphi_j) = ([L_k, \bar{L}_j]\varphi_k, \varphi_j) + (\bar{L}_j L_k \varphi_k, \varphi_j)$$

$$= (a_{kj} L_n \varphi_k, \varphi_j) - (L_k \varphi_k, L_j \varphi_j)$$

$$+ O(\|\varphi\|_{\bar{z}} \|\varphi\|),$$

where again no boundary term appears and we use the fact that

$$(4.23)\qquad (fL_k \varphi_k, \varphi_j) = O(\|\varphi\|_{\bar{z}} \|\varphi\|)$$

since the boundary term appears only when $k = n$ and $\varphi_n = 0$ on $b\Omega$. Observe that

$$(4.24)\qquad \|\vartheta\varphi\|^2 = \sum (L_k \varphi_k, L_j \varphi_j) + O(\|\varphi\|_{\bar{z}} \|\varphi\|).$$

and that

$$(4.25)\qquad (c_{kj} L_n \varphi_k, \varphi_j) = \int_{b\Omega} c_{kj} \varphi_k \bar{\varphi}_j dS + O(\|\varphi\|_{\bar{z}} \|\varphi\|).$$

Summing on k and j and combining the above we obtain, for $\varphi \in \dot{\mathcal{D}}^{0,1} \cap C_0^\infty(U \cap \Omega)$

$$(4.26)\qquad \|\bar{\partial}\varphi\|^2 + \|\vartheta\varphi\|^2 = \|\varphi\|_{\bar{z}}^2 + \int_{b\Omega} c_{kj} \varphi_k \bar{\varphi}_j dS$$

$$+ O(\|\varphi\|_{\bar{z}} \|\varphi\|),$$

which is the desired formula (4.14) when $p = 0$, $\lambda = 0$ and $q = 1$.

The following theorems give estimates which lead to the solution of the $\bar{\partial}$-Neumann problem.

Theorem. If $P \in b\Omega$ then the following are equivalent

(a) There exists a neighborhood U of P such that

$$(4.27)\quad \int_{b\Omega} |\varphi|^2 dS + \|\varphi\|_z^2 \leq C(\|\bar{\partial}\varphi\|^2 + \|\vartheta\varphi\|^2 + \|\varphi\|^2)$$

for all $\varphi \in \dot{\mathcal{D}}^{p,q}$ with coefficients in $C_0^\infty(\bar{\Omega} \cap U)$.

(b) The Levi form at P has either at least $n-q$ positive eigenvalues or at least $q+1$ negative eigenvalues.

In case Ω is strongly pseudo-convex, i.e. the Levi form has $n-1$ positive eigenvalues, the inequality (4.27) for $q \geq 1$ is an immediate consequence of (4.14) with $\lambda = 0$. For $q = n$ being in $\dot{\mathcal{D}}^{p,n}$ means that φ vanishes on the boundary and hence again (4.27) is a consequence of (4.14). In case $q = 0$ we have $\dot{\mathcal{D}}^{p,q} = \mathcal{A}^{p,0}$ and $\vartheta\varphi = 0$, thus (4.14) with $\lambda = 0$ is replaces by

$$(4.28)\quad \|\bar{\partial}\varphi\|^2 = \|\varphi\|_z^2 + O(\|\varphi\|_z \|\varphi\|)$$

for $\varphi \in \mathcal{A}^{p,0}$. Condition (b) for $q = 0$ says that the Levi form has at least one negative eigenvalue so that for some $i < n$ we have $c_{ii}(P) < 0$, we can suppose that

$c_{11}(P) < 0$. For any function $u \in C_0^\infty(\bar{\Omega} \cap U)$ we have

$$(4.29)\quad \|\bar{L}_1 u\|^2 = -([L_1, \bar{L}_1]u, u) + \|L_1 u\|^2 + O(\|u\|_{\bar{z}}\|u\|)$$

$$- \int_{b\Omega} c_{11} |u|^2 dS + O(\|u\|_{\bar{z}}\|u\|)$$

taking U small enough so that $c_{11} < \frac{1}{2} c_{11}(P)$ in U we obtain

$$(4.30)\quad \int_{b\Omega} |u|^2 dS \leq \text{const.}(\|u\|_{\bar{z}}^2 + \|u\|^2)$$

and from this (4.27) follows in case $q = 0$. As a final illustration of how (4.27) is derived consider the case $q = 1$ assuming that the Levi form has at least two negative eigenvalues. Choose a basis $L_1, \ldots, L_n$ such that the Levi form is diagonal at P and assume that $c_{ii}(P) < 0$ for $1 \leq i \leq m$, $c_{ii}(P) > 0$ for $m < i < m'$ and $c_{ii}(P) = 0$ for $m' \leq i \leq n-1$ where $m \geq 2$. Choose U small enough so that $|c_{ii} - c_{ii}(P)| < \delta$ on U. Then

$$(4.31)\quad \|\bar{L}_i \varphi_j\|^2 \geq \varepsilon_{ij} \|\bar{L}_i \varphi_j\|^2 - (1 - \varepsilon_{ij}) \int_{b\Omega} c_{ii} |\varphi_j|^2 dS$$

$$+ O(\|\varphi\|_{\bar{z}}\|\varphi\|),$$

We use this formula when $1 \leq i$, $j < m'$ and set

$\varepsilon_{ii} = \varepsilon$ and $\varepsilon_{ij} = \frac{1}{2}$ when $i \neq j$ substituting in (4.26) we then obtain

$$(4.32)\quad \|\bar{\partial}\varphi\|^2 + \|\vartheta\varphi\|^2 \geq \text{const.}\|\varphi\|_{\bar{z}}^2 + 0\,(\|\varphi\|_{\bar{z}}\|\varphi\|)$$

$$+ \int_{b\Omega} \Big\{ \sum_{j=1}^{m'} \big(-\tfrac{1}{2} \sum_{i\neq j<m} c_{ii} + \varepsilon c_{jj}\big)|\varphi_j|^2$$

$$+ \sum_{j\geq m'} c_{jj}|\varphi_j|^2 + \sum_{i\neq j} c_{ij}\varphi_i\bar{\varphi}_j \Big\}\, dS$$

By choosing ε and δ small enough we get the coefficients of $|\varphi_j|^2$ in the boundary integral positive and bounded away from zero in U. On the other hand, the last term in the boundary integral is bounded by const. $\sum|\varphi_j|^2$ where the constant can be choosen as small as we please by making U small. Thus we obtain the inequality (4.27) in case $p = 0$, $q = 1$, and the Levi form has at least two negative eigenvalues. To establish (4.27) for arbitrary q when the Levi form has at least $q + 1$ negative eigenvalues we proceed in the same way.

The proof that (4.27) implies (b) is given in Hörmander [10].

<u>Proposition</u>. If condition (b) of the above theorem is satisfied then there exists a constant $C > 0$ and a neighborhood W of $b\Omega$

$$(4.33)\qquad \tau\int_{W\cap\Omega}|\varphi|^2 e^{-\tau\lambda}\,dV \le C(\|T^*\varphi\|^2_{(\tau\lambda)} + \|S\varphi\|^2_{(\tau\lambda)} + \|\varphi\|^2_{(\tau\lambda)}$$

for all $\tau \ge 0$, and $\varphi \in \dot{\mathcal{D}}^{p,q} \cap C_0^\infty(\bar{\Omega}\cap U)$,

where $\lambda = e^{sr}$ for some fixed large s.

For the proof of this proposition see Hörmander [11]. To establish (4.33) in the case that Ω is strongly pseudo-convex, choose s sufficiently large so that $(\lambda_{z_i\bar{z}_j})$ is positive definite as in (2.10) of section 2; it will then be positive definite in a neighborhood W of bM. Then from (4.16) we see that $\lambda_{[jk]}$ is positive definite and then (4.33) follows from (4.14) by use of a partition of unity.

Finally the same argument also yields the following result.

Proposition. If $\lambda \in C^\infty(\bar{\Omega})$ is strongly pluri-subharmonic in a neighborhood W of $b\Omega$ and if Ω is pseudo-convex then (4.33) holds for all $\tau \ge 0$ and $\varphi \in \dot{\mathcal{D}}^{p,q} \cap C_0^\infty(\bar{\Omega}\cap U)$ when $q \ge 1$.

In case there exists a $\lambda \in C^\infty(\bar{\Omega})$ which is strongly pluri-subharmonic throughout Ω, then the integral in (4.33) can be all taken over Ω so that we have

$$(4.34) \qquad \|\varphi\|^2_{(\tau\lambda)} \le \frac{C}{\tau - C}\left(\|T^*\varphi\|^2_{(\tau\lambda)} + \|S\varphi\|^2_{(\tau\lambda)}\right)$$

Hence by the results of the previous lecture we obtain the following result.

<u>Theorem</u>. If there exists a strongly pluri-subharmonic function in $C^\infty(\overline{\Omega})$ and if Ω is pseudo-convex with a smooth boundary then the $\bar{\partial}$-Neumann problem $L_2^{p,q}(\Omega, \tau\lambda)$ has a solution for $q \geq 1$ and τ sufficiently large.

The condition that there exists $\lambda \in C^\infty(\overline{\Omega})$ which is pluri-subharmonic is satisfied in $\mathbb{C}^n$, by taking

$$\lambda = |z|^2 = \sum |z_j|^2$$

for example. It is also satisfied for Ω in a Stein manifold, by taking

$$\lambda = \sum_1^N |h_k|^2,$$

where $h_1, \ldots, h_n$ are holomorphic functions that separate points. Thus we obtain the operator N, for $\tau > C$, and for each τ and L_2 solution of the $\bar{\partial}$-problem given by $T^*N\alpha$. Now we have, if $S\alpha = 0$:

$$(4.35) \qquad \|T^*N\alpha\|^2_{(\tau\lambda)} = (TT^*N\alpha, N\alpha)_{(\tau\lambda)}$$

$$= (\alpha, N\alpha)_{(\tau\alpha)}$$

(continued on next page)

$$\leq \frac{C}{\tau - C} \|\alpha\|^2_{(\tau\lambda)}.$$

If σ is the minimum of the least eigenvalue of $(\lambda_{z_i \bar{z}_j})$ we can choose $C = q\sigma^{-1}$. In $\mathbb{C}^n$ we can select the coordinates so that the origin lies in Ω then setting δ = diameter of Ω (i.e. $\delta = \sup |P-Q|$, $P, Q \in \Omega$) choosing $\lambda = |z|^2$, we have $C = q$ and

(4.36) $$e^{-\tau\delta^2} < e^{-|z|^2} \leq 1 \quad \text{for } z \in \Omega.$$

Choosing $\tau = q + \delta^{-2}$ we obtain

(4.37) $$\|T^* N\alpha\| \leq e\,\delta\|\alpha\|.$$

Hence in $\mathbb{C}^n$ we obtain the result of Hörmander ([10]) that there exists an L_2 solution of the $\bar{\partial}$-problem which is bounded as in (4.37).

The above theorem, and also estimate (4.37) can then be generalized to pseudo-convex domains whose boundary is not smooth by constructing an appropriate sequence corresponding to the exhaustion of $\Omega = U\Omega_c$ (see Hormander [10]).

J. J. Kohn

Lecture 5. Pseudo-differential operators.

In this lecture we will give a brief review of Sobolev spaces and pseudo-differential operators. If $\Omega \subset\subset \mathbb{R}^n$ and if s is a non-negative integer then for $u \in C^\infty(\bar{\Omega})$ we define

$$(5.1) \qquad \|u\|_s^2 = \sum_{|\alpha| \leq s} \|D^\alpha u\|^2$$

and let $H_s(\Omega)$ denote the completion of $C^\infty(\bar{\Omega})$ under this norm. $H_s(\Omega)$ is called a Sobolev space and $\|u\|_s$ is called the Sobolev s-norm.

These spaces have the following properties:

A. $H_s(\Omega) \subset H_t(\Omega)$ if $s > t$,

further if $u \in H_s$ we have

$$(5.2) \qquad \|u\|_s \geq \|u\|_t$$

and if $t_0 < t < s$ then for any $\varepsilon > 0$ there exists a constant $C(\varepsilon)$ such that

$$(5.3) \qquad \|u\|_t \leq \varepsilon \|u\|_s + C(\varepsilon) \|u\|_{t_0}$$

for all $u \in H_s(\Omega)$.

B. If $u_\nu \in H_s(\Omega)$, $\nu = 1,2,\ldots$

and if $\|u_\nu\| < C$ independently of ν then there exists a subsequence $\{u_{\nu_j}\}$ which converges in $H_t(\Omega)$ whenever $t < s$.

C. If Ω has a smooth boundary and if $s > k + \frac{1}{2}n$ then

$$H_s(\Omega) \subset C^k(\bar{\Omega}),$$

this follows from the inequality

$$\text{(5.4)} \qquad \sup \sum_{|\alpha| \leq k} |D^\alpha u| \leq C \|u\|_s$$

for all $u \in C^\infty(\bar{\Omega})$. A corollary of this property is that $u \in C^\infty(\bar{\Omega})$ if and only if $u \in \bigcap H_s(\Omega)$.

By duality we define $\| \ \|_s$ and H_s for negative integers as follows. If $u \in C^\infty(\bar{\Omega})$ and if $s < 0$ then we define $\|u\|_s$ by:

$$\text{(5.5)} \qquad \|u\|_s = \sup \frac{(u,v)}{\|v\|_{-s}} ,$$

where the sup is taken over all $v \in C^\infty(\bar{\Omega})$. From (5.5) follows the generalized Schwarz inequality

$$\text{(5.5')} \qquad |(u,v)| \leq \|u\|_s \|v\|_{-s} ,$$

which is very useful.

$H_s(\Omega)$ is then again defined as the completion of

$C^\infty(\bar{\Omega})$ under $\|u\|_s$ and the properties A, B still hold as well as (5.2) and (5.3).

For $u \in C_0^\infty(\Omega)$ the above norms can be expressed in terms of the Fourier transform as follows:

(5.6) $$\|u\|_s \sim \|\Lambda^s u\|,$$

where $\Lambda^s : C_0^\infty(\Omega) \longrightarrow C^\infty(\mathbb{R}^n)$ is defined by

(5.7) $$\widehat{\Lambda^s u}(\xi) = (1 + |\xi|^2)^{\frac{s}{2}} \hat{u}(\xi) ,$$

where $\hat{v}$ denotes the Fourier transform of v which is given by

(5.8) $$\hat{v}(\xi) = \int e^{-ix\cdot\xi} v(x)dx .$$

Formula (5.6) can be used as a definition of s-norms for arbitrary $s \in \mathbb{R}$, of course it then applies only to compactly supported functions.

Formulas (5.7) and (5.8) make sense for distribution with compact support. If u is a distribution on Ω then we say that u is locally in $H_s(\Omega)$ if

$$\Lambda^s(\zeta u) \in L_2(\Omega) \quad \text{for all } \zeta \in C_0^\infty(\Omega).$$

We denote by $H_s^{loc}(\Omega)$ the set of all $u \in \mathcal{D}'(\Omega)$ which are locally in $H_s(\Omega)$. Then if $u \in H_s^{loc}(\Omega)$ we have

$\zeta u \in H_s(\Omega)$ for all $\zeta \in C_0^\infty(\Omega)$.

If $P: C_0^\infty(\mathbb{R}^n) \longrightarrow C^\infty(\mathbb{R}^n)$ is a linear operator and if $U \subset \mathbb{R}^n$ we say that P is of order m on U if for every $s \in \mathbb{R}$ and every $\zeta \in C_0^\infty(\mathbb{R}^n)$ there exists C such that

(5.9) $$\|\zeta P u\|_s \le C\|u\|_{s+m} \quad \text{for all } u \in C_0^\infty(U) .$$

If P is of order m for every m then we say P is of order $-\infty$.

It then follows that $\sum_{|\alpha|\le m} a_\alpha D^\alpha$ with $a_\alpha \in C^\infty(\mathbb{R}^n)$ is of order m and that Λ^m is of order m.

Definition. A function $p \in C^\infty(\mathbb{R}^n \times \mathbb{R}^n)$ is called symbol order m if for every compact $K \subset\subset \mathbb{R}^n$ and for every pair of multi-indices α, β there exists a constant C (depending on K, α and β) such that

(5.10) $$\sup_K \left| D_x^\alpha D_\xi^\beta p(x,\xi)\right| \le C(1 + |\xi|)^{m-|\beta|} .$$

The operator $P: C_0^\infty(\mathbb{R}^n) \longrightarrow C^\infty(\mathbb{R}^n)$ defined by

(5.11) $$Pu(x) = \int_{\mathbb{R}^n} e^{ix\cdot\xi} p(x,\xi)\, \hat{u}(\xi)\, d\xi ,$$

where $d\xi = d\xi_1, \ldots, d\xi_n$ and $x\cdot\xi = \sum x_j \xi_j$, is called the pseudo-differential operator with symbol p.

In the above example $\sum_{|\alpha|\leq m} a_\alpha D^\alpha$ is the pseudo-differential operator with symbol $\sum_{|\alpha|=m} a_\alpha(x)\xi^\alpha$, where $\xi^\alpha = \xi_1^{\alpha_1}, \ldots \xi_n^{\alpha_n}$ and Λ^m is the pseudo-differential operator with symbol $(1+|\xi|^2)^{m/2}$.

The following are the properties of pseudo-differential operators that will be most useful to us.

(A) If p is a symbol of order m then P defined by (5.11) is an operator of order m.

(B) If P is a pseudo-differential operator of order m then its adjoint P^* is also a pseudo-differential operator of order m.

(C) If P and Q are pseudo-differential operators of orders a and b then $P + Q$ is a pseudo-differential operator of order max(a,b). Further the composition PQ, modulo operators of order-∞, is a pseudo-differential operator of order $a + b$ and the commutator:

$$[P,Q] = PQ - QP$$

is of order $a + b - 1$.

The meaning of "PQ modulo operators of order $-\infty$" should be interpreted as follows. If W is any proper neighborhood of the diagonal in $\mathbb{R}^n \times \mathbb{R}^n$, (proper means that if $U \subset \mathbb{R}^n$

is relatively compact then

$$U' = \{(x,y) \in W \mid x \in U\}$$

and

$$U'' = \{(x,y) \in W \mid y \in U\},$$

are relatively compact) , then there exists an operator P_0 of order $-\infty$ such that if $U \subset \mathbb{R}^n$ is relatively compact, then setting

$$\tilde{U} = \{x \in \mathbb{R}^n \mid (x,y) \in U'\}$$

then for any $u \in C_0^\infty(\mathbb{R}^n)$ such that

$$u = 0 \text{ on } \tilde{U}, \quad (P + P_0)(u) = 0 \text{ on } U.$$

Thus $P + P_0$ has a unique extension to an operator of

$$C^\infty(\mathbb{R}^n) \to C^\infty(\mathbb{R}^n)$$

so that $(P + P_0)Q$ makes sense. Further, if P'_0 is another such operator then

$$(P + P'_0)Q - (P + P_0)Q$$

is of order $-\infty$.

(D) The inequality (5.9) and the discussion above show that a pseudo-differential operator P can be extended, modulo operators of order $-\infty$ to $\cup H_s^{loc}(\mathbb{R}^n)$. A crucial property of pseudo-differential operators is that they are pseudo-local. We say that an operator P is _pseudo-local_ if, whenever u is distribution in the domain of P then

(5.12) $$\text{sing supp}(Pu) \subset \text{sing supp}(u),$$

where the complement of sing supp(v) is defined as those points which have neighborhoods on which v is in C^∞.

<u>Definition</u>. A pseudo-differential operator P of order m is called elliptic if for every compact K there exists C such that

(5.13) $$|p(x,\xi)| \geq C|\xi|^m \quad \text{for } |\xi| \text{ large},$$

where p is the symbol of P.

<u>Theorem</u>. If P is an elliptic pseudo-differential operator of order m then there exists a unique (modulo operators of order $-\infty$) pseudo-differential operator Q such that both $PQ - I$ and $QP - I$ are of order $-\infty$. Further, this Q is elliptic of order $-m$.

In view of (5.12) we have the following:

<u>Corollary</u>. If P is elliptic and u is a distribution such that $Pu = f$ then sing supp(u) $\subset$ sing supp(f), in particular $u \in C^\infty(\mathbb{R}^n)$ whenever $f \in C^\infty(\mathbb{R}^n)$.

<u>Proposition</u>. If P is elliptic of order m and if

$$u \in H_{-N}(\mathbb{R}^n) \text{ and } Pu \in H_s(\mathbb{R}^n) \text{ then } u \in H_{s+m}(\mathbb{R}^n).$$

Further we have

(5.14) $$\|u\|_{s+m} \leq C(\|Pu\|_s + \|u\|_{-N})$$

for all $u \in H_{s+m}(\mathbb{R}^n)$.

Conversely, if (5.14) holds for all $u \in C_0^\infty(\mathbb{R}^n)$ and some fixed s and N with $-N<s<m$ then P is elliptic of order m.

The estimate (5.14) follows immediately from the theorem and (5.9) since

$$u = QPu + Ku ,$$

with Q of order -m and K of order $-\infty$.

Proposition. If P is a pseudo-differential operator of order $m>0$ and if

(5.15) $$\|u\|^2_{\frac{1}{2}m} \leq C(Pu,u) + \|u\|^2$$

then P is elliptic.

Proof: It suffices to prove (5.14) with $s = 0$ and $N = 0$. We have

(5.16) $$\|u\|^2_m = \|\Lambda^{m/2} u\|^2_{m/2} \leq C(P\Lambda^{m/2} u, \Lambda^{m/2} u) + \|u\|^2_{m/2}$$

$$\leq C\Big\{ |([P,\Lambda^{m/2}]u, \Lambda^{m/2} u)| + \|Pu\|\|u\|_m + \|u\|^2_{m/2} \Big\}$$

cont. next page

$$\leq C\{\|u\|^2_{m-\frac{1}{2}} + \text{large const.}\|Pu\|^2$$

$$+ \text{small const.}\|u\|^2_m + \|u\|^2_{m/2}\}.$$

The desired estimate is then obtained by use of (5.3).

In case $P = \sum_{|\alpha|\leq m} a_\alpha D^\alpha$ ellipticity is equivalent to the condition

$$\sum_{|\alpha|=m} a_\alpha \xi^\alpha \neq 0 \quad \text{if} \quad \xi \neq 0 \ . \tag{5.17}$$

All the above results can be generalized for determined systems. If P represents a $k \times k$ matrix of pseudo-differential operators and u represents a k-tuple of functions. Defining $\|u\|_s$ by

$$\|u\|^2_s = \sum_{j=1}^{k} \|u_j\|^2_s \tag{5.18}$$

Then ellipticity can be defined by requiring (5.14) to hold for some $-N < s < m$. For a matrix of differential operators this is equivalent to requiring that the matrix of the principal parts be non-singular. All the above theorems then hold for the case of such operators.

We conclude with the observation that all the above results can be "localized" in a natural manner i.e. the notion of ellipticity can be defined on any open subset of R^n and the natural analogues of the above results hold. In particular we have:

<u>Theorem</u>. If P is a differential operator defined on $U \subset \mathbb{R}^n$ which is elliptic and of order m, i.e.

$$P = \sum_{|\alpha| \leq m} a_\alpha D^\alpha \quad \text{with} \quad a_\alpha \in C^\infty(U)$$

and (5.17) holds then for

$$\zeta, \zeta' \in C_0^\infty(U) \quad \text{with} \quad \zeta = 1$$

in a neighborhood of the support of ζ we have

(5.19) $$\|\zeta u\|_{s+m} \leq C(\|\zeta' Pu\|_s + \|\zeta' u\|_{-N})$$

for all $u \in C^\infty(U)$, where C depends on ζ, ζ', s and N. It then follows that sing supp $(u) \subset$ sing supp Pu.

<u>Proof</u>. Let $\zeta_1 \in C_0^\infty(U)$ be such that $\zeta_1 = 1$ in a neighborhood of the support of ζ and $\zeta' = 1$ in a neighborhood of the support of ζ_1. Then let P' be an elliptic differential operator on R^n which is equal to P in a neighborhood of the support of ζ'_0. Let Q be a pseudo-differential operator such that $QP' = I + K$ where K is of order $-\infty$ and Q of order $-m$. Then

$$\zeta u = QP\zeta u + K(\zeta u)$$

taking $s + m$ norms we obtain

$$(5.20)\qquad \|\zeta u\|_{s+m} \le \|P\zeta u\|_s + \text{const.}\|u\|_{-N}$$

$$\le \|\zeta Pu\|_s + \|[P,\zeta]u\|_s + \text{const.}\|\zeta u\|_{-N} ,$$

since the support of $[P,\zeta]$ is contained in the support of ζ and since $[P,\zeta]$ is an operator of order m-1 we have

$$(5.21)\qquad \|[P,\zeta]u\|_s = \|[P,\zeta]\zeta_1 u\|_s \le \text{const.}\|\zeta_1 u\|_{s+m-1}$$

so we obtain

$$(5.22)\qquad \|\zeta u\|_{s+m} \le \|\zeta_1 u\|_{s+m-1} + \|\zeta Pu\|_s + \text{const.}\|\zeta u\|_{-N}.$$

Repeating the same argument with s replaced by s -1 and ζ by ζ_1 and ζ_1 by ζ_2 where

$$\zeta_2 \in C_0^\infty(U) \text{ with } \zeta_2 = 1$$

in a neighborhood of the support of ζ_1 and $\zeta' = 1$ in a neighborhood of the support of ζ_2. After k steps we obtain

$$(5.23)\qquad \|\zeta u\|_{s+m} \le \text{const.}(\|\zeta_k Pu\|_s + \|\zeta_k u\|_{s+m-k}$$

$$+ \|\zeta_k u\|_{-N}.$$

Choosing k = N + s + m concludes the proof of (5.16).

It remains to show that if $Pu|_V$ is C^∞ then $u|_V$ is C^∞, where V is an open subset of V. To do this we choose the ζ and ζ' in $C_0^{\infty}(V)$. Then using a standard smoothing operator in (5.19) it can be shown that $\zeta u \in H_{s+m}$ for every s and hence in C^∞.

This same result holds for when P is an elliptic system.

J.J. Kohn

Lecture 6. Interior regularity and existence theorems.

As in Lecture 3 we set

$$(6.1)\qquad q_\lambda(\varphi,\psi) = (T^*\varphi, T^*\psi)_{(\lambda)} + (S\varphi, S\psi)_{(\lambda)}$$

for $\varphi, \psi \in \mathcal{D}$. From (4.14) we deduce the following

Lemma. There exists $C > 0$ such that

$$(6.2)\qquad \|\varphi\|_1^2 \leq C\, q_\lambda(\varphi,\varphi) + \|\varphi\|^2 ,$$

for all $\varphi \in \mathcal{a}^{p,q}$ with $\varphi = 0$ on $b\Omega$, where C depends on λ.

Proof. Since $\varphi = 0$ on $b\Omega$ we have

$$(6.3)\qquad \|\varphi\|_z^2 = \sum \|L_i \varphi_{IJ}\|^2 + \|\varphi\|^2 \leq \text{const.} \|\varphi\|^2_{(\lambda)\bar{z}}$$

also

$$\|\varphi\|_{\bar{z}}^2 \leq \text{const.} \|\varphi\|^2_{(\lambda)\bar{z}}$$

and

$$(6.4)\qquad \|\varphi\|_1^2 \leq \text{const.} \left(\|\varphi\|_z^2 + \|\varphi\|_{\bar{z}}^2 \right) \leq \text{const.} \|\varphi\|^2_{(\lambda)\bar{z}} .$$

Since the boundary integral is zero we obtain (6.1) from (6.4) and (4.14).

From this lemma we see that the operator $T^*T + SS^*$ is elliptic of second order and hence we conclude that whenever the $\bar\partial$-Neumann problem has a solution N on $L_2^{p,q}(\Omega,\lambda)$ then $N\alpha$ is smooth on any open subset on which α is smooth. Hence also the solution of the $\bar\partial$-problem given by $T^*N\alpha$, when $\alpha \perp \mathcal{H}^{p,q}$ and $S\alpha = 0$, is smooth where α is smooth. Under these circumstances if ζ and $\zeta' \in C_0^\infty(\Omega)$ we have, by (5.16) with

$$u = N\alpha \quad \text{and} \quad P = T^*T + SS^*$$

$$\|\zeta N\alpha\|_{s+2} \leq C(\|\zeta'\alpha\|_s + \|\alpha\|). \tag{6.5}$$

__Theorem.__ If Ω is pseudo-convex and if in a neighborhood W of $b\Omega$ there is a strongly pluri-subharmonic function λ and $q \geq 1$; or if the Levi form has either at least $n - q$ positive eigenvalues or $q + 1$ negative eigenvalues then for sufficiently large τ the $\bar\partial$-Neumann problem has a solution on $L_2^{p,q}(\Omega,\tau\lambda)$, and the space $\mathcal{H}^{p,q}$ is finite dimensional and consists of elements which are in C^∞ on $\bar\Omega$. Furthermore, if $W = \Omega$ then $\mathcal{H}^{p,q} = 0$ for $q \geq 1$.

__Proof:__ The case $W = \Omega$ follows from (4.34). To establish the general case it suffices to show (by the results of Lecture 3) that if $\|\varphi_\nu\|$ is bounded, $\varphi_\nu \perp \mathcal{H}^{p,q}$ and that if

$$\lim q_{\tau\lambda}(\varphi_\nu, \varphi_\nu) = 0$$

then there is a subsequence φ_{ν_j} such that $\lim \varphi_{\nu_j} = 0$

in $L_2^{p,q}$. To do this, choose a function $\zeta_1 \in C^\infty(\bar{\Omega})$ such that $\zeta_1 = 1$ in a neighbrohood V of $b\Omega$ and is zero in Ω - W. Then, if τ is sufficiently large, we obtain the following inequality as a consequence of (4.33).

$$\|\zeta_1\varphi\|^2 \leq \text{const.} q_{(\tau\lambda)}(\zeta_1\varphi, \zeta_1\varphi) \tag{6.6}$$

$$\leq \text{const.}(q_{(\tau\lambda)}(\varphi,\varphi) + \|\zeta_0\varphi\|_1^2$$

for $\varphi \in \dot{\mathcal{D}}^{p,q}$ where $\zeta_0 \in C_0^\infty(\Omega)$ and $\zeta_0 = 1$ on Ω - V. The second inequality is obtained by noting that

$$\|S\zeta_1\varphi\| \leq \|\zeta_1 S\varphi\| + \|[S,\zeta_1]\varphi\| \tag{6.7}$$

$$\leq \text{const.}(\|S\varphi\| + \|\zeta_0\varphi\|)$$

since $[S,\zeta_1]$ is a matrix whose components have supports in Ω - V and hence are bounded by const. $|\zeta_0|$. A similar calculation for $T^*\zeta_1$ establishes (6.6). Further, from (6.2) we obtain

$$\|\zeta_0\varphi\|_1^2 \leq \text{const. } q_{(\tau\lambda)}(\varphi,\varphi) + \|\zeta_0'\varphi\|^2 . \tag{6.8}$$

where $\zeta_0' \in C_0^\infty(\Omega)$ and $\zeta_0' = 1$ on the support of ζ_0. Combining (6.6) and (6.8) yields

$$\|\zeta_1\varphi\|^2 \leq \text{const.}\left(q_{(\tau\lambda)}(\varphi,\varphi) + \|\zeta_0'\varphi\|^2\right) . \tag{6.9}$$

Applying (6.8) to the sequence $\{\varphi_\nu\}$ we have to conclude that $\|\zeta_0\varphi_\nu\|_1$ is bounded and thus there exists a subsequence φ_{ν_j} such that $\zeta_0\varphi_{\nu_j}$ converges in $L_2^{p,q}(\Omega)$ thus we see that for every $\zeta \in C_0^\infty(\Omega)$ there exists a subsequence φ_{ν_j} such that $\zeta\varphi_{\nu_j}$ converges in $L_2^{p,q}(\Omega)$. Setting $\zeta = \zeta_0'$ and applying (6.9) to the differences $\varphi_{\nu_i} - \varphi_{\nu_j}$ we see that $\{\zeta_1\varphi_{\nu_j}\}$ is a Cauchy sequence and hence converges in $L_2^{p,q}(\Omega)$. Therefore φ_{ν_j} converges to $\theta \in L_2^{p,q}(\Omega)$ and so it also converges to θ in $L_2^{p,q}(\Omega, \tau\lambda)$. Since $q_{\tau\lambda}(\theta,\theta) = 0$ and $\theta \perp \mathcal{H}^{p,q}$ we conclude that $\theta = 0$ as required.

This solution of the $\bar\partial$-Neumann problem can be applied to prove existence of holomorphic functions as indicated in the first lecture. It gives the following result.

<u>Theorem</u>. Under the hypothesis of the previous theorem and if there exists $P \in b\Omega$ such that at P there exists a local holomorphic separating function, then there exists a holomorphic function on Ω which cannot be continued to any domain which contains P in the interior.

<u>Proof</u>: As in the first lecture we construct a function F which is not in $L_2(\Omega)$ but such that $\alpha = \bar\partial F$ is in

$L_2(\Omega)$, what is more the construction can be so arranged that if F_ν is a small translation of F in the direction normal to $b\Omega$ then $\alpha = \lim \bar\partial F_\nu$. Thus we see that $S\alpha = 0$ and that $\alpha \perp \mathcal{H}^{0,1}$. Then

$$\alpha = T(T^*N\alpha) \quad \text{and} \quad T^*N\alpha \in L_2(\Omega) \quad \text{thus} \quad h = F - T^*N\alpha$$

is the desired holomorphic function.

Another application of the $\bar\partial$-Neumann problem is to prove the Newlander-Nirenberg theorem. To do this, we first define almost complex structure.

Definition. If Ω is a differentiable manifold and $\mathbb{C}T$ denotes the complexified tangent bundle then an almost-conplex structure on Ω is given by sub-bundle $T^{1,0}$ of $\mathbb{C}T$ with the following property. If $T^{0,1}$ denotes the conjugate of $T^{1,0}$ then

$$\mathbb{C}T = T^{1,0} \oplus T^{0,1}, \text{ i.e. } T^{1,0} \cap T^{0,1} = 0$$

$$\mathbb{C}T = T^{1,0} + T^{0,1}.$$

If Ω is a conplex manifold then the underlying almost-complex structure is given by the vectors of type (1,0). It is clear that if there is a complex structure associated with a given almost-complex structure then it is unique.

Definition. If Ω is almost-complex with ther almost-complex structure given by $T^{1,0}$ is called integrable if for any two local vector fields L and L' with values in

$T^{1,0}$ the commutator $[L, L']$ also has values in $T^{1,0}$.

Clearly if $T^{1,0}$ is the almost-complex structure associated to a complex structure then it is integrable. Conversely we have

<u>Theorem</u>. (Newlander, Nirenberg [25]). If $T^{1,0}$ gives an integrable almost-complex structure on Ω then Ω has a complex structure such that $T^{1,0}$ gives the associated almost-complex structure.

First observe that this theorem is strictly local, that is, it suffices to show that given a point $P \in \Omega$ there exists function $h_1, \ldots, h_n$ defined in a neighborhood U of P, such that $dh_1, \ldots, dh_n$ are linearly independent and such that for every vector field L with values in $T^{1,0}$ we have

(6.10) $\qquad \bar{L}(h_j) = \overline{L(\bar{h}_j)} = 0 \quad$ for $\; j = 1, \ldots, n.$

Then $h_1, \ldots, h_n$ are a complex coordinate system and it is clear that a function u defined on an open subset of U satisfies the equations $\bar{L}(u) = 0$ for all L with values in $T^{1,0}$ if and only if u satisfies the Cauchy-Riemann equations with respect to the coordinates $h_1, \ldots, h_n$.

On an almost-complex manifold the space of form $\mathcal{A}$ has a natural bigradation

$$\mathcal{A} = \mathcal{A}^{p,q}$$

induced by the decomposition $\mathbb{C}T = T^{1,0} + T^{0,1}$. Denote by $\Pi_{p,q} : \mathcal{A} \to \mathcal{A}^{p,q}$ the corresponding projection mappings.

We define the operators ∂ and $\bar{\partial}$ by

$$(6.11) \qquad \partial = \sum_{p,q} \Pi_{p+1,q}\, d\, \Pi_{p,q}$$

$$\bar{\partial} = \sum_{p,q} \Pi_{p,q+1}\, d\, \Pi_{p,q}$$

Proposition. For an almost complex structure the following are equivalent.

(a) integrable

(b) For every form φ of type (0,1) we have $\Pi_{2,0} d\varphi = 0$

(c) If φ is a (0,1)-form then $d = \partial\varphi + \bar{\partial}\varphi$

(d) If u is a function $\partial^2 u = 0$

(e) $d = \partial + \bar{\partial}$, $\partial^2 = \bar{\partial}^2 = 0$ and $\partial\bar{\partial} = -\bar{\partial}\partial$.

Proof: By a standard formula for an exterior derivative we have

$$(6.12) \qquad \langle d\varphi, L \wedge L' \rangle = L(\langle \varphi, L' \rangle) - L'(\langle \varphi, L \rangle) - \langle \varphi, [L, L'] \rangle ,$$

choosing L and L' to be any two vector vields with values in $T^{1,0}$ we have

$$\langle\!\langle \varphi, L' \rangle\!\rangle = \langle \varphi, L \rangle = 0$$

since φ is of degree (0,1). Thus we see that $\Pi_{2,0} d = 0$ if and only of L, L' is of type (1,0); which shows that (a) is equivalent to (b). Clearly (c) is equivalent to (b) since for forms φ of degree (0,1) we have by (6.11) $\partial\varphi = \Pi_{1,1} d\varphi$ and $\bar\partial\varphi = \Pi_{0,2} d\varphi$ so that (c) holds if and only if $\Pi_{2,0} d\varphi = 0$. By (6.11) we have $\partial^2 u = \Pi_{2,0} d u$ hence if L and L' are vector fields in $T^{1,0}$ then

$$(6.13) \qquad \langle \partial^2 u, L \wedge L' \rangle = \langle d\partial u, L \wedge L' \rangle = L\langle u, L' \rangle - L'\langle u, L \rangle - \langle u, [L, L'] \rangle$$

since $\partial u = \Pi_{1,0} du$ we obtain

$$(6.14) \qquad \langle \partial^2 u, L \wedge L' \rangle = [L, L'] u - \langle \partial u, [L, L'] \rangle$$

this is zero if and only if $[L, L']$ is of type (1,0). So that (e) is equivalent to (a). Finally the conditions (c) and (e) and their conjugates extend to forms of all types and hence are equivalent to (f).

<u>Corollary</u>. On an integrable almost-complex manifold a function u satisfies the equation $\bar\partial u = 0$ if and only if $du = \partial u$ and this in turn is equivalent to $\bar{L} u = 0$ for all L of type (1,0).

If Ω is a domain in an integrable almost-complex manifold then we can put this manifold a hermitian metric, i.e. a hermitian metric whose inner product induces a hermitian inner product on $\mathbb{C}T$ such that $T^{1,0}$ is orthogonal to $T^{0,1}$. Then all our results on the $\bar{\partial}$-Neumann problem generalize with exactly the same proofs.

Let $U \subset \mathbb{R}^{2n}$ be an open subset on which is defined an integrable almost-complex structure. We choose real coordinates $x^1,\ldots,x^{2n}$ such that the ball

$$B = \left\{ x \in \mathbb{R}^{2n} \;\middle|\; \sum_1^{2n} |x_j|^2 \leq 1 \right.$$

is contained in U. For each $t \in [0,1]$ we define the map $\Phi_t : \mathbb{R}^n \to \mathbb{R}^n$ by $\Phi_t(x) = tx$. Let $\omega^1,\ldots,\omega^n$ be a basis for the forms of type $(1,0)$ i.e.

$$\omega^j = \sum_{k=1}^{2n} a_k^j(x)dx^k \tag{6.15}$$

for each $t \in (0,1]$ we set

$$\omega_t^j = t^{-1}\Phi_t^* \omega^j$$

these define an integrable almost-complex structure on B for each $t \in [0,1]$ we shall denote this by B_t. In terms of coordinates we have

$$\omega_t^j = \sum_{k=1}^{2n} a_k^j(tx)dx^k \tag{6.16}$$

We denote by $\partial_t, \bar{\partial}_t$ the operators associated with B_t. Note that B_0 has a complex structure, in fact setting

$$z_0^j = \sum a_k^j(0)x^k \tag{6.17}$$

we have $\omega_0^j = dz_0^j = \partial_0 z_0^j$ so that $\bar{\partial}_0 z_0^j = 0$ and the dz_0^j are independent. Now B_0 is strongly pseudo-convex and the function $\lambda = \sum |z_0^j|^2$ is strongly plurisubharmonic on B_0. Hence for small t, B_t is strongly pseudo-convex and λ is strongly plurisubharmonic (in the sense that $\lambda_{[jk]}$ is positive definite). Choosing a hermitian metric which varies smoothly with t we conclude that there exists a $\delta < 0$ a fixed τ such that (4.34) holds in $\dot{\mathcal{D}}_t^{1,0}$ when $t \leq \delta$ with a constant independent of t i.e.

$$_t\|\varphi\|^2_{(\tau\lambda)} \leq \text{const.}\left({}_t\|T^*_t\varphi\|^2_{(\tau\lambda)} + {}_t\|S_t\varphi\|^2_{(\tau\lambda)}\right) \tag{6.18}$$

for all $\varphi \in \dot{\mathcal{D}}_t^{1,0}$, where ${}_t\|\ \|_{(\tau\lambda)}$ denotes the norm in the space $L_2^{0,1}(B_t, \tau\lambda)$ with respect to the hermitian metric on B_t. Hence, when $t \leq \delta$, the $\bar{\partial}$-Neumann problem on $L_2^{0,1}(B_t, \tau\lambda)$ has a solution and $\mathcal{H}_t^{0,1} = 0$. Hence if $\mathfrak{I}$ and $\mathfrak{I}'$ have supports in the interior of B and $\mathfrak{I}' = 1$

on support of ζ then (6.5) holds independently of t, i.e.

$$\|N_t \alpha\|_{s+2} \leq \text{const.}(\|\zeta'\alpha\|_s + \|\alpha\|) \tag{6.19}$$

for all $\alpha \in L_2^{0,1}(B_t)$ and $t \leq \delta$. Further, we have

$$\bar{\partial}_t u(x) = \sum b_i^k(tx) \frac{\partial u}{\partial x^k} \bar{\omega}^i , \tag{6.20}$$

where

$$\sum_k b_i^k(tx)\, a_k^j(tx) = \delta_i^j \tag{6.21}$$

Hence if $u \in C^\infty(B)$ we have for any α

$$\lim_{t \to 0} D^\alpha \bar{\partial}_t u = D^\alpha \bar{\partial}_0 u \tag{6.22}$$

uniformly. Now we define the function z_t^j by

$$z_t^j = z_0^j - T^* N_t \bar{\partial}_t z_0^j \tag{6.23}$$

and we have $\bar{\partial}_t z_t^j = 0$.

In (6.19) choose ζ_0, ζ and ζ' with supports in the interior of B such that $\zeta_0 = 1$ in a neighborhood of the origin, $\zeta = 1$ in a neighborhood of the support of ζ_0 and $\zeta' = 1$ in a neighborhood of the support of ζ ; then

(6.24) $$\|\zeta_0(z_t^j - z_0^j)\|_s \le \|T^*_t \zeta_0 N_t \bar\partial_t z_0^j\|_s$$

$$+ \|[T^*_t, \zeta_0] N_t \bar\partial_t z_0^j\|_s$$

$$\le \|\zeta N_t \bar\partial_t z_0^j\|_{s+1} \le \text{const.}\ \|\zeta' \bar\partial_t z_0^j\|_s + \|\bar\partial_t z_0^j\| .$$

Now for s large we have

(6.25) $$\sup_V \sum_{|\alpha| \le 1} |D^\alpha(z_t^j - z_0^j)| \le \text{const.} \|\zeta_0(z_t^j - z_t^j)\|_s ,$$

hence, since the right of (6.24) goes to zero as $t \to 0$ we see that if $t = t_0$ is sufficiently small then on V the gradients of $z_{t_0}^1, \ldots, z_{t_0}^n$ are linearly independent. Thus B_{t_0} admits holomorphic coordinates in a neighborhood of the origin and hence so does B_1, in fact the coordinates on B_1 are given by $z_{t_0}^j(xt_0^{-1})$, which proves the theorem.

J. J. Kohn

Lecture 7. Boundary regularity.

In this lecture we will assume Ω is a domain in a complex manifold with a C^∞ boundary which is pseudo-convex and such that there exists a strongly plurisubharmonic function in a neighborhood of $b\Omega$. We wish to discuss smoothness of solutions of the $\bar\partial$-problem and of the $\bar\partial$-Neumann problem in the closed domain $\overline{\Omega}$, i.e. up to and including the boundary. We will restrict our attention to the $\bar\partial$-problem for functions or equivalently the $\bar\partial$-Neumann problem on (0,1)-forms. A nautral question to ask is, given a (0,1)-form $=\bar\partial v$ when does there exist a solution u of $\alpha = \bar\partial$ such that

(7.1) sing supp (ω) sing supp ( ).

It is easy to see that every solution u has this property in Ω ; however, we wish to interpret the above in $\overline{\Omega}$. The problem is more delicate there for if u is a solution and h is a holomorphic function then u + h is also a solution which in general will not be smooth on $b\Omega$.

The following example shows that it is not always possible to find a solution satisfying (7.1). Let $\Omega \subset \mathbb{C}^2$ such that in a neighborhood U of (0,0) we have

$$\Omega \cap U = (z_1, z_2) \in U \mid \operatorname{Re}(z_2) < 0 \quad . \tag{7.2}$$

Let $\rho \in C_0^\infty(U')$, with U' a neighborhood of (0,0) and $\overline{U}'$ U such that $\rho = 1$ on a neighborhood V of (0,0) and let

$$\alpha = \bar\partial\left(\frac{\rho}{z_2}\right) = \frac{\bar\partial\rho}{z_2} \quad . \tag{7.3}$$

Then the support of α is contained in $W = (U' - V)$.
We will show that every solution of $\bar{\partial} u = \alpha$ is unbounded in $\bar{\Omega} - W'$. where $W"$ is a neighborhood of $\bar{W}$. For simplicity we suppose that U and U' are so chosen that for some positive numbers A and δ_0 the set

(7.4) $\{(z_1, z_2) \mid |z_1| \leq A, \operatorname{Im}(z_2) = 0, -\delta_0 \leq \operatorname{Re}(z_2) \leq 0\}$

is contained in $\bar{\Omega}$ and the set

(7.5) $\{(z_1, z_2) \mid |z_1| = A, \operatorname{Im}(z_2) = 0, -\delta_0 \leq \operatorname{Re}(z_2) \leq 0$

does not intersect U'. Let $h = u - \frac{\rho}{z_2}$, then h is holomorphic; consider the restriction of h to the line $z_2 = -\delta$ with $0 < \delta < \delta_0$ we have

$$h(z_1, -\delta) = u(z_1, -\delta) + \frac{(z_1, -\delta)}{\delta}$$

If u is bounded by K in $\bar{\Omega}$ -W' then $h(z_1, -\delta)$ is bounded by K on the circle $z_1 = A$, $\operatorname{Re}(z_2) = -\delta$; on the other hand

$$h(0, -\delta) \geq \frac{1}{\delta} - K$$

which is a contradiction, since $h(0, -\delta)$ is an average of $h(z_1, -\delta)$ on $|z_1| = A$.

The theory of interior regularity of elliptic operators extends to boundary regularity for the so-called coercive boundary value problems. Suppose Q is an integer-differential form on k-tuples of functions on Ω expressed by:

$$(7.6) \qquad Q(u,v) = \sum_{j,|\alpha|,|\beta|\le 1} \sum_{i,j} \int_\Omega a^{ij}_{\alpha\beta} D^\alpha u_i \overline{D^\beta v_j}\, dV ,$$

where $u = (u_1,\dots,u_k)$, $v = (v_1,\dots,v_k)$ and $u_j, v_j, a^{ij} \in C^\infty(\bar\Omega)$. Suppose that B is a $k \times h$ matrix of C^∞ functions on $b\Omega$ such that B is of constant rank. Then if

$$(7.7) \qquad \sum_{j=1}^{k} \|u_j\|_1^2 \le \text{const.}\left(Q(u,u) + \sum_{j=1}^{k} \|u_j\|^2\right)$$

for all u with $Bu = 0$ on $b\Omega$, we say that the Q restricted to the space of k-tuples satisfying $Bu = 0$ is coercive.

Theorem. Suppose Q with the boundary condition $Bu = 0$ is coercive. Let

$$\mathcal{B} = \{u \mid Bu = 0 \text{ on } b\Omega\}$$

and let $\tilde{\mathcal{B}}$ denote the completion of $\mathcal{B}$ under the norm given by the right side of (7.7). Then if $u \in \tilde{\mathcal{B}}$ and satisfies

$$(7.8) \qquad Q(u,v) = (f,v)$$

for all $v \in \tilde{\mathcal{B}}$. Then sing supp(u) $\subset$ sing supp(f) and if

$$\zeta, \zeta' \in C_0^\infty(\bar\Omega),\ \zeta' u\ L_2(\Omega) \text{ and } \zeta' f \in H_s(\Omega)$$

then

$$\zeta u \in H_{s+2}(\Omega) \text{ and}$$

(7.9) $$\|\zeta u\|_{s+2} \leq \text{const.}\left(\|\zeta'' f\|_s + \|\zeta' u\|\right).$$

The $\bar{\partial}$-Neumann problem is coercive (i.e. (7.7) holds when $Q = q$ acting on $\dot{\mathcal{D}}$) only when $q = n-1$. To see this note that (4.14) implies that

(7.10) $$q(\varphi,\varphi) \leq \text{const.}\left(\|\varphi\|_{\bar{z}}^2 + \int_{b\Omega} |\varphi|^2 \, dS + \|\varphi\|^2\right),$$

hence if (7.7) holds then

(7.11) $$\|\varphi\|_1^2 \leq \text{const.}\left(\|\varphi\|_{\bar{z}}^2 + \int_{b\Omega} |\varphi|^2 \, dS + \|\varphi\|^2\right)$$

and this is true if and only if $\varphi = 0$ on $b\Omega$.

We do however, have the following global regularity result (see Kohn [17]).

<u>Theorem</u>. Suppose $b\Omega$ is C^∞, Ω pseudo-convex and if there exists a function λ which is C^∞ in a neighborhood of $b\Omega$. Then for every integer k there exists a number τ_k such that if N gives the solution of the $\bar{\partial}$-Neumann problem in

$$L_2^{p,q}(\Omega,\tau\lambda) \text{ with } q \geq 1$$

then

$$N(\mathcal{A}^{p.q}) \subset C^k(\bar{\Omega}) \text{ if } \tau \geq \tau_k .$$

<u>Corollary</u>. Under the same hypotheses on Ω if $\alpha \in \mathcal{A}^{p.q}$ $q \geq 1$, and if $\bar{\partial} u = \alpha$ in L_2 then for each k there ex-

ists $u_k \in C^k(\bar{\Omega})$ such that $\bar{\partial} u_k = \alpha$.

Here we give a brief outline of the proof of the above theorem. First we set

$$(7.12) \qquad Q_\tau(\varphi,\psi) = q_{\tau\lambda}(\varphi,\psi) + (\varphi,\psi)_{(\tau\lambda)}$$

Now for each $\tau \geq 0$ and each $\alpha \in L_2^{p,q}(\Omega)$ there exists a unique $\varphi_\tau \in \mathcal{D}^{p,q}$ such that

$$(7.13) \qquad Q_\tau(\varphi_\tau,\psi) = (\alpha,\psi)_{(\tau\lambda)} \quad \text{for each } \psi \in \mathcal{D}^{p,q} .$$

We wish to show that given an integer s there exists a number A_s such that $\varphi_\tau \in H_s(\Omega)$ whenever $\tau \geq A_s$ and that in fact

$$(7.14) \qquad \|\varphi_\tau\|_s \leq \text{const.} \|\alpha\|_s .$$

To prove that $\varphi_\tau \in H_s(\Omega)$, we first prove (7.14) under the assumption that φ_τ is smooth. For $P \in b\Omega$ choose a "boundary coordinate system" in a neighborhood U of P ; that is, we choose functions $x^1,\dots,x^{2n-1}$ such that $dx^1,\dots,dx^{2n-1},dr$ are linearly independent throughout U we say that $x^1,\dots,x^{2n-1}$ are the tangential coordinates and $x^{2n} = r$ is the normal coordinate. Note that if $\varphi \in \dot{\mathcal{D}}$ and the support of φ lies in $U \cap \bar{\Omega}$ then for any index α $D_b^\alpha \varphi \in \dot{\mathcal{D}}$, where

$$(7.15)\qquad D_b^{\alpha} = (-i)^{|\alpha|} \frac{\partial^{|\alpha|}}{x_1^{\alpha_1}, \dots, x_{2n-1}^{\alpha_{2n-1}}}$$

and

$$D_b^{\alpha}\varphi = \sum_{IJ} D_b^{\alpha}\varphi_{IJ}\, \omega^{IJ} ,$$

where

$$= \sum \varphi_{IJ}\, \omega^{IJ} .$$

Starting with the inequality (4.33)

$$(7.16)\qquad \tau \|\varphi\|^2_{(\tau\lambda)} \leq C\, Q_\tau(\varphi,\varphi) ,$$

substituting $D_b^{\alpha}\zeta\varphi_\tau$, for φ in (7.16) we obtain

$$(7.17)\qquad \tau \sum_{|\alpha|\leq s} \|D_b^{\alpha}\zeta\varphi_\tau\|^2_{(\tau\lambda)} \leq \text{const.} \sum_{|\alpha|=s} Q\, (D_b^{\alpha}\zeta\varphi_\tau , D_b^{\alpha}\zeta\varphi_\tau) .$$

We wish to estimate the right-hand side of (7.17) by using (7.13) with an appropriate $\psi \in \dot{\mathcal{D}}$. Integration by parts gives

$$(7.18)\qquad Q_\tau(D_b^{\alpha}\zeta\varphi_\tau,\, D_b\zeta\varphi_\tau) = Q_\tau(\varphi_\tau,\, \zeta D_b^{\alpha} D_b^{\alpha}\zeta\varphi_\tau)$$

$$+ O\Big(\Big\{ \sum_{|\beta|=s} \|D_b^{\beta}\zeta\varphi_\tau\|_{(\tau\lambda)} + \sum_{|\beta|=s-1} \|\frac{\partial}{\partial r} D_b^{\beta}\zeta\varphi_\tau\|_{(\tau\lambda)}$$

$$+ C_\tau \sum_{|\gamma|<s} \|D^{\gamma}\zeta'\varphi_\tau\|_{(\tau\lambda)} \Big\} \sqrt{Q_\tau(D_b^{\alpha}\zeta\varphi_\tau, D_b^{\alpha}\zeta\varphi_\tau)}$$

$$+ \{\ldots\}^2 \Big),$$

where the γ runs through m-tuples, $\zeta' \in C_0^{\infty}(U \cap \bar{\Omega})$ with $\zeta' = 1$

on support of ζ . Further, we have by (7.12) and integration by parts

(7.19) $$Q_\tau(\varphi_\tau, \zeta D^\alpha_b D^\alpha_b \zeta\varphi_\tau) = (\alpha, \zeta D^\alpha_b D^\alpha_b \zeta\varphi_\tau)_{(\tau\lambda)}$$

$$= (D^\alpha_b \zeta\alpha, D^\alpha_b \zeta\varphi_\tau) + O(c_\tau \|\alpha\|_s \|\varphi_\tau\|_{s-1}) .$$

The fact, $b\Omega$ is non-characteristic (which is a consequence of the ellipticity of the "laplacian") implies:

(7.20) $$\sum_{|\beta|=s-1} \|\tfrac{\partial}{\partial r} D^\beta_b \zeta\varphi_\tau\|^2_{(\tau\lambda)} \leq \text{const.}\Big\{\sum_{|\alpha|=s} Q_\tau(D^\alpha\zeta\varphi_\tau, D^\alpha\zeta\varphi_\tau) + \sum_{|\alpha|=s} \|D^\alpha\zeta\varphi_\tau\|^2_{(\tau\lambda)}\Big\} .$$

Assuming (7.14) is true for $s - 1$ in place of s, we obtain from (7.18), (7.19) and (7.20) with τ sufficiently large

(7,21) $$\sum_{|\alpha|\leq s} \|D^\alpha_b \zeta\varphi_\tau\|^2_{(\tau\lambda)} + \sum_{|\gamma|\leq s-1} \|\tfrac{\partial}{\partial r} D_b \zeta\varphi_\tau\|^2_{(\tau\lambda)} \leq \text{const.}\|\alpha\|^2_s$$

The fact that φ_τ and α are connected by an elliptic system i.e. $E_\tau(\varphi_\tau) = \alpha$ to solve for the normal derivates in U as follows

(7.22) $$\frac{\partial^2}{\partial r^2}\varphi_{IJ\tau} = \sum a^{KL}_{IJ}\alpha_{KL} + \sum b^{KLij}_{IJ} \frac{\partial^2}{\partial x^i \partial x^j}\varphi_{KL\tau} + \sum c^{KLi}_{IJ} \frac{\partial^2}{\partial x^i \partial r}\varphi_{KL\tau} + F_{IJ}(\varphi_\tau),$$

where F_{IJ} is an operator of first order. Differentiation of (7.22) and use of (7.21) the yield

(7.23) $$\|\zeta\varphi_\tau\|^2_{(\tau\lambda)} \leq \text{const.} \|\alpha\|^2_s \, .$$

The desired estimate (7.14) is then obtained using a partition of unity with (7.23) for neighborhoods of boundary points and (6. 5) for neighborhoods of interior points. To actually prove that $\varphi_\tau \in H_s(\Omega)$ we use the method of "elliptic regularization" (see Kohn and Nirenberg [19]). Consider the form

(7.24) $$Q^\delta_t(\varphi,\psi) = Q_t(\varphi,\psi) + \delta \sum_{|\alpha|\leq 1} (D^\alpha \zeta_v \varphi, D^\alpha \zeta_v \varphi)_{(\tau\lambda)}$$

where $\delta > 0$, the $\{\zeta_v\}$ are a partition of unity and $\varphi, \psi \in \dot{\mathcal{D}}^{p,q}$. Then, clearly Q_τ is coericive i.e. (7.7) is satisfied. Hence, if we denote By $\tilde{\mathcal{D}}^{p,q}$ the completion of $\dot{\mathcal{D}}^{p,q}$ under Q^δ_τ then the unique $\varphi^\delta_\tau \in \tilde{\mathcal{D}}^{p,q}$ which satisfies

(7.25) $$Q^\delta_\tau(\varphi^\delta_\tau, \psi) = (\alpha, \psi)_{(\tau\lambda)}$$

for all $\psi \in \dot{\mathcal{D}}^{p,q}$, is in $C^\infty(\bar{\Omega})$ and hence $\varphi^\delta_\tau \in \dot{\mathcal{D}}^{p,q}$. Now, the same derivation is that of (7.14) yields

(7.26) $$\|\varphi^\delta_\tau\|_s \leq \text{const.} \|\alpha\|_s \, ,$$

where the constant is independent of δ when δ is small.

Definition. We say that the $\bar\partial$-Neumann problem in $L_2^{p,q}(\Omega)$ is ε-subelliptic at $P \in b\Omega$ if there exists a neighborhood U of P such that

(7.29) $$\|\Lambda_b^\varepsilon \varphi\|^2 \leq \text{const. } Q(\varphi,\varphi)$$

for all $\varphi \in \dot{D}^{p,q} \cap C_0(U \cap \bar\Omega)$,

where

$$\|\Lambda_b^\varepsilon \varphi\|^2 = \sum_{IJ} \int_{\mathbb{R}^{2n}} |\Lambda_b^\varepsilon \varphi_{IJ}|^2 dx\, dr$$

and $Q = Q_0$ in (7.12).

The estimate (7.29) is equivalent to

(7.30) $$\sum_{j=1}^{2n} \|\Lambda_b^{\varepsilon-1} \frac{\partial}{\partial x_j} \zeta\varphi\|^2 \leq \text{const. } Q(\varphi,\varphi),$$

in particular if $\varepsilon = 1$ then Q is coercive (i.e. (7.7) holds). If ε-ellipticity holds and if $\varepsilon < 1$ then necessarly $\varepsilon \leq \frac{1}{2}$.

Theorem. If Ω is such that at each $P \in b\Omega$ the $\bar\partial$-Neumann on $L^{p,q}(\Omega)$ is ε-subelliptic then it is soluable and (7.27) holds.

From the theory of Sobolev spaces it follows that the norm given on the left side of (7.30) is compact in $L_2^{p,q}(U \cap \bar\Omega)$ from this it follows that the $\bar\partial$-Neumann problem has a solution in $L_2^{p,q}(\Omega)$. The pseudo-localness depends

It then follows that there is a sequence $\delta_v \to 0$ such that the arithemetic means of the sequence $\{\varphi_\tau^\delta v\}$ are a Cauchy sequence in H_s and since they converge to φ_τ in L_2 this implies that $\varphi_\tau \in H_s$. To pass from the smoothness of φ_τ to the smoothness of $N_\tau \alpha$ requires a functional analytic argument given in Kohn [17].

We now return to the question when the $\bar{\partial}$-Neumann problem has boundary regularity. More precisely, we want to show that, under certain circumstances, the operator N is pseudo-local; in the sense that

(7.27) $$\text{sing supp } N\alpha \subset \text{sing supp } \alpha$$

where the singular support is considered in $\bar{\Omega}$.

If U is a boundary coordinate neighborhood as above, with boundary coordinates $x^1,\dots,x^{2n-1}$, $x^{2n} = r$ then $s \in \mathbb{R}$ and $u \in C_0^\infty(U \cap \bar{\Omega})$ we define $\Lambda_b^s u \in C^\infty(\mathbb{R}_-^{2n})$, where $\mathbb{R}_-^{2n} = \{(x^1,\dots,x^{2n}) \mid x^{2n} \le 0\}$.

(7.28) $$\widetilde{\Lambda_b^s u}(\xi,r) = (1+|\xi|^2)^{s/2}\tilde{u}(\xi,r),$$

where $\xi = (\xi_1,\dots,\xi_{2n-1})$ and $\tilde{v}(\xi,r)$ denotes the partial Fourier transform defined by

$$\tilde{v}(\xi,r) = \int_{\mathbb{R}^{2n-1}} e^{-ix\cdot\xi} v(x,r)dx$$

where $x\cdot\xi = \sum_1^{2n-1} x^j\xi_j$ and $dx = dx^1,\dots,dx^{2n-1}$,

on the following: suppose $\varphi \in \mathcal{D}$ is the solution of

(7.31) $$Q(\varphi,\psi) = (\alpha,\psi) \text{ for } \psi \in \mathcal{D}.$$

Then if $\zeta, \zeta' \in C_0^\infty(U \cap \overline{\Omega})$ with $\zeta' = 1$ on the support of ζ and if $\zeta'\alpha \in H_s(\Omega)$ then $\Lambda_b^{2\varepsilon-2}\zeta\varphi \in H_{s+2}(\Omega)$ and we have

(7.32) $$\|\Lambda_b^{2\varepsilon-2}\zeta\varphi\|_{s+2} \leq \text{const.}\|\zeta'\alpha\|_s$$

There ~~estimates are obtained by replacing in (7.30) with~~ $\Lambda_b^{m\varepsilon}$ then proceeding by induction on m with the same type of argument as in the previous theorem.

__Theorem.__ If the Levi-form at $P \in b\Omega$ either has at least $n-q$ positive eigenvalues or $q+1$ negative eigenvalues then the $\bar\partial$-Neumann problem in $L_2^{p,q}(\Omega)$ is $\frac{1}{2}$-subelliptic at P.

__Proof:__ Due to the results of the previous lecture it suffices to show that

(7.33) $$\|\Lambda_b^{\frac{1}{2}} u\|^2 \leq \text{const.}\left(\|u\|_{\bar z}^2 + \int_{b\Omega} |u|^2 dS + \|u\|^2\right)$$

for all $$u \in C_0^\infty(U \cap \overline{\Omega}).$$

Let $L_1,\dots,L_n$ be a basis of the vectorfields in $T^{1,0}$ on U such that $L_j(r) = 0$ and $L_n(r) = 1$ throughout U. It is then easily verified that if for each r, P_r is pseudo-differential operator on $\mathbb{R}^{2n-1}$ which varies smoothly with r (i.e. its symbol $p_r(x,\xi) \in C^\infty$ as a function r, x and ξ) then on $U \subset \mathbb{R}^{2n}$ the operator $[L_n, P_r]$ acts as a pseudo-differential operator on $r = \text{const.}$ of the same order at P_r. Now it suffices to bound $\|\Lambda_b^{-\frac{1}{2}} \frac{\partial u}{\partial x_j}\|^2$ by the right-hand side of (7.33) with $j < 2n$. Since these $\frac{\partial}{\partial x_j}$ are linear combinations the vectors $L_1, \ldots, L_{n-1}, \bar{L}_1, \ldots, \bar{L}_{n-1}$ and $N = L_n - \bar{L}_n$ it therefore suffices to bound

$$\|\Lambda_b^{-\frac{1}{2}} L_j u\|^2, \ \|\Lambda_b^{-\frac{1}{2}} \bar{L}_j u\|^2$$

and for $j < n$ and $\|\Lambda^{-\frac{1}{2}} N u\|^2$. The second of these is immediate.

$$(7.34) \quad \|\Lambda^{-\frac{1}{2}} L_j u\|^2 = (L_j u, P^0 u) = (u, P^0 L_j u) + (u, Q^0 u)$$

$$\leq \text{const.}(\|u\| \, \|u\|_{\mathbb{Z}} + \|u\|^2)$$

where P^0, Q^0 are operators whose restrictions to $r = \text{const.}$ are pseudo-differential operators of order 0. Finally:

$$(7.35) \quad \|\Lambda^{-\frac{1}{2}} N u\|^2 = ((L_n - \bar{L}_n) u, P^0 u)$$

$$= \int_{b\Omega} u \overline{P^0 u}\, dS + (u, \bar{L}_n P_n^0) + (\bar{L}_n u, P_n^0),$$

in view of the above remark this is bounded by the right side of (7.33).

J. J. Kohn

The following theorem is proven in Kohn [16].

<u>Theorem</u>. If Ω is pseudo-convex and if in a neighborhood of $P \in b\Omega$ each non-zero vector field of degree (1,0) and values in $T^{1,0}(b\Omega)$ on $b\Omega$ is of finite type at P and if the Levi form is diagonalizable in a neighborhood of P (i.e. in C) the $c_{ij} = c_{ii}\delta_{ij}$ throughout a neighborhood) then there exists $\varepsilon > 0$ such that the $\bar{\partial}$-Neumann problem on $L^{p,q}(\Omega)$ with $q \geq 0$ is ε-subelliptic at P.

J. J. Kohn

Lecture 8. The induced Cauchy-Riemann equations.

In this lecture we will take up systematically the euqations on $b\Omega$ which arise in the extension problem discussed in lecture 1. We define

$$\mathcal{C}^{p,q} = \{\varphi \in \mathcal{A}^{p,q} \mid \bar{\partial} r \wedge \varphi = 0 \text{ on } b\Omega\} \tag{8.1}$$

Observe that if $q = 0$ then $\varphi \in \mathcal{C}^{p,q}$ is equivalent to

$$\varphi = \bar{\partial} r \wedge \psi + r\theta \text{ near } b\Omega, \tag{8.2}$$

where $\psi \in \mathcal{A}^{p,q-1}$ and $\theta \in \mathcal{A}^{p,q}$. From (8.2) it follows that

$$\bar{\partial}(\mathcal{C}^{p,q}) \qquad {}^{p,q+1} \;. \tag{8.3}$$

Therefore we have the following diagram

$$\begin{array}{ccccccccc} 0 & \rightarrow & \mathcal{C}^{p,q+1} & \longrightarrow & \mathcal{A}^{p,q+1} & \longrightarrow & \mathcal{B}^{p,q+1} & \longrightarrow & 0 \\ & & \bar{\partial}\uparrow & & \bar{\partial}\uparrow & & \uparrow\bar{\partial}_b & & \\ 0 & \rightarrow & \mathcal{C}^{p,q} & \longrightarrow & \mathcal{A}^{p,q} & \longrightarrow & \mathcal{B}^{p,q} & \longrightarrow & 0 \end{array} \tag{8.4}$$

where

$$\mathcal{B}^{p,q} = \mathcal{A}^{p,q}/\mathcal{C}^{p,q} \tag{8.5}$$

and $\bar{\partial}_b$ is the mapping induced by $\bar{\partial}$. In case $p = q = 0$ the space $\mathcal{B}^{0,0}$ is naturally identified with $C^\infty(b\Omega)$ and the $\bar{\partial}_b f = 0$ is necessary for extending f to a holomorphic function.

We define the following cohomology groups:

$$(8.6)\qquad H^{p,q}(\mathcal{A}) = \frac{\{\varphi\in\mathcal{A}^{p,q} \mid \bar\partial\varphi=0\}}{\bar\partial\mathcal{A}^{p,q-1}}$$

$$H^{p,q}(\mathcal{C}) = \frac{\{\varphi\in\mathcal{C}^{p,q} \mid \bar\partial\varphi=0\}}{\bar\partial\mathcal{C}^{p,q-1}}$$

$$H^{p,q}(\mathcal{B}) = \frac{\{\varphi\in\mathcal{B}^{p,q} \mid \bar\partial_b\varphi=0\}}{\bar\partial_b\mathcal{B}^{p,q-1}}$$

then (8.4) induces a sequence

$$(8.7)\qquad H^{p,q}(\mathcal{C}) \longrightarrow H^{p,q}(\mathcal{A}) \longrightarrow H^{p,q}(\mathcal{B}) \longrightarrow H^{p,q+1}(\mathcal{C})$$

To illustrate, we define the map $H^{0,0}(\mathcal{B}) \longrightarrow H^{0,1}(\mathcal{C})$. If $f\in H^{0,0}(\mathcal{B})$, f is represented by a function in $C^\infty(b\Omega)$ which we will also denote by f and which has the property if $\tilde f$ is an extension of f then $\bar\partial\tilde f\in\mathcal{C}^{0,1}$ the cohomology class of $\bar\partial\tilde f$ in $H^{0,1}(\mathcal{C})$ then gives the image of f under the map $H^{0,0}(\mathcal{B}) \longrightarrow H^{0,1}(\mathcal{C})$. It isclear that this image is independent of the extension $\tilde f$ and that f can be extended to a holomorphic function on Ω if and only if $\bar\partial\tilde f$ is cohomologous to 0 in $H^{0,1}(\mathcal{C})$; that is there exists $g\in\mathcal{C}^{0,0}$ such that $\bar\partial\tilde f = \bar\partial g$. The desired extension then is $\tilde f - g$.

<u>Proposition</u>. $H^{p,q}(\mathcal{C}) \approx H_0^{p,q}(\mathcal{A})$, where $H_0^{p,q}(\mathcal{A})$ is the cohomology of forms which vanish on $b\Omega$.

<u>Proof</u>: If $\varphi\in\mathcal{C}^{p,q}$ then from (8.2) we see that

$$\varphi = \bar\partial(r\psi) + r(\theta - \bar\partial\psi)$$

and that

(8.11) $F(\mathcal{H}^{p,q}) \subset \mathcal{C}^{n-p,n-q}$.

From (8.9) it follows that F is an isomorphism and from (8.10) it follows that if $\varphi \in \mathcal{H}^{p,q}$ then $\bar\partial F\varphi = 0$. Thus we have proved that F induces an isomorphism between $H^{p,q}(\mathcal{Q})$ and $H^{n-p,n-q}(\mathcal{C})$.

<u>Theorem</u>. If $b\Omega$ is connected and if the Levi form has at least one positive eigenvalue at each $P \in b\Omega$ and if $f \in C^\infty(b\Omega)$ with $\bar\partial_b f = 0$ then there exists a function $h \in C^\infty(\bar\Omega)$ such that $h = f$ on $b\Omega$ and h is holomorphic in Ω.

<u>Proof</u>: Let $\tilde f \in C^\infty(\Omega)$ be an extension of f, as we have seen the desired h exists if and only if $\bar\partial\tilde f$ is cohomoolgous to zero in $H^{0,1}(\mathcal{C})$. By the previous theorem $H^{0,1}(\mathcal{C}) \approx \mathcal{H}^{n,n-1}$. Let $\psi_1,\dots,\psi_k$ be a basis for $\mathcal{H}^{n,n-1}$ so $\bar\partial\tilde f$ is cohomologous to zero if and only if

(8.12) $$\int_\Omega \bar\partial\tilde f \wedge \psi_j = \int_{b\Omega} f\psi_j = 0 \quad \text{for } j = 1,\dots,k.$$

Thus we see that the space

(8.13) $$\mathcal{E} = \frac{\{f \in C^\infty(b\Omega) \mid \bar\partial_b f = 0\}}{\{f = h \text{ on } b\Omega,\ h\ C^\infty(\bar\Omega) \text{ hol.}\}}$$

has dimension $\leq k$. Thus $f, f^2, \dots, f^{k+1}$ have linearly dependent cosets in $\mathcal{E}$ so that if

showing that φ is cohomologous to a form that vanishes on $b\Omega$. Further, $\bar\partial\varphi = \bar\partial(r(\bar\partial\psi + \theta))$ which concludes the proof.

<u>Proposition</u>. If φ represents a cohomology class in $H^{p,q}(\mathcal{C})$ and ψ represents a cohomology class in $H^{n-p,\, n-q}(\mathcal{Q})$ then $\int_\Omega \varphi \wedge \psi$ depends only on the cohomology classes of φ and ψ.

<u>Proof</u>: If $\varphi = \bar\partial\theta$ with $\theta \in \mathcal{C}^{p,q-1}$ then we have

$\bar\partial\theta \wedge \psi = d(\theta \wedge \psi)$ and hence

$$(8.8) \qquad \int_{b\Omega} \bar\partial\theta \wedge \psi = \int_{b\Omega} \theta \wedge \psi = 0 \quad ,$$

since $\theta \wedge \psi \in \mathcal{C}^{n,n-1}$ the same calculation yields the result when $\psi = \bar\partial\xi$.

<u>Theorem</u>. If at each point $P \in b\Omega$ the Levi-form has either $n-q$ positive or $q+1$ negative eigenvalues then $H^{n-p,n-1}(\mathcal{C})$ is finite dimensional and is isomorphic to $H^{p,q}(\mathcal{Q})$.

<u>Proof</u>: By our previous results the condition on the Levi-form implies that problem is $\frac{1}{2}$-subelliptic and hence the space $\mathcal{H}^{p,q}$ is finite dimensional and all its elements are in C^∞ on $\bar\Omega$. We define a map $F : \mathcal{Q}^{p,q} \to \mathcal{Q}^{n-p,n-q}$ by

$$(8.9) \qquad F\varphi \wedge \psi = \langle \varphi, \psi \rangle dV, \quad \text{for } \psi \in \mathcal{Q}^{p,q} .$$

It is easy to verify that

$$(8.10) \qquad F\vartheta = \bar\partial F$$

$$P(f) = \sum_{j=1}^{m} a_j f^j \quad \text{with } a_j \in \mathbb{C},\ a_m \neq 0, \tag{8.14}$$

then $P(f)$ has a holomorphic extension F to Ω. Similarly, $Ff, F^2 f, \ldots, F^{k+1} f$ are linearly dependent and hence there is a polygonial Q so that $Q(F)f$ has an holomorphic extension G. If f is constant there is nothing to prove. If f is not constant then it takes on infinitely many values and thus F and $Q(F)$ take on infinitely many values and hence the set of zeroes of $Q(F)$ is thin. Then from

$$P(f) = P\left(\frac{G}{Q(F)}\right) \tag{8.15}$$

on $b\Omega - \{\text{zeroes of } Q(F)\}$ we conclude that $F = \frac{G}{Q(F)}$ holds on $\Omega - \{\text{zeroes of } Q(F)\}$. Thus we know that $G/Q(F)$ is locally bounded and hence can be extended to holomorphic function on Ω which gives the desired extension of f.

From (8.7) and from the identification of $H^{p,q}(\mathcal{O})$ with $H^{n-p,n-q}(\zeta)$ we see that $H^{p,q}(\mathcal{B})$ is finite dimensional whenever $H^{p,q}(\mathcal{O})$ and $H^{n-p,n-q-1}(\mathcal{O})$ are finite dimensional. Thus the following condition implies the finite dimensionality of $H^{p,q}(\mathcal{B})$: The Levi-form has either at least $\max(n-q, q+1)$ non-zero eigenvalues of the same sign or at least $\min(n-q, q+1)$ non-zero eigenvalues of opposite signs. In fact this depends on the "internal structure" of $b\Omega$ which can be defined abstractly as follows.

__Definition__. If X is a real C^∞ manifold, we say that X

has a <u>partially almost-complex structure of codimension</u> k if there exists a <u>sub-bundle</u> $T^{1,0}(X) \subset \mathbb{C}T(X)$ such that; setting $T^{0,1}(X) = \overline{T^{1,0}}(X)$, we have

(a) $T^{1,0}(X) \cap T^{0,1}(X) = 0$

(b) $\dim T(X) = 2 \dim T^{1,0}(X) + k$

(c) If L and L' are local vector fields with values in $T^{1,0}(X)$ then $[L, L']$ has also values in $T^{1,0}(X)$

Under these circumstances we define the space of exterior forms $\mathcal{A}_b$ on $T^{1,0}(X) + T^{0,1}(X)$, on this space this sum induces a bigradation

$$\mathcal{A}_b = \mathcal{A}_b^{p,q} \tag{8.16}$$

and we define the operators

$$\partial_b : \mathcal{A}_b^{p,q} \longrightarrow \mathcal{A}_b^{p+1,q} \quad \text{and} \quad \bar{\partial}_b : \mathcal{A}_b^{p,q} \longrightarrow \mathcal{A}_b^{p,q+1} \tag{8.17}$$

by setting

$$\langle \partial_b u, L \rangle = Lu, \quad u \text{ a function } L \in T^{1,0}(X) \tag{8.18}$$

and extending this to forms in the usual way. These operators have the same formal properties as ∂ and $\bar{\partial}$. Restricting ourselves to the case of co-dimension 1 we can define the Levi-form at $P \in X$ as follows. Take a real 1-form defined in a neighborhood of P, which annihilates

$T^{1,0}(X) + T^{0,1}(X)$ the Levi-form, then is given by

(8.19) $$L \mapsto i\langle d\gamma, L\wedge\bar{L}\rangle = -i\langle\gamma, [L,\bar{L}]\rangle$$

for $L \in T_p^{1,0}(X)$. Since at each point the space of such is one-dimensional the numbers of non-zero eigenvalues and the numbers of eigenvalues of the same and opposite signs are invariants. Now we can put a hermitian metric on $\mathbb{C}T(X)$ which makes $T^{1,0}(X)$ orthogonal to $T^{0,1}(X)$ and we obtain the adjoint of ϑ_b of $\bar{\partial}_b$. By methods analogous as for the $\bar{\partial}$-Neumann problem we can prove the following result (see Kohn [15]).

<u>Theorem</u>. Let X be a compact partially complex manifold of co-dimension 1 with dim X = 2n-1 and if the Levi-form (8.19) has either at least max (n-q, q+1) eigenvalues of the same sign or at least min(n-q,q+1) eigenvalues of opposite signs. Then, setting

(8.20) $$\mathcal{H}_b^{p,q} = \{\varphi \in \mathcal{A}^{p,q} \mid \bar{\partial}_b\varphi = 0, \vartheta_b\varphi = 0\},$$

where ϑ_b is the adjoint of $\bar{\partial}_b$ and

(8.21) $$\Box_b = \vartheta_b\bar{\partial}_b + \bar{\partial}_b\vartheta_b ,$$

then there exists a completely continuous self-adjoint operator

$N_b : L_2^{p,q}(X) \to L_2^{p,q}(X)$ such that N_b

is pseudo-local and

(8.22) $$\square_b N_b = I + H_b ,$$

where H_b is the othogonal projection of $L_2^{p,q}(X)$ on $\mathcal{H}_b^{p,q}$. Furthermore, if $\alpha \in L^{p,q}(X)$, $\bar\partial_b \alpha = 0$ and $\alpha \perp \mathcal{H}^{p,q}$ then $\alpha = \bar\partial_b (\vartheta_b N_b \alpha)$.

The theory indicated in the above theorem is based on the estimate

(8.23) $$\|\varphi\|_{\frac{1}{2}}^2 \leq \text{const.}\left(\|\bar\partial_b \varphi\|^2 + \|\vartheta_b \varphi\|^2 + \|\varphi\|^2\right)$$

for $\varphi \in \mathcal{A}_b^{p,q}$.

Observe that dim x = 3 and q = 1 the conditions in the above theorem cannot be satisfied because then the Levi-form is 1×1. In this case $\bar\partial_b$ can be represented locally by single first order operator. In fact, if $X \subset \mathbb{C}^2$ is given by

(8.24) $$r = \operatorname{Im}(z_1) + |z_2|^2$$

then to the equation $\bar\partial_b u = f$ is the Lewy equation which has no solutions for "most" functions f.

The following natural problem was posed by H. Lewy.

Suppose L is a complex vector field in $\mathbb{R}^3$, i.e.

$$L = \sum_{j=1}^{3} a_j \frac{\partial}{\partial x_j} \qquad a_j \in C^\infty(\mathbb{R}^3) \tag{8.25}$$

do there exist non-trivial local solutions of the equation $Lu = 0$. Recently, L. Nirenberg (see [32]) has found an example of such a vector field for which the only local solution of $Lu = 0$ are $u = \text{const}$. What is more, in Nirenberg's example the vector fields L, $\bar{L}$ and $[L,\bar{L}]$ are linearly independent. It is still an open question whether on X which satisfies the conditions on the Levi-form for $q = 1$ given in the above theorem the equation $\bar{\partial}_b u = 0$ has non-trivial local solutions.

To conclude these lectures we wish to point out an application of these results due to Kerzman (see [30]). Namely, if $H : L_2(\Omega) \to \mathcal{H}^{0,0}$ denotes the orthogonal projection map onto the space of holomorphic functions, and if N is pseudo-local then H is also pseudo-local. Now H can be expressed as

$$Hu(z) = \int_\Omega K(z,\omega)u(\omega)\, dV_\omega . \tag{8.25}$$

Then the pseudo-locality of H implies that

$$K \in C^\infty(\bar{\Omega} \times \bar{\Omega} - \textbf{diag}\,(b\Omega \times b\Omega)\} .$$

REFERENCES

[1] A. ANDREOTTI, and C.D. HILL, several articles to appear in Ann. Scuola Norm. Sup. Pisa.

[2] BOCHNER, S. "Analytic and meromorphic continuation by means of Green's formula," Ann. Math. (2) 44, 652-673 (1943).

[3] EHRENPREIS, L. "A new proof and extension of Hartog's theorem," Bull. Amer. Math. Soc. 67, 5007-509 (1961).

[4] FOLLAND, G.B. and KOHN, J. J. "The Neumann problem for the Cauchy-Riemann complex," Ann. of Math. Study Vol. 75, Princeton Univ. Press, 1972.

[5] FOLLAND, G.B. and STEIN, E.M. "Parametrices and estimates for the $\bar{\partial}_b$ on strongly pseudo-convex boundaries," Bull. Amer. Math. Soc. (to appear).

[6] GAFFNEY, M.P., "Hilbert space methods in the theory of harmonic integrals," trans. Amer. Math. Soc. 78(1955) 426 - 444.

[7] GRAUERT, H. "Bemerkenswerte pseudo konvexe Mannigfaltigkeiten," Math. Z. 81 (1963), 377 - 391.

[8] ______ and LIEB I., "Das Ramirezsche Integral und die Gleichung $\bar{\partial} f =$ im Bereich der beschrankten Formen, Rice University Studies, (to appear).

[9] HENKIN, G"Integral representations of holomorphic functions in strongly pseudocomvex domains and certain applications," Mat. Sbornik 78 (120): 4(1969), 611-632 (Russian), English translation in Math. of the U.S.S.R. April 1969, 7(4), 597-616.

[10] HÖRMANDER, L., "Estimates and existence theorems for the $\bar{\partial}$ operator,"Acta Math. 113 (1965), 89 - 152.

[11] ______ , "An introduction to complex analysis in several variables, Van Nostrand, Princeton, 1966.

[12] ______ , and WERMER, J., "Uniform approximation on compact subsets in ^{n}", Math. Scand. 23(1968), 5-21.

[13] KERZMAN, N., "Holder and L^p estimates for solutions of $\bar{\partial}u = f$ in strongly pseudoconvex domains, Comm. Pure Appl. Math. 24(1971), 301 - 379.

[14] KOHN, J.J. "Harmonic integrals on strongly pseudo-convex manifolds," I, Ann. of Math. 78(1963), 112-148; ibid. 79 (1964), 450-472.

[15] ______ , "Boundaries of complex manifolds," Proc. Conference on Complex Manifolds (Minneapolis), Springer-Verlag, New York, 1965.

[16] ______ , "Boundary behavior of $^-$ on weakly pseudo-convex manifolds of dimension two," J. Diff. Geom. Vol 6, 523 - 542 (1972).

[17] KOHN, J.J., "Global regularity for $\bar{\partial}$ on weakly pseudo-convex manifolds," Trans. Amer. Math. Soc. (to appear).

[18] ______, and NIRENBERG, L. "An algebra of pseudo-differential operators," Comm. Pure Appl. Math. 18 (1965), 269 - 305.

[19] ______ , and NIRENBERG, L., "Non-coercive boundary value problems, Comm. Pure Appl. Math. 18 (1965) 443 - 492.

[20] ______ , and NIRENBERG, L.,"A pseudo-convex domain not admitting a holomorphic support function," Math. Ann. 201, 265 - 268 (1973).

[21] ______ , and ROSSI, H., "On the extension of holomorphic functions from the boundary of a complex manifold, Ann. of Math. 81 (1965), 451 - 472.

[22] LEWY, H., "On the local character of the solutions of an atypical linear differential equation in three variables and a related theorem for regular functions of two complex variables," Ann. of Math. 64 (1956), 514 - 522.

[23] ______ , "An example of a smooth linear partial differential equation without solution, Ann. of Math. 66 (1957), 155 - 158.

[24] LIEB, I., "Ein Approximationssatz auf streng pseudokonvexen Gebieten, Math. Annalen 184 (1969), 55-60.

[25] NEWLANDER, A. and NIRENBERG, L., "Complex analytic coordinates in almost-complex manifolds," Ann. of Math. 65 (1957), 391 - 404.

[26] NIRENBERG, R., "On the H. Lewy extension phenomenon," Trans. Amer. Math. Soc., (to appear).

[27] ______ , and WELLS, R.O., "Approximation theorems on differentiable submanifolds of a complex manifold," Trans. Amer. Math. Soc. 142 (1969), 15 -36.

[28] ØVRELID, N., "Integral representation formulas and L^p estimates for the equation $\bar{\partial}u = f$," Math. Scand. 29 (1971), 137 - 160.

[29] RAMIREZ, E., "Ein Divisionsproblem in der komplexen Analysis mit einer Anwendung auf Randintegraldarstellung," Math. Annalen 184 (1970), 172 - 187.

[30] KERZMAN, N., "The Bergmann kernel function: differentiability at the boundary," Math. Ann. 195 (1972) 149-158.

[31] LAX, P.D. and PHILLIPS, R.S., "Local boundary conditions for dissipative symmetric operators." Comm. Pure Appl. Math. 13 (1960), 427 - 455.

[32] NIRENBERG, L., "Lectures on linear partial differential equations," Mimeographed notes. Courant Institute (1973).

CENTRO INTERNAZIONALE MATEMATICO ESTIVO
(C. I. M. E.)

THE MIXED CASE OF THE DIRECT IMAGE THEOREM AND ITS APPLICATIONS

YUM-TONG SIU

Corso tenuto a Bressanone dal 3 al 12 giugno 1973

THE MIXED CASE OF THE DIRECT IMAGE THEOREM AND ITS APPLICATIONS

Yum-Tong Siu [1]

§ 0. Introduction

In these lectures we will discuss the so-called mixed case of the direct image theorem and its applications. The starting point of the direct image theorem is the following finiteness theorem.

(0.1) Theorem (Cartan-Serre). *If $\mathcal{F}$ is a coherent analytic sheaf on a compact complex space X, then $H^\nu(X,\mathcal{F})$ is finite-dimensional over $\mathbb{C}$ for $\nu \geq 0$.*

The proof is obtained by Schwartz's finiteness theorem on the perturbation of a surjective operator between Fréchet spaces by a compact operator (see [8, VIII.A.19] and [6]).

In 1960 Grauert proved the following parametrized version of the above theorem which is known as the proper case of the direct image theorem [6].

(0.2) Theorem (Grauert). *If $\pi: X \longrightarrow S$ is a proper*

[1] Partially supported by a National Science Foundation Grant and a Sloan Fellowship

<u>holomorphic map of complex spaces and</u> $\mathcal{F}$ <u>is a coherent analytic sheaf on</u> X , <u>then the</u> ν^{th} <u>direct image</u> $R^{\nu}\pi_{*}\mathcal{F}$ <u>of</u> $\mathcal{F}$ <u>under</u> π <u>is coherent on</u> S <u>for</u> $\nu \geqq 0$.

Grauert's proof uses the power series method. The idea is to expand a ν-dimensional cohomology class in a power series in the variables of S (after reducing the general case to the case where S is an open subset of a number space). The coefficients in the power series expansion may not be cocycles, but, by using descending induction on ν , Grauert showed that they can be approximated by cocycles. Then he used induction on dim S and applied the induction hypothesis to the approximating cocycles to get the coherence of the νth direct image.

About ten years later Knorr [13] and Narasimhan [18] gave simplified presentations of Grauert's original proof.

Recently Kiehl [10] used nuclear and homotopy operators and a form of Schwartz's finiteness theorem to obtain a new proof for an important special case of Grauert's theorem. Then Forster-Knorr [3] and Kiehl-Verdier [12] succeeded in obtaining new proofs of Grauert's theorem along such lines. Their proofs make use of descending induction on ν , but does not use induction on dim S . This opens the way to generalizing Grauert's theorem to relative-analytic spaces and such generalizations were carried out by Kiehl [11], Forster-Knorr [4] and Houzel [9].

Y-T. Siu

In 1962 Andreotti-Grauert [1] generalized the theorem of Cartan-Serre in another direction. They introduced the concepts of strongly pseudoconvexity and pseudoconcavity and proved finiteness theorems for spaces which are strongly pseudoconvex or pseudoconcave. A function φ on a complex space X is said to be *strongly p-pseudoconvex* if for every $x \in X$ there exists an embedding τ of an open neighborhood U of x onto a complex subspace of an open subset G of $\mathbb{C}^N$ and there exists a real-valued C^2 function ψ on G such that $\varphi = \psi \circ \tau$ and the $N \times N$ hermitian matrix

$$\left(\frac{\partial^2 \psi}{\partial z_i \partial \bar{z}_j}\right)$$ has at least $N - p + 1$ positive eigenvalues at every point of G (where $z_1, \ldots, z_N$ are the coordinates of $\mathbb{C}^N$).

(0.3) *Theorem* (Andreotti-Grauert). *Suppose* X *is a complex space and* $\varphi : X \longrightarrow (a_*, b_*) \subset \mathbb{R}$ *is a proper* C^2 *map. Suppose* $a_* < a_\# < b_\# < b_*$ *such that* φ *is strongly* p-*pseudoconvex on* $\{\varphi > b_\#\}$ *and is strongly* q-*pseudoconvex on* $\{\varphi < a_\#\}$ *and* $\{\varphi \leq b\} = \{\varphi < b\}^-$ *for* $b_\# < b < b_*$. *Suppose* $\mathcal{F}$ *is a coherent analytic sheaf on* X *such that* $\operatorname{codh} \mathcal{F} \geq r$ *on* $\{\varphi < a_\#\}$. *Then, for* $p \leq \nu < r-q$, $H^\nu(X, \mathcal{F})$ *is finite-dimensional over* $\mathbb{C}$ *and* $H^\nu(X, \mathcal{F}) \longrightarrow H^\nu(\{a < \varphi < b\}, \mathcal{F})$ *is an isomorphism for* $a_* \leq a < a_\#$ *and* $b_\# < b \leq b_*$.

(For the definition of $\operatorname{codh} \mathcal{F}$, see (A.1) of the Appendix.)

A holomorphic map $\pi: X \longrightarrow S$ of complex spaces is called *strongly* *(p,q)-pseudoconvex-pseudoconcave* if there exists a C^2 map $\varphi: X \longrightarrow (a_*, b_*) \subset \mathbb{R}$ and there exists $a_* < a_\# < b_\# < b_*$ such that

i) $\pi \,\big|\, \{a \leqq \varphi \leqq b\}$ is proper for $a_* < a < b < b_*$.

ii) $\{\varphi \leqq b\} = \{\varphi < b\}^-$ for $b_\# < b < b_*$.

iii) φ is strongly p-pseudoconvex on $\{\varphi > b_\#\}$.

iv) φ is strongly q-pseudoconvex on $\{\varphi < a_\#\}$.

We introduce the following notations. For $a_* \leqq a < b \leqq b_*$,

$$X_a^b = \{a < \varphi < b\}$$

$$\pi_a^b = \pi \,\big|\, X_a^b .$$

For a coherent analytic sheaf $\mathcal{F}$ on X, $R^\nu (\pi_a^b)_* \mathcal{F}$ denotes $R^\nu (\pi_a^b)_* (\mathcal{F} \,\big|\, X_a^b)$.

The so-called mixed case of the direct image theorem is the following parametrized version of the theorem of Andreotti-Grauert.

(0.4) *Conjecture*. *Suppose* $\pi: X \longrightarrow S$ *is a strongly (p,q)-pseudoconvex-pseudoconcave holomorphic map (given with* φ *and* $a_* < a_\# < b_\# < b_*$) *and* $\mathcal{F}$ *is a coherent analytic sheaf on* X *such that* $\dim S \leqq n$ *and* $\operatorname{codh} \mathcal{F} \geqq r$ *on* $\{\varphi < a_\#\}$. *Then, for* $p \leqq \nu < r-q-n$, $R^\nu \pi_* \mathcal{F}$ *is coherent on* S *and* $R^\nu \pi_* \mathcal{F} \longrightarrow R^\nu (\pi_a^b)_* \mathcal{F}$ *is an isomorphism for*

$a_* \leqq a < a_\#$ and $b_\# < b < b_*$.

This conjecture so far has not been completely proved. The special case $\{\varphi < a_\#\} = \emptyset$ is called the pure pseudoconvex case. The special case $\{\varphi > b_\#\} = \emptyset$ is called the pure pseudoconcave case. Partial results for these two pure cases were obtained by Knorr [14] and Siu [24,25]. Recently the pure pseudoconvex case was completely proved by Siegfried [21] by using the methods of the new proofs of Grauert's theorem and the pure pseudoconcave case was completely proved by Ramis-Ruget [19] by using the methods of the new proofs of Grauert's theorem together with duality. Unfortunately these methods cannot be applied to the mixed case, because any induction on the dimension of the direct image is impossible. A partial result on the mixed case was obtained by Siu [26]. In these lectures we will prove the following improved partial result of the mixed case which is good enough for the known applications.

(0.5) Main Theorem. Suppose $\pi: X \longrightarrow S$ is a strongly (p,q)-pseudoconvex-pseudoconcave holomorphic map (given with φ and $a_* < a_\# < b_\# < b_*$) and $\mathcal{F}$ is a coherent analytic sheaf on X such that $\dim S \leqq n$ and $\operatorname{codh} \mathcal{F} \geqq r$ on $\{\varphi < a_\#\}$. Suppose $\operatorname{codh}_{\mathcal{O}_{S,\pi(x)}} \mathcal{F}_x \geqq n$ for $x \in X$. Let $a_* < a' < a < a_\#$ and $b_\# < b < b' < b_*$. Then the following conclusions hold.

i) $(R^\nu(\pi_a^b)_* \mathcal{F})_s$ is finitely generated over $\mathcal{O}_{S,s}$ for

$s \subseteq S$ <u>and</u> $p \leq \nu < r-q-n$.

ii) $R^\nu \pi_* \mathcal{F} \longrightarrow R^\nu (\pi_a^b)_* \mathcal{F}$ <u>is an isomorphism for</u> $p < \nu < r-q-n$.

iii) $R^p(\pi_{a'}^{b'})_* \mathcal{F} \longrightarrow R^p(\pi_a^b)_* \mathcal{F}$ <u>is an isomorphism</u>.

iv) <u>If</u> $p < r-q-2n$, <u>then</u> $R^\nu (\pi_a^b)_* \mathcal{F}$ <u>is coherent on</u> S <u>for</u> $p \leq \nu < r-q-n-1$.

For the applications, only conclusions i) and iii) of the Main Theorem are needed. The Main Theorem will be proved by the power series method. If we couple the power series method with the methods of new proofs of Grauert's theorem and duality, we can improve conclusions iii) and iv) to the following, but we will not discuss it in these lectures.

(0.6) <u>Theorem</u>. <u>Under the assumptions of</u> (0.5), <u>if</u> $p < r-q-2n$, <u>then, for</u> $p \leq \nu < r-q-n$, $R^\nu \pi_* \mathcal{F}$ <u>is coherent on</u> S <u>and</u> $R^\nu \pi_* \mathcal{F} \longrightarrow R^\nu (\pi_a^b)_* \mathcal{F}$ <u>is an isomorphism for</u> $a_* \leq a < a_\#$ <u>and</u> $b_\# < b \leq b_*$.

The Main Theorem will be applied to the following:

i) extending coherent analytic sheaves

ii) blowing down strongly 1-pseudoconvex maps

iii) blowing down relative exceptional sets

iv) obtaining a criterion for the projectivity of a map

v) extending families of complex spaces.

(For applications ii), iii) and iv), the pure pseudoconvex case of the direct image theorem suffices.)

For coherent sheaf extension, we will not obtain the best known result of extension from Hartogs' figures [23]. We will only obtain the result of extension from ring domains [22] (which implies the extension across subvarieties [29,5]). The proof of coherent sheaf extension by means of the direct image theorem is not the simplest approach. A very simple proof of the extension from Hartogs' figures was given in [27] which does not use the power series method of Grauert and does not use the method of privileged sets of Douady.

The smooth case of the local result on blowing down strongly 1-pseudoconvex maps was obtained by Markoe-Rossi [17] and the general case of the complete result was obtained in [25]. The results on relative exceptional sets and projectivity criterion were due to Knorr-Schneider [15]. The special case of the result on extending families of complex spaces where the parameter space is a single point was obtained by Rossi [20] and the general case was due to Ling [16].

Now we give here a brief sketch of the main ideas of the proof of the Main Theorem. In the actual proof, for technical reasons, we use sheaf systems to construct complexes of Banach bundles to calculate the direct image sheaves, but, here in this sketch, for simplicity, we compromise the accuracy by calculating the direct image sheaves by the usual Čech cochain complex. In this sketch there are also other compromises of accuracy in some minor points for the sake of simplicity. The proof of the Main Theorem has three key steps.

The first key step is the existence of privileged sets for a coherent sheaf, i.e. if $\theta : {}_n\mathcal{O}^p \longrightarrow {}_n\mathcal{O}^q$ is a sheaf-homomorphism on an open neighborhood G of 0 in $\mathbb{C}^n$ which is part of a finite resolution of the given sheaf (where ${}_n\mathcal{O}$ is the structure sheaf of $\mathbb{C}^n$), then there exists an open polydisc neighborhood P of 0 in G satisfying the following. If $s \in \Gamma(P, {}_n\mathcal{O}^q)$ is bounded (in a suitable sense) and the germ of s at 0 belongs to the image of θ, then s is the image of a bounded section v of ${}_n\mathcal{O}^p$ over P and v can be so chosen that $s \longmapsto v$ is continuous $\mathbb{C}$-linear. The existence of privileged sets can be proved in three ways (some of which are valid only for certain senses of boundedness). The first proof by Cartan uses the Weierstrass preparation and division theorems and it is usually used in the proof of the Closure-of-Modules Theorem [8, II.D]. The second proof by Grauert uses power series expansion and it is used in Grauert's original proof of the proper case of the direct image theorem. The third proof by Douady uses holomorphic Banach bundles and it is used in his solution of the module problem [2]. In these lectures, we present Douady's proof, because it works for all senses of boundedness needed for our purpose and because the idea of Grauert's proof occurs in the second key step of the proof of the Main Theorem and we will see it there anyway. The existence of privileged sets gives rise to Theorem B with bounds and Leray's theorem with bounds, which are used, together with the bumping techniques of Andreotti-Grauert [1], to construct complexes of Banach bundles to calculate the direct image sheaves.

The second key step is the analog of the Hauptlemma of Grauert's original proof of the proper case. Suppose S is an open neighborhood of 0 in $\mathbb{C}^n$. Let $\Delta(\rho)$ be the open polydisc in $\mathbb{C}^n$ centered at 0 with polyradius ρ. Roughly the analog of the Hauptlemma can be described as follows. Suppose $\mathfrak{U} = \{U_i\}$, $\mathfrak{V} = \{V_j\}$ be suitable collections of Stein open subsets of X such that each V_j is relatively compact in some U_i. Let

$$\mathfrak{U}(\rho) = \{U_i \cap \pi^{-1}(\Delta(\rho))\}$$

$$\mathfrak{V}(\rho) = \{V_j \cap \pi^{-1}(\Delta(\rho))\}.$$

Then there exist

$$\xi^{(1)}, \ldots, \xi^{(k)} \in Z^{\nu}(\mathfrak{U}(\rho^{o}), \mathcal{F})$$

for some ρ^{o} satisfying the following. For ρ sufficiently small (in a suitable sense), every $\xi \in Z^{\nu}(\mathfrak{U}(\rho), \mathcal{F})$ can be written as

$$\xi = \sum_{i=1}^{k} a_i \xi^{(i)} + \delta\eta$$

when restricted to $\mathfrak{V}(\rho)$, where

$$a_i \in \Gamma(\Delta(\rho), {}_n\mathcal{O})$$

$$\eta \in C^{\nu-1}(\mathfrak{U}(\rho), \mathcal{F}) .$$

Moreover the bounds (in a suitable sense) of a_i and η are dominated by a constant times the bound of ξ . We will look upon this as a generalization of the existence of privileged sets. Instead of lifting bounded sections in the map

$$\Gamma(P, {}_n\mathcal{O}^p) \longrightarrow \Gamma(P, {}_n\mathcal{O}^q)$$

induced by θ , in this case we consider the lifting of bounded sections in the map

$$\Gamma(\Delta(\rho), {}_n\mathcal{O}^k) \oplus C^{\nu-1}(\mathfrak{U}(\rho), \mathcal{F}) \longrightarrow Z^{\nu}(\mathfrak{U}(\rho), \mathcal{F})$$

defined by

$$(a_1, \ldots, a_k) \oplus \eta \longrightarrow \sum_{i=1}^{k} a_i \xi^{(i)} + \delta\eta$$

(restricted to $\mathfrak{U}(\rho)$) .

It turns out that the ideas of Grauert's proof of the existence of privileged sets can be generalized to this case when we apply Leray's theorem with bounds. For this we have to use induction on dim S, but we do not need any descending induction on ν . The reason why we can avoid this descending induction on ν which is so essential in Grauert's original proof of the proper case is that we look upon the analog of the Hauptlemma as the generalization of the existence of privileged sets and we use both the surjectivity and the injectivity statements of Leray's theorem with bounds, whereas Grauert did not use the technique for proving the existence of privileged sets in his proof of the Hauptlemma and he used only the surjectivity statement of Leray's theorem with bounds. Our approach is simpler and gives the best possible result in the finite generation of the stalks of the direct image sheaves.

After the above two key steps there is still one obstacle to proving the coherence of the direct image sheaves. Suppose S is an open neighborhood of 0 in $\mathbb{C}^n$. To get the coherence of the direct image sheaves by induction on n, we need the following statement on global isomorphism. There exists an open polydisc neighborhood U of 0 in S such that

$$H^{\nu}(\pi^{-1}(U), \mathcal{G}) \longrightarrow \Gamma(U, R^{\nu}\pi_{*}\mathcal{G})$$

is an isomorphism for all sheaves $\mathcal{G}$ of the form $\mathcal{F}/(t_n - c)^m\mathcal{F}$, where $t_1, \ldots, t_n$ are the coordinates of $\mathbb{C}^n$. The third key step is to obtain this isomorphism statement. Suppose $\mathfrak{U} = \{U_i\}$ is a Stein open covering of X. For any open subset D of S let $\mathfrak{U}(D) = \{U_i \cap \pi^{-1}(D)\}$. The presheaves

$$D \longmapsto B^{\nu}(\mathfrak{U}(D), \mathcal{G})$$

$$D \longmapsto C^{\nu}(\mathfrak{U}(D), \mathcal{G})$$

define sheaves $\mathcal{B}^{\nu}(\mathcal{G})$, $\mathcal{C}^{\nu}(\mathcal{G})$ on S. We derive the isomorphism statement by constructing a sheaf-homomorphism

$\mathcal{B}^{\nu}(\mathcal{F}) \longrightarrow \mathcal{C}^{\nu-1}(\mathcal{F})$ on U which is a right inverse of δ. This right inverse gives rise to a right inverse $\mathcal{B}^{\nu}(\mathcal{G}) \longrightarrow \mathcal{C}^{\nu-1}(\mathcal{G})$ of δ for sheaves $\mathcal{G}$ of the form $\mathcal{F}/(t_n-c)^m\mathcal{F}$. The existence of a right inverse $\mathcal{B}^{\nu}(\mathcal{G}) \longrightarrow \mathcal{C}^{\nu-1}(\mathcal{G})$ of δ implies the vanishing of $H^1(U, \mathcal{B}^{\nu}(\mathcal{G}))$ which, together with the coherence of $R^{\nu}\pi_*\mathcal{G}$, $R^{\nu+1}\pi_*\mathcal{G}$, ..., $R^{\mu-1}\pi_*\mathcal{G}$, yields the isomorphism

$$H^{\mu}(\pi^{-1}(U), \mathcal{G}) \longrightarrow \Gamma(U, R^{\mu}\pi_*\mathcal{G}).$$

The construction of a right inverse $\mathcal{B}^{\nu}(\mathcal{F}) \longrightarrow \mathcal{C}^{\nu-1}(\mathcal{F})$ of δ is based on the generalization of the following observation. For an open polydisc G in $\mathbb{C}^n$, a continuous $\mathbb{C}$-linear map

$$\alpha: \Gamma(G, {}_n\mathcal{O}^p) \longrightarrow \Gamma(G, {}_n\mathcal{O}^q)$$

is induced by a sheaf-homomorphism ${}_n\mathcal{O}^p \longrightarrow {}_n\mathcal{O}^q$ on G if and only if α is linear over the polynomial ring $\mathbb{C}[t_1, \ldots, t_n]$. We show that, under the additional assumption of the vanishing of $(R^{\nu+1}\pi_*\mathcal{F})_0, \ldots, (R^{\nu+n-1}\pi_*\mathcal{F})_0$, the lifting

$$\xi \longmapsto (a_1, \ldots, a_k) \oplus \eta$$

in the analog of Grauert's Hauptlemma can be done in such a way that it is linear over the polynomial ring $\mathbb{C}[t_1, \ldots, t_n]$. When we have the finite generation of $(R^{\nu}\pi_*\mathcal{F})_0, \ldots, (R^{\nu+n}\pi_*\mathcal{F})_0$ over ${}_n\mathcal{O}_0$, we can apply the above argument to a complex of the form $\mathcal{C}^{\lambda}(\mathcal{F}) \oplus {}_n\mathcal{O}^{p_\lambda}$ instead of to $\mathcal{C}^\lambda(\mathcal{F})$ and obtain the isomorphism

$$H^\mu(\pi^{-1}(U),\mathcal{G}) \longrightarrow \Gamma(U, R^\mu\pi_*\mathcal{G})$$

for $\nu \leq \mu \leq \nu + n$. It is the third key step that makes the additional assumption of $p < r - q - 2n$ necessary, because we need some room to get a right inverse of δ. It is also the arguments of this step that necessitate the introduction of complexes of Banach bundles for the calculation of the direct image sheaves, although such an introduction streamlines the presentation elsewhere as well.

In these lectures some tedious details, which are obvious and can easily be filled in, are left out, especially in §3, §4, and §5. Details of this nature can be found in [13, 18, 24]. There is an appendix at the end which deals with homological codimension, flatness, and gap-sheaves. Consult the appendix when these concepts are mentioned or their properties are used.

Now we list the notations we will use in these lectures.

$\mathbb{N}$ = the set of all positive integers

$\mathbb{N}_0$ = $\mathbb{N} \cup \{0\}$

$\mathbb{N}_*$ = $\mathbb{N} \cup \{\infty\}$

$\mathbb{R}_+$ = the set of all positive numbers

${}_n\mathcal{O}$ = the structure sheaf of $\mathbb{C}^n$.

The components of $a \in \mathbb{C}^m$ are denoted by $a_1, \cdots, a_m$. For $a,b \in \mathbb{R}^m$, by $a < b$ we mean $a_i < b_i$ for $1 \leq i \leq m$; and by $a \leq b$ we mean $a_i \leq b_i$ for $1 \leq i \leq m$.

n occupies a special position in these lectures. The coordinates of $\mathbb{C}^n$ are always denoted by $t = (t_1, \cdots, t_n)$. $\mathbb{C}[t]$ denotes the polynomial ring $\mathbb{C}[t_1, \cdots, t_n]$. Suppose $t^o \in \mathbb{C}^n$, $\rho \in \mathbb{R}^n_+$, and $\lambda \leq \nu$ in $\mathbb{N}^n_o$.

$\Delta(t^o, \rho)$ = the open polydisc in $\mathbb{C}^n$ with center t^o and polyradius ρ

$$\left(\frac{t-t^o}{\rho}\right)^\nu = \left(\frac{t_1-t_1^o}{\rho_1}\right)^{\nu_1} \cdots \left(\frac{t_n-t_n^o}{\rho_n}\right)^{\nu_n}$$

$$|\nu| = \nu_1 + \cdots + \nu_n$$

$$D^\nu = \frac{\partial^{|\nu|}}{\partial t_1^{\nu_1} \cdots \partial t_n^{\nu_n}}$$

$$\binom{\nu}{\lambda} = \binom{\nu_1}{\lambda_1} \cdots \binom{\nu_n}{\lambda_n} .$$

When $t^o = 0$, we denote $\Delta(t^o, \rho)$ simply by $\Delta(\rho)$. When $\rho = (1, \cdots, 1)$, we denote $\Delta(\rho)$ simply by Δ. In general, for $b \in \mathbb{R}^N_+$, $\Delta^N(b)$ denotes the open polydisc in $\mathbb{C}^N$ with center 0 and polyradius b. For $0 \leq a < b$ in $\mathbb{R}^N$,

$$G^N(a,b) = \{z \in \Delta^N(b) \,\big|\, |z_i| > a_i \text{ for some } 1 \leq i \leq N\} .$$

The closure of a set G is denoted by G^{-}. $\|\cdot\|_G$ denotes the sup norm on G. If G is an open subset of $\mathbb{C}^N$, $\Gamma_{L^2}(G, {}_N\mathcal{O})$ denotes the set of all L^2 holomorphic functions on G.

The stalk of a sheaf $\mathcal{F}$ at a point s is denoted by $\mathcal{F}_s$. If U is an open neighborhood of s and $f \in \Gamma(U, \mathcal{F})$, then f_s denotes the germ of f at s.

A complex space may have nonzero nilpotent elements in its structure sheaf. The structure sheaf of a complex space X is denoted by $\mathcal{O}_X$. For $x \in X$, $\mathfrak{m}_{X,x}$ means the maximum ideal of the local ring $\mathcal{O}_{X,x}$ and sometimes (when no confusion can arise) it also means the ideal sheaf for the subvariety $\{x\}$ of X. Suppose $\pi: X \longrightarrow S$ is a holomorphic map of complex spaces and $s \in S$. Then $\mathfrak{m}_{S,s}$ means also the ideal sheaf on X generated by the inverse image of $\mathfrak{m}_{S,s}$ when no confusion can arise. For a coherent analytic sheaf $\mathcal{F}$ on X, $R^{\nu}\pi_*\mathcal{F}$ denotes the νth direct image of under π. If Y is an open subset of X and $\sigma = \pi | Y$, then $R^{\nu}\sigma_*(\mathcal{F}|Y)$ is simply denoted by $R^{\nu}\sigma_*\mathcal{F}$.

Suppose $\mathfrak{U} = \{U_i\}$ and $\mathfrak{V} = \{V_j\}$ are collections of open subsets of a complex space X and $\mathcal{F}$ is a coherent analytic sheaf on X. $\mathfrak{U} \ll \mathfrak{V}$ means that each U_i is relatively compact in some $V_{\tau(i)}$. For every $\xi \in C^{\nu}(\mathfrak{V}, \mathcal{F})$, we can define $\tau^*\xi \in C^{\nu}(\mathfrak{U}, \mathcal{F})$ by means of the index map τ. $\tau^*\xi$ is also denoted by $\xi | \mathfrak{U}$. For $\xi, \eta \in C^{\nu}(\mathfrak{V}, \mathcal{F})$, we say that $\xi = \eta$ on $\mathfrak{U}$ if $\tau^*\xi = \tau^*\eta$. $|\mathfrak{U}|$ denotes $\bigcup_i U_i$. For a complex space Y , $Y \times \mathfrak{U}$ denotes $\{Y \times U_i\}$.

When we have a sheaf-homomorphism $\theta : \mathcal{F} \longrightarrow \mathcal{G}$ of analytic sheaves on a complex space X , we sometimes use the same symbol θ to denote also the maps

$$\Gamma(X, \mathcal{F}) \longrightarrow \Gamma(X, \mathcal{G})$$

$$\mathcal{F}/\mathcal{I}\mathcal{F} \longrightarrow \mathcal{G}/\mathcal{I}\mathcal{G}$$

(where $\mathcal{I}$ is an ideal sheaf on X) and other similar maps induced by θ when no confusion can arise.

Y.-T. Siu

Table of Contents

Y-T. Siu

PART I CONSTRUCTION OF COMPLEXES OF BANACH BUNDLES

§1 Privileged Polydiscs

(1.1) Suppose $D = \Delta(t^0,\rho)$ and E_0 is a Banach space. Define $B(D,E_0)$ as one of the following:

i) the set of all E_0-valued uniformly bounded holomorphic functions on D,

ii) the set of all E_0-valued uniformly continuous holomorphic functions on D,

iii) the set of all E_0-valued holomorphic functions

$$f = \sum_\nu f_\nu \left(\frac{t-t^0}{\rho}\right)^\nu \quad (f_\nu \in E_0)$$

on D with

$$\sup_\nu \|f_\nu\|_{E_0} < \infty$$

where $\|\cdot\|_{E_0}$ is the norm of E_0.

In any of these three cases $B(D,E_0)$ is a Banach space. $B(D,\mathbb{C})$ is simply denoted by $B(D)$. If U is an open neighborhood of D^- and E is the trivial bundle on U with fiber E_0, we denote $B(D,E_0)$ also by $B(D,E)$. Suppose $\mathcal{F}$ is a coherent analytic sheaf on an open neighborhood of D^-. There exists an exact sequence

$$0 \longrightarrow {}_n\mathcal{O}^{p_m} \longrightarrow \cdots \longrightarrow {}_n\mathcal{O}^{p_1} \longrightarrow {}_n\mathcal{O}^{p_0} \longrightarrow \mathcal{F} \longrightarrow 0$$

on an open neighborhood of D^-.

Definition. D is an $\mathcal{F}$-privileged neighborhood if

(a) the induced sequence

$$0 \longrightarrow B(D)^{p_m} \longrightarrow \cdots \longrightarrow B(D)^{p_1} \xrightarrow{\alpha} B(D)^{p_0}$$

is split exact,

(b) $\operatorname{Coker} \alpha \longrightarrow \mathcal{F}_{t^0}$ is injective.

When (a) is satisfied, one defines $B(D, \mathcal{F})$ as $\operatorname{Coker} \alpha$.

This privilegedness is said to be in the sense of Cartan, Douady, or Grauert according as $B(D)$ has the meaning of i), ii), or iii).

The definition of privilegedness and $B(D, \mathcal{F})$ is independent of the choice of the resolution of $\mathcal{F}$, because, by using Theorem B of Cartan-Oka, we can easily prove that any two finite free resolution of $\mathcal{F}$ on a Stein open neighborhood of D^- become isomorphic finite free resolutions after we apply to each of them a finite number of modifications [8, Def. VI.F.1], i.e. after we apply to each of them a finite number of times the process of replacing it by its direct sum with some finite free resolution of the zero sheaf which has only two nonzero terms (cf. [8, p.202, VI.F.3]).

(1.2) For Banach spaces E_0, F_0 we denote by $L(E_0, F_0)$ the Banach space of all continuous linear maps from E_0 to F_0.

Suppose S is an open subset of $\mathbb{C}^n$ and E is a holomorphic Banach bundle on S with fiber E_0. For $s \in S$ we denote by E_s the fiber of E at s. For any open subset U of S we denote by $E|U$ the restriction of E to U.

Suppose F is a holomorphic Banach bundle on S with fiber F_0. A map $\gamma: E \longrightarrow F$ is called a <u>bundle-homomorphism</u> if for every open subset U of S for which there are trivializations

$$\alpha: E|U \xrightarrow{\approx} U \times E_0$$

$$\beta: F|U \xrightarrow{\approx} U \times F_0$$

there exists a holomorphic map $A(\cdot)$ from U to $L(E_0,F_0)$ such that

$$(\beta\gamma\alpha^{-1})(s,x) = (s,A(s)x)$$

for $s \in U$ and $x \in E_0$. For any open subset U of S we denote by $\gamma|U$ the bundle-homomorphism

$$E|U \longrightarrow F|U$$

induced by γ. For $s \in S$ we denote by γ_s the map

$$E_s \longrightarrow F_s$$

induced by γ .

Suppose

$$0 \longrightarrow E^{(m)} \xrightarrow{\theta} E^{(m-1)} \longrightarrow \cdots \longrightarrow E^{(0)}$$

is a complex of bundle-homomorphisms of holomorphic Banach bundles on S . <u>If for some</u> $s_0 \in S$ <u>the sequence</u>

$$0 \longrightarrow E^{(m)}_{s_0} \xrightarrow{\theta_{s_0}} E^{(m-1)}_{s_0} \longrightarrow \cdots \longrightarrow E^{(0)}_{s_0}$$

<u>is split exact, then there exists an open neighborhood</u> U <u>of</u> s_0 <u>in</u> S <u>such that the sequence</u>

$$0 \longrightarrow E^{(m)}\big|U \longrightarrow E^{(m-1)}\big|U \longrightarrow \cdots \longrightarrow E^{(0)}\big|U$$

<u>is split exact</u>.

To prove this, it suffices to prove the case where $m = 1$ and $E^{(1)}$, $E^{(0)}$ are both trivial bundles. Let H be the closed subspace of $E^{(0)}_{s_0}$ which complements $\operatorname{Im} \theta_{s_0}$. Let

$$\sigma : E^{(1)} \oplus (S \times H) \longrightarrow E^{(0)}$$

be the bundle-homomorphism induced by θ and the inclusion map

$$S \times H \hookrightarrow S \times \left(H \oplus \operatorname{Im} \theta_{s_0}\right) = E^{(0)} .$$

σ_{s_0} is an isomorphism. Since the invertible elements of

$L\left(E^{(1)}_{s_0} \oplus H, E^{(0)}_{s_0}\right)$ form an open subset, there exists an open neighborhood U of s_0 such that $\sigma | U$ is a bundle-isomorphism (i.e. $(\sigma|U)^{-1}$ is a bundle-homomorphism).

(1.3) Suppose S (respectively Ω) is an open subset of $\mathbb{C}^n$ (respectively $\mathbb{C}^N$) and $\mathcal{F}$ is a coherent analytic sheaf on $S \times \Omega$. For

$$s = (s_1, \dots, s_n) \in S$$

we denote by $\mathcal{F}(s)$ the sheaf

$$\mathcal{F} \Big/ \sum_{i=1}^{k} (t_i - s_i) \mathcal{F}$$

where $t_1, \dots, t_n$ are the coordinates of $\mathbb{C}^n$. $\mathcal{F}(s)$ can be regarded in a natural way as a sheaf on Ω.

For $p \geq 1$, we denote by $B(\Omega, {}_{n+N}\mathcal{O}^p)$ the trivial bundle on $\mathbb{C}^n$ whose fiber is $B(\Omega)^p$.

Let $\pi: S \times \Omega \longrightarrow S$ be the natural projection. Suppose $s \in S$ and $\mathcal{F}$ is π-flat at $\{s\} \times \Omega$ and $\mathcal{F}$ admits a finite free resolution

(*)

$$0 \longrightarrow {}_{n+N}\mathcal{O}^{p_m} \longrightarrow \dots \longrightarrow {}_{n+N}\mathcal{O}^{p_1} \longrightarrow {}_{n+N}\mathcal{O}^{p_0} \longrightarrow \mathcal{F} \longrightarrow 0$$

on $S \times \Omega$. *Suppose* $z \in \Omega$ *and* $G \subset\subset \Omega$ *is an open polydisc centered at* z *which is an* $\mathcal{F}(s)$*-privileged neighborhood of* z. *Then there exists an open neighborhood* U *of* s *in* S *such that, for any open polydisc* $D \subset U$ *centered at* s,

$D \times G$ <u>is an $\mathcal{J}$-privileged neighborhood of</u> (s,z) .

To prove this, consider the following sequence of bundle-homomorphisms induced by (*):

(#)

$$0 \longrightarrow B(G, {}_{n+N}\mathcal{O}^{p_m}) \longrightarrow \cdots \longrightarrow B(G, {}_{n+N}\mathcal{O}^{p_1}) \xrightarrow{\alpha} B(G, {}_{n+N}\mathcal{O}^{p_0})$$

Since G is $\mathcal{J}(s)$-privileged and since by the π-flatness of $\mathcal{J}$ at $\{s\} \times \Omega$ the sequence

$$0 \longrightarrow {}_{n+N}\mathcal{O}^{p_m}(s) \longrightarrow \cdots \longrightarrow {}_{n+N}\mathcal{O}^{p_1}(s) \longrightarrow \mathcal{J}(s) \longrightarrow 0$$

induced by (*) is exact, we conclude that the sequence (#), when restricted to the singleton $\{s\}$, is split exact. By (1.2), on some open neighborhood of s in S , (#) is split exact and Coker α is a holomorphic Banach bundle. Observe that, for any bounded open polydisc D centered at s , $B(D, B(G, {}_{n+N}\mathcal{O}^{p_i}))$ is naturally topologically isomorphic to $B(D \times G, {}_{n+N}\mathcal{O}^{p_i})$ $(0 \leq i \leq m)$. Hence, when D is contained in a sufficiently small neighborhood U of s in S ,

$$0 \longrightarrow B(D \times G, {}_{n+N}\mathcal{O}^{p_m}) \longrightarrow \cdots$$
$$\longrightarrow B(D \times G, {}_{n+N}\mathcal{O}^{p_1}) \longrightarrow B(D \times G, {}_{n+N}\mathcal{O}^{p_0})$$

is split exact and $B(D \times G, \mathcal{J})$ is topologically isomorphic to $B(D, \text{Coker } \alpha)$. To show the injectivity of

$$B(D \times G, \mathcal{F}) \longrightarrow \mathcal{F}_{(s,z)} ,$$

it suffices to show the injectivity of

$$\beta: (\mathcal{O}(\mathrm{Coker}\ \alpha))_s \longrightarrow \mathcal{F}_{(s,z)}$$

(where $\mathcal{O}(\mathrm{Coker}\ \alpha)$ is the sheaf of germs of holomorphic sections of the bundle $\mathrm{Coker}\ \alpha$) and, by induction on n, it suffices to do the case $n = 1$. Take

$$f \in \mathrm{Ker}\ \beta .$$

Suppose f is nonzero. There exists a maximum nonnegative integer k such that

$$f = (t - s)^k g$$

with

$$g \in (\mathcal{O}(\mathrm{Coker}\ \alpha))_s .$$

By the π-flatness of $\mathcal{F}$,

$$g \in \mathrm{Ker}\ \beta .$$

Since G is an $F(s)$-privileged neighborhood of z, it follows that $g(s) = 0$, contradicting the maximality of k.

(1.4) Suppose $\mathcal{F}$ is a coherent analytic sheaf on an open subset Ω of $\mathbb{C}^N$. Identify $\mathbb{C}^N$ with the subset $0 \times \mathbb{C}^N$ of $\mathbb{C}^{N+1}$ and extend $\mathcal{F}$ trivially to a coherent analytic sheaf $\tilde{\mathcal{F}}$ on $\mathbb{C} \times \Omega$. <u>If</u> $G \subset\subset \Omega$ <u>is an open polydisc centered at</u> z <u>and if</u> G <u>is an</u> $\mathcal{F}$<u>-privileged neighborhood of</u> z, <u>then</u>,

<u>for any bounded open disc</u> $D \subset \mathbb{C}$ <u>centered at</u> 0, $D \times G$ <u>is an</u> $\tilde{\mathcal{F}}$-<u>privileged neighborhood of</u> $(0,z)$.

To prove this, we can assume without loss of generality that there exists an exact sequence

$$0 \longrightarrow {}_N\mathcal{O}^{p_m} \xrightarrow{\alpha_m} \cdots \longrightarrow {}_N\mathcal{O}^{p_1} \xrightarrow{\alpha_1} {}_N\mathcal{O}^{p_0} \longrightarrow \mathcal{F} \longrightarrow 0$$

on Ω. Define

$$0 \longrightarrow {}_{N+1}\mathcal{O}^{p_m} \oplus {}_{N+1}\mathcal{O}^{p_{m-1}} \xrightarrow{\tilde{\alpha}_m} \cdots$$

$$\longrightarrow {}_{N+1}\mathcal{O}^{p_2} \oplus {}_{N+1}\mathcal{O}^{p_1} \xrightarrow{\tilde{\alpha}_2} {}_{N+1}\mathcal{O}^{p_1} \oplus {}_{N+1}\mathcal{O}^{p_0} \xrightarrow{\tilde{\alpha}_1} {}_{N+1}\mathcal{O}^{p_0} \longrightarrow \tilde{\mathcal{F}} \longrightarrow 0$$

by

$$\tilde{\alpha}_1 = (\alpha_1, w)$$

$$\tilde{\alpha}_j = \begin{pmatrix} \alpha_j & (-1)^{j-1} w \\ 0 & \alpha_{j-1} \end{pmatrix} \quad (1 < j \leqq m)$$

where w is the coordinate of $\mathbb{C}$, an element of ${}_{N+1}\mathcal{O}^p$ is represented as a column vector, and α_j is regarded as a matrix of holomorphic functions which are considered as functions on $\mathbb{C} \times \Omega$ independent of w. The sequence is exact on $\mathbb{C} \times \Omega$. Let

$$\beta_j \colon B(G, {}_N\mathcal{O}^{p_{j-1}}) \longrightarrow B(G, {}_N\mathcal{O}^{p_j}) \quad (1 \leqq j \leqq m)$$

be a continuous linear map, which, when composed with the map

$$\alpha_j' : B(G, {}_N\mathcal{O}^{p_j}) \longrightarrow B(G, {}_N\mathcal{O}^{p_{j-1}})$$

induced by α_j, gives rise to a projection

$$B(G, {}_N\mathcal{O}^{p_{j-1}}) \longrightarrow \operatorname{Im} \alpha_j' .$$

A corresponding

$$\tilde{\beta}_j : B\left(D \times G, {}_{N+1}\mathcal{O}^{p_{j-1}} \oplus {}_{N+1}\mathcal{O}^{p_{j-2}}\right) \longrightarrow B\left(D \times G, {}_{N+1}\mathcal{O}^{p_j} \oplus {}_{N+1}\mathcal{O}^{p_{j-1}}\right)$$

(where $p_{-1} = 0$) can be given by

$$\tilde{\beta}_j(a' + ta'', b) = \left(\beta_{j+1}a' + t\beta_{j+1}(a'' - (-1)^j \beta_j b), \beta_j b\right)$$

where a' is independent of t and one denotes also by β_j the map

$$B\left(D \times G, {}_{N+1}\mathcal{O}^{p_{j-1}}\right) \longrightarrow B\left(D \times G, {}_{N+1}\mathcal{O}^{p_j}\right)$$

it induces. Hence

$$B(D \times G, \tilde{\mathcal{F}}) \approx B(G, \mathcal{F})$$

and the result follows.

(1.5) Suppose D is a bounded open polydisc in $\mathbb{C}^n$ centered at t^0 and suppose

$$0 \longrightarrow \mathcal{F}' \longrightarrow \mathcal{F} \longrightarrow \mathcal{F}'' \longrightarrow 0$$

is an exact sequence of coherent analytic sheaves on an open neighborhood of D^-. _If_ D _is an_ $\mathcal{F}'$_-privileged and_ $\mathcal{F}''$_-privileged neighborhood of_ t^0, _then_ D _is an_ $\mathcal{F}$_-privileged_ neighborhood of t^0.

To prove this, we take finite free resolutions on D^-:

$$0 \longrightarrow {}_n\mathcal{O}^{p'_m} \xrightarrow{\alpha'_m} \cdots \longrightarrow {}_n\mathcal{O}^{p'_1} \xrightarrow{\alpha'_1} {}_n\mathcal{O}^{p'_0} \longrightarrow \mathcal{F}' \longrightarrow 0$$

$$0 \longrightarrow {}_n\mathcal{O}^{p''_m} \xrightarrow{\alpha''_m} \cdots \longrightarrow {}_n\mathcal{O}^{p''_1} \xrightarrow{\alpha''_1} {}_n\mathcal{O}^{p''_0} \longrightarrow \mathcal{F}'' \longrightarrow 0 .$$

We can construct the following commutative diagram

$$\begin{array}{ccccccccccc}
 & & 0 & & & & 0 & & 0 & & 0 & & \\
 & & \downarrow & & & & \downarrow & & \downarrow & & \downarrow & & \\
0 & \longrightarrow & {}_n\mathcal{O}^{p'_m} & \longrightarrow & \cdots & \longrightarrow & {}_n\mathcal{O}^{p'_1} & \longrightarrow & {}_n\mathcal{O}^{p'_0} & \longrightarrow & \mathcal{F}' & \longrightarrow & 0 \\
 & & \downarrow & & & & \downarrow & & \downarrow & & \downarrow & & \\
0 & \longrightarrow & {}_n\mathcal{O}^{p_m} & \xrightarrow{\alpha_m} & \cdots & \longrightarrow & {}_n\mathcal{O}^{p_1} & \xrightarrow{\alpha_1} & {}_n\mathcal{O}^{p_0} & \longrightarrow & \mathcal{F} & \longrightarrow & 0 \\
 & & \downarrow & & & & \downarrow & & \downarrow & & \downarrow & & \\
0 & \longrightarrow & {}_n\mathcal{O}^{p''_m} & \longrightarrow & \cdots & \longrightarrow & {}_n\mathcal{O}^{p''_1} & \longrightarrow & {}_n\mathcal{O}^{p''_0} & \longrightarrow & \mathcal{F}'' & \longrightarrow & 0 \\
 & & \downarrow & & & & \downarrow & & \downarrow & & \downarrow & & \\
 & & 0 & & & & 0 & & 0 & & 0 & &
\end{array}$$

where

i) ${}_n\mathcal{O}^{p_j} = {}_n\mathcal{O}^{p'_j} \oplus {}_n\mathcal{O}^{p''_j}$ and, except in the last column, the vertical maps are the natural injections and projections.

ii) α_j is of the form $\begin{pmatrix} \alpha'_j & \gamma_j \\ 0 & \alpha''_j \end{pmatrix}$ (γ_j: ${}_n\mathcal{O}^{p''_j} \longrightarrow {}_n\mathcal{O}^{p'_j}$ being a sheaf-homomorphism).

Let

$$\beta'_j: B(D, {}_n\mathcal{O}^{p'_{j-1}}) \longrightarrow B(D, {}_n\mathcal{O}^{p'_j})$$

be a continuous linear map which, when composed with the map

$$\tilde{\alpha}'_j: B(D, {}_n\mathcal{O}^{p'_j}) \longrightarrow B(D, {}_n\mathcal{O}^{p'_{j-1}})$$

induced by α'_j, gives rise to a projection

$$B(D, {}_n\mathcal{O}^{p'_j}) \longrightarrow \operatorname{Im} \tilde{\alpha}'_j .$$

Let

$$\beta''_j: B(D, {}_n\mathcal{O}^{p''_{j-1}}) \longrightarrow B(D, {}_n\mathcal{O}^{p''_j})$$

be a similar map. Then a corresponding

$$\beta_j: B(D, {}_n\mathcal{O}^{p_{j-1}}) \longrightarrow B(D, {}_n\mathcal{O}^{p_j})$$

can be defined by

$$B_j(a,b) = (\beta'_j(a - \beta''_j b), \beta''_j b) .$$

Since clearly

$$0 \longrightarrow B(D, \mathcal{F}') \longrightarrow B(D, \mathcal{F}) \longrightarrow B(D, \mathcal{F}'') \longrightarrow 0$$

is exact, the result follows.

Before we state the principal theorem on privileged polydiscs, we have to introduce a terminology. Suppose S_ρ is a statement depending on $\rho \in \mathbb{R}^n_+$. We say that S_ρ holds for ρ <u>sufficiently strictly small</u>, if there exist

$\omega_1 \in \mathbb{R}_+$ and positive-valued functions $\omega_i(\rho_1, \ldots, \rho_{i-1})$ on $\mathbb{R}_+^{i-1}$ $(1 < i \leqq n)$ such that S_ρ holds for ρ satisfying

$$\begin{cases} \rho_1 < \omega_1 \\ \rho_i < \omega_i(\rho_1, \ldots, \rho_{i-1}) \qquad (1 < i \leqq n) . \end{cases}$$

(1.6) Theorem. Suppose $\mathcal{F}$ is a coherent analytic sheaf on an open neighborhood U of 0 in $\mathbb{C}^n$. Then, for ρ sufficiently strictly small, $\Delta(\rho)$ is an $\mathcal{F}$-privileged neighborhood of 0.

Proof. Use induction on n. By shrinking U, we can assume that there exists a nonnegative integer d such that t_n is not a zero-divisor of $t_n^d \mathcal{F}_{t^0}$ for $t^0 \in U$. Then $\mathcal{G} := t_n^d \mathcal{F}$ is π-flat at $U \cap \{t_n = 0\}$, where $\pi : \mathbb{C}^n \longrightarrow \mathbb{C}$ is the projection onto the last coordinate. By induction hypothesis, when $(\rho_1, \ldots, \rho_{n-1})$ is sufficiently strictly small, the polydisc $G \subset \mathbb{C}^{n-1}$ with polyradius $(\rho_1, \ldots, \rho_{n-1})$ and centered at 0 is relatively compact in $U \cap \{t_n = 0\}$ and is a privileged neighborhood of 0 for the coherent analytic sheaf

$$\left(\bigoplus_{j=0}^{d-1} t_n^j \mathcal{F} \big/ t_n^{j+1} \mathcal{F} \right) \oplus \mathcal{G} / t_n \mathcal{G}$$

on $U \cap \{t_n = 0\}$. By (1.3), when an open disc $D \subset \mathbb{C}$ centered at 0 is sufficiently small, $D \times G \subset\subset U$ and $D \times G$ is a $\mathcal{G}$-privileged neighborhood of 0. By using (1.4) and applying (1.5) to the exact sequences

$$0 \longrightarrow {t_n^j \mathcal{F}}\big/{t_n^{j+1}\mathcal{F}} \longrightarrow \mathcal{F}\big/{t_n^{j+1}\mathcal{F}} \longrightarrow \mathcal{F}\big/{t_n^{j}\mathcal{F}} \longrightarrow 0 \quad (1 \leqq j < d)$$

$$0 \longrightarrow \mathcal{G} \longrightarrow \mathcal{F} \longrightarrow \mathcal{F}\big/{t_n^{d}\mathcal{F}} \longrightarrow 0 ,$$

we conclude that $D \times G$ is an $\mathcal{F}$-privileged neighborhood of

§2 Semi-norms on Unreduced Spaces

(2.1) Lemma. Suppose $\mathcal{F}$ is a coherent analytic sheaf on a complex space X and $D \subset\subset X$ is an open set. Then there exists a nonnegative integer p such that, if $f \in \Gamma(U, \mathcal{F})$ for some open subset U of D and $f_x \in \mathfrak{m}_{X,x}^{p+1} \mathcal{F}_x$ for $x \in U$, then $f = 0$.

Proof. Take $x_0 \in X$. Let $0 = \bigcap_i Q_i$ be the primary decomposition of the zero submodule of $\mathcal{F}_{x_0}$. Let k_i be the dimension of the radical of Q_i. Q_i is the stalk at x_0 of a coherent analytic subsheaf $\mathcal{Q}_i$ of $\mathcal{F}$ defined on some open neighborhood of x_0. Then

$$\dim_x \operatorname{Supp} \mathcal{F}/\mathcal{Q}_i \leq k_i .$$

$$\left((\mathcal{Q}_i)_{[k_i-1]}\mathcal{F}\right)_x = (\mathcal{Q}_i)_x$$

for $x = x_0$ (see (A.7) of the Appendix). Hence these two conditions hold for x in some open neighborhood of x_0. Since we need only prove the lemma for each $\mathcal{F}/\mathcal{Q}_i$, we can assume without loss of generality that X has pure dimension k and

$$0_{[k-1]}\mathcal{F} = 0 .$$

We can also assume the following:

i) X is a subspace of Δ.

ii) The projection $\tilde{\pi}: \Delta \longrightarrow D$ onto the first k coordinates makes X an analytic cover over D.

iii) The reduction of X is defined by the ideal sheaf $\mathcal{I}$ generated by holomorphic functions $g_{k+1}, \cdots, g_n$ on Δ.

iv) The unreduced structure of X is defined by $\mathcal{I}^{\ell}$.

We are going to prove that it suffices to set $p = \ell$. Let $\pi = \tilde{\pi}|X$. Let S be the subvariety of X outside which $t_1, \cdots, t_k, g_{k+1}, \cdots, g_n$ form a local coordinates system. Let T be the set where $R^0\pi_* \mathcal{F}$ is not locally free. Since

$$0_{[k-1]}\mathcal{F} = 0 ,$$

it suffices to prove that $f = 0$ on $U - \pi^{-1}(T \cup \pi(S))$. There

$$f_x \in \mathfrak{m}_{X,x}^{\ell+1} \mathcal{F}_x$$

means that

$$f_x = \sum_{|\alpha|=\ell} t_1^{\alpha_1} \cdots t_k^{\alpha_k} g_{k+1}^{\alpha_{k+1}} \cdots g_n^{\alpha_n} h_\alpha$$

in $\mathcal{F}_x$ with

$$h_\alpha \in \mathcal{F}_x .$$

Since

$$g_{k+1}^{\alpha_{k+1}} \cdots g_n^{\alpha_n} = 0$$

in $\mathcal{O}_X$ for $\alpha_1 = \cdots = \alpha_k = 0$, one has

$$f_x = \sum_{\alpha_1+\dots+\alpha_k \geq 1} t_1^{\alpha_1} \dots t_k^{\alpha_k} g_{k+1}^{\alpha_{k+1}} \dots g_n^{\alpha_n} h_\alpha$$

in $\mathcal{F}_x$. Let W be an open neighborhood of $\pi(x)$ in D such that $\pi^{-1}(W) \cap U$ is closed in $\pi^{-1}(W)$. Let f' be the element of $\Gamma(\pi^{-1}(W), \mathcal{F})$ which equals f on $\pi^{-1}(W) \cap U$ and equals 0 on $\pi^{-1}(W) - U$. The element f^* of $\Gamma(W, R^0\pi_* \mathcal{F})$ corresponding to f' satisfies

$$f_y^* \in \mathfrak{m}_{\mathbb{C}^k, y} (R^0\pi_* \mathcal{F})_y \qquad (y \in W).$$

It follows that $f^* = 0$ on $W - T$. Q.E.D.

(2.2) Suppose V is a subvariety of an open subset G of $\mathbb{C}^n$ and p is a nonnegative integer. Define $\mathcal{J}_V(p)$ as the sheaf of germs of holomorphic functions on G whose partial derivatives of order $\leq p$ vanish identically on V. Then $\mathcal{J}_V(p)$ <u>is a coherent ideal sheaf on</u> G.

We prove by induction on p. The case $p = 0$ is the theorem of Cartan-Oka. We can assume without loss of generality that

$$\mathcal{J}_V(p-1) = \sum_{i=1}^{k} {}_n\mathcal{O} f_i$$

for some holomorphic functions $f_1, \dots, f_k$ on G. Let ℓ be the number of n-tuples of nonnegative integers α with $|\alpha| = p$. Define

$$\varphi : {}_n\mathcal{O}^k \longrightarrow {}_n\mathcal{O}^\ell$$

by the matrix

$$(D^\alpha f_i)_{|\alpha|=p,\ 1 \leq i \leq k}$$

and define

$$\psi : {}_n\mathcal{O}^k \longrightarrow {}_n\mathcal{O}$$

by the row vector

$$(f_1, \ldots, f_k) .$$

Let

$$\eta : {}_n\mathcal{O}^\ell \longrightarrow {}_n\mathcal{O}^\ell / \mathcal{I}_V(0)\, {}_n\mathcal{O}^\ell$$

be the quotient map. Then

$$\mathcal{I}_V(p) = \psi(\operatorname{Ker} \eta\varphi) .$$

For, every $g \in \mathcal{I}_V(p-1)_x$ can be written as

$$g = \sum_{i=1}^k a_i\ (f_i)_x \quad (a_i \in {}_n\mathcal{O}_x)$$

and, for $|\alpha| = p$,

$$D^\alpha g = \sum_{i=1}^k a_i\ (D^\alpha f_i)_x .$$

It follows that $g \in \mathcal{I}_V(p)_x$ if and only if, when restricted to V,

$$\sum_{i=1}^k a_i\ (D^\alpha f_i)_x = 0 \quad (|\alpha| = p) .$$

(2.3) Suppose V is a subvariety of an open subset G of $\mathbb{C}^n$, p is a nonnegative integer, and

$$f \in \Gamma(V, {}_n\mathcal{O}/\mathcal{J}_V(p)) .$$

Suppose L is an open subset of U. We define a semi-norm ${}_p\|f\|_L$ as follows. For $x \in U$, there exists

$$\tilde{f} \in \Gamma(D, {}_n\mathcal{O})$$

for some open neighborhood D of x such that $\tilde{f}$ induces $f|D \cap V$. Let

$$f_\alpha(x) = D^\alpha \tilde{f}(x) \quad (|\alpha| \leqq p) .$$

Then

$$\{f_\alpha(x)\}_{|\alpha| \leqq p}$$

is independent of the choice of $\tilde{f}$. Define

$${}_p\|f\|_L = \sup\{|f_\alpha(x)| \,\big|\, |\alpha| \leqq p, x \in L\} .$$

<u>The semi-norms</u> ${}_p\|\cdot\|_L$ <u>on</u> $\Gamma(V, {}_n\mathcal{O}/\mathcal{J}_V(p))$ <u>define a Fréchet space structure when</u> L <u>runs through all relatively compact open subsets of</u> V.

We prove this by induction on p. Let $\{f_\nu\}$ be a Cauchy sequence in $\Gamma(V, {}_n\mathcal{O}/\mathcal{J}_V(p))$. It suffices to show that every point of V admits an open neighborhood U in V such that $\{f_\nu|U\}$ converges to some element of $\Gamma(U, {}_n\mathcal{O}/\mathcal{J}_V(p))$. We can assume that G is Stein and

$$\mathcal{I}_V(p-1) = \sum_{i=1}^{k} {}_n\mathcal{O}\, s_i$$

for some holomorphic functions $s_1, \ldots, s_k$ on G. Consider the following commutative diagram of quotient maps

$$\begin{array}{ccc} {}_n\mathcal{O} & \xrightarrow{\ \varphi\ } & {}_n\mathcal{O}/\mathcal{I}_V(p) \\ & \psi \searrow \quad \swarrow \eta & \\ & {}_n\mathcal{O}/\mathcal{I}_V(p-1) & \end{array} .$$

By induction hypothesis, $\eta(f_\nu)$ converges to some element f' of $\Gamma(V, {}_n\mathcal{O}/\mathcal{I}_V(p-1))$. By applying the open mapping theorem to the map

$$\Gamma(G, {}_n\mathcal{O}) \longrightarrow \Gamma\big(V, {}_n\mathcal{O}/\mathcal{I}_V(p-1)\big)$$

induced by ψ, we can find

$$g_\nu, g \in \Gamma(G, {}_n\mathcal{O})$$

such that

$$\psi(g_\nu) = \eta(f_\nu)$$

$$\psi(g) = f'$$

$$g_\nu \longrightarrow g .$$

Since it suffices to show that $f_\nu - \varphi(g_\nu)$ converges in $\Gamma\big(V, {}_n\mathcal{O}/\mathcal{I}_V(p)\big)$, we can assume without loss of generality that $\eta(f_\nu) = 0$. For some holomorphic functions $a_{\nu i}$ on G

$$f^{(\nu)} = \varphi\Big(\sum_{i=1}^{k} a_{\nu i} s_i\Big) .$$

From the definition of $_p\|\cdot\|_L$, it follows that, for $|\alpha| = p$,

$$\Big(\sum_{i=1}^{k} a_{\nu i} D^{\alpha} s_i\Big)\Big| V$$

is a Cauchy sequence in ν . By the Closure-of-Modules Theorem [8, II.D.3] there exist holomorphic functions a_i on G such that, for $|\alpha| = p$,

$$\Big(\sum_{i=1}^{k} a_{\nu i} D^{\alpha} s_i\Big)\Big| V$$

converges to

$$\Big(\sum_{i=1}^{k} a_i D^{\alpha} s_i\Big)\Big| V$$

as $\nu \longrightarrow \infty$. Then f_ν converges to $\varphi\Big(\sum_{i=1}^{k} a_i s_i\Big)$ in $\Gamma\big(V, {}_n\mathcal{O}/\mathcal{I}_V(p)\big)$.

(2.4) Suppose X is a complex space. Define the <u>reduction order</u> of X as the smallest nonnegative integer p_0 such that, if

$$f \in \Gamma(U, \mathcal{O}_X)$$

for some open subset U of X and

$$f_x \in \mathfrak{m}_{X,x}^{p_0+1}$$

for all $x \in U$, then $f = 0$. A complex space is reduced if and only if its reduction order is zero. By (2.1) a relatively compact open subset of a complex space has finite reduction order.

Suppose the reduction order of X is $\leqq p < \infty$. Let U be a relatively compact open subset of a Stein open subset $\tilde{U}$ of X. For an element f of $\Gamma(U, \mathcal{O}_X)$ define $\|f\|_U$ as follows. Imbed $\tilde{U}$ as a subspace V of an open subset G of $\mathbb{C}^N$ by a holomorphic map Φ. Let f' be the element of $\Gamma(\Phi(U), \mathcal{O}_V)$ corresponding to f. Define

$$\|f\|_U = \inf\{{}_p\|f^*\|_{\Phi(U)} \mid f^* \in \Gamma(\Phi(U), {}_n\mathcal{O}/\mathcal{J}_V(p)) \text{ induces } f'\}.$$

Two different choices of $\tilde{U}$ or its embedding give rise to equivalent semi-norms. (However, semi-norms defined by different p's may not be equivalent.) When we write $\|\cdot\|_U$, we assume that a fixed U and a fixed Φ (and a fixed p) are chosen. Let $\Gamma_b(U, \mathcal{O}_X)$ denote the set of all $f \in \Gamma(U, \mathcal{O}_X)$ with $\|f\|_U < \infty$.

Suppose $f \in \Gamma\left(\Delta(t^0, \rho) \times U, \mathcal{O}_{\mathbb{C}^n \times X}\right)$. Let

$$f = \Sigma f_\nu \left(\frac{t-t^0}{\rho}\right)^\nu$$

be its power series expansion in t. The <u>Grauert norm</u> $\|f\|_{U,t^0,\rho}$ of f is defined as $\sup_\nu \|f_\nu\|_U$. Suppose $\mathfrak{U} = \{U_i\}$ is a collection of open subsets of X such that U_i is relatively compact in some Stein open subset of X.

For

$$\xi = \{\xi_{i_0 \cdots i_k}\} \in C^k(\Delta(t^0,\rho)\times \mathfrak{U}, \mathcal{O}_{\mathbb{C}^n \times X})$$

define

$$\|\xi\|_{\mathfrak{U},t^0,\rho} = \sup_{i_0,\ldots,i_k} \|\xi_{i_0\cdots i_k}\|_{U_{i_0}\cap\cdots\cap U_{i_k},t^0,\rho}$$

When $n = 0$, $\|\xi\|_{\mathfrak{U},t^0,\rho}$ is simply denoted by $\|\xi\|_{\mathfrak{U}}$. Let $C^k_b(\mathfrak{U}, \mathcal{O}_X)$ denote the set of all $\xi \in C^k(\mathfrak{U}, \mathcal{O}_X)$ with $\|\xi\|_{\mathfrak{U}} < \infty$.

For a holomorphic function g on $\Delta(t^0,\rho)$ with power series expansion

$$g = \sum g_\nu \left(\frac{t-t^0}{\rho}\right)^\nu ,$$

define

$$|g|_{t^0,\rho} = \sup_\nu |g_\nu| .$$

§3. Theorem B with Bounds

(3.1) Lemma. Suppose X is a complex subspace of an open subset Ω of $\mathbb{C}^N$ and $G_1 \subset\subset G_2 \subset\subset \Omega$ are open subsets with G_2 Stein. Let $H_i = X \cap G_i$ $(i = 1,2)$. Then there exists a continuous map

$$\psi : \Gamma(\Delta(t^0,\rho) \times H_2, \mathcal{O}_{\mathbb{C}^n \times X}) \longrightarrow \Gamma(\Delta(t^0,\rho) \times G_1, {}_{n+N}\mathcal{O})$$

linear over $\mathbb{C}[t]$ such that

i) $\theta\psi$ is the restriction map from $\Gamma(\Delta(t^0,\rho) \times H_2, \mathcal{O}_{\mathbb{C}^n \times X})$ to $\Gamma(\Delta(t^0,\rho) \times H_1, \mathcal{O}_{\mathbb{C}^n \times X})$ where $\theta : {}_{n+N}\mathcal{O} \longrightarrow \mathcal{O}_{\mathbb{C}^n \times X}$ is the quotient map.

ii) $\|\psi(\xi)\|_{G_1,t^0,\rho} \leq C\|\xi\|_{H_2,t^0,\rho}$ where C is a constant independent of ξ, t^0, and ρ.

Proof. By considering the power series expansion, we observe that, it suffices to prove the special case where $n = 0$.

Take open subsets $G_1 \subset\subset G' \subset\subset G'' \subset G_2$ with G'' Stein. Let $H' = X \cap G'$ and $H'' = X \cap G''$. Take $\xi \in \Gamma(H_2, \mathcal{O}_X)$. There exists $\eta \in \Gamma(G', \mathcal{O}_X)$ such that

$$\theta\eta = \xi | H'$$

$$\|\eta\|_{G'} \leq C \|\xi\|_{H''}$$

where C is independent of ξ and comes from applying the open mapping theorem to

$$\Gamma(G'', {}_N\mathcal{O}) \longrightarrow \Gamma(H'', \mathcal{O}_X) .$$

Let η' be the projection of η onto the orthogonal complement of the kernel of

$$\Gamma_{L^2}(G', {}_N\mathcal{O}) \longrightarrow \Gamma(H', \mathcal{O}_X) .$$

Define $\psi(\xi) = \eta'$. Then $\psi(\xi)$ is independent of the choice of η and satisfies the required conditions. Q.E.D.

(3.2) Lemma. Suppose X is a Stein complex space and $U_1 \subset\subset U_2 \subset\subset X$ are open subsets with U_2 Stein. Then the restriction map $r : \Gamma_b(U_2, \mathcal{O}_X) \longrightarrow \Gamma_b(U_1, \mathcal{O}_X)$ factors through a Hilbert space.

Proof. U_2 can be identified with a complex subspace of a Stein open subset Ω of some $\mathbb{C}^N$. Take open subsets $\overline{U_1} \subset G_1 \subset\subset G_2 \subset\subset G_3 \subset\subset \Omega$ with G_3 Stein. Let $H_3 = U_2 \cap G_3$. From (3.1) we obtain a continuous linear map

$$\psi : \Gamma(H_3, \mathcal{O}_X) \longrightarrow \Gamma(G_2, {}_N\mathcal{O})$$

such that the composite map of

$$\Gamma_b(U_2, \mathcal{O}_X) \xrightarrow{\text{restr.}} \Gamma(H_3, \mathcal{O}_X) \xrightarrow{\psi} \Gamma(G_2, {}_N\mathcal{O})$$

$$\xrightarrow{\text{restr.}} \Gamma_{L^2}(G_1, {}_N\mathcal{O}) \xrightarrow{\text{quot.}} \Gamma_b(U_1, \mathcal{O}_X)$$

is r. Hence r factors through the Hilbert space $\Gamma_{L^2}(G_1, {}_N\mathcal{O})$. Q.E.D.

(3.3) Lemma. Suppose X is a Stein complex space and $G_1 \subset\subset G_2 \subset\subset X$ are open subsets with G_2 Stein. Let $\mathfrak{U}_\nu = \{U_i^{(\nu)}\}$ be a finite Stein open covering of G_ν ($\nu = 1,2$) with $U_i^{(1)} \subset\subset U_i^{(2)}$. Then for $k \geq 1$ there exists a continuous map

$$\varphi : Z^k(\Delta(t^0,\rho) \times \mathfrak{U}_2, \mathcal{O}_{\mathbb{C}^n \times X}) \longrightarrow C^{k-1}(\Delta(t^0,\rho) \times \mathfrak{U}_1, \mathcal{O}_{\mathbb{C}^n \times X})$$

linear over $\mathbb{C}[t]$ such that $\delta\varphi$ agrees with the restriction map and $\|\varphi(\xi)\|_{\mathfrak{U}_1, t^0, \rho} \leq C\|\xi\|_{\mathfrak{U}_2, t^0, \rho}$ where C is a constant independent of ξ, t^0, and ρ.

Proof. By considering the power series expansion in t, we observe that it suffices to prove the special case where $n = 0$.

Take Stein open subsets $G_1 \subset\subset G' \subset\subset G'' \subset\subset G_2$. Take a finite Stein open covering $\mathfrak{U}' = \{U_i'\}$ (respectively $\mathfrak{U}'' = \{U_i''\}$) of G' (respectively G'') such that $U_i^{(1)} \subset\subset U_i' \subset\subset U_i'' \subset\subset U_i^{(2)}$.

Take $\xi \in Z^k(\mathfrak{U}_2, \mathcal{O}_X)$. There exists $\eta \in C^{k-1}(\mathfrak{U}', \mathcal{O}_X)$ such that

$$\delta\eta = \xi | \mathfrak{U}'$$

$$\|\eta\|_{\mathfrak{U}'} \leq C\|\xi\|_{\mathfrak{U}''}$$

where C is independent of ξ and comes from applying the open mapping theorem to

$$C^{k-1}(\mathfrak{U}'', \mathcal{O}_X) \longrightarrow Z^k(\mathfrak{U}'', \mathcal{O}_X) .$$

By (3.2) the restriction map

$$C_b^{k-1}(\mathfrak{U}', \mathcal{O}_X) \longrightarrow C_b^{k-1}(\mathfrak{U}_1, \mathcal{O}_X)$$

factors through a Hilbert space H. Let η' be the projection of η onto the orthogonal complement of the kernel of the composite map

$$H \longrightarrow C_b^{k-1}(\mathfrak{U}_1, \mathcal{O}_X) \xrightarrow{\delta} Z^k(\mathfrak{U}_1, \mathcal{O}_X) .$$

Define $\varphi(\xi) = \eta' | \mathfrak{U}_1$. Then $\varphi(\xi)$ is independent of the choice of η and satisfies the requirements. Q.E.D.

(3.4) <u>Proposition</u>. <u>Suppose</u> X <u>is a Stein complex space and</u> $G_1 \subset\subset G_2 \subset\subset X$ <u>are open subsets with</u> G_2 <u>Stein. Let</u> $\mathfrak{U}_\nu = \{U_i^{(\nu)}\}$ <u>be a finite Stein open covering of</u> G_ν $(\nu = 1,2)$ <u>with</u> $U_i^{(1)} \subset\subset U_i^{(2)}$. <u>Let</u> $\alpha : \mathcal{O}_{\Delta \times X}^{p_1} \longrightarrow \mathcal{O}_{\Delta \times X}^{p_2}$ <u>be a sheaf-homomorphism such that</u> Coker α <u>is flat with respect to</u> $\Delta \times X \longrightarrow \Delta$. <u>Then there exists an open neighborhood</u> Ω <u>of</u> 0 <u>in</u> Δ <u>satisfying the following</u>:

(a) <u>for</u> $\Delta(t^0, \rho) \subset \Omega$, <u>there exists</u>

$$\psi : \Gamma(\Delta(t^0,\rho) \times G_2, \operatorname{Im} \alpha) \longrightarrow \Gamma(\Delta(t^0,\rho) \times G_1, \mathcal{O}_{\Delta \times X}^{p_1})$$

<u>linear over</u> $\mathbb{C}[t]$ <u>such that</u> $\alpha\psi$ <u>agrees with the restriction map and</u> $\|\psi(\eta)\|_{G_1, t^0, \rho} \leq C \|\eta\|_{G_2, t^0, \rho}$;

(b) *for* $\Delta(t^0,\rho) \subset \Omega$ *and* $k \geqq 1$ *there exists*

$$\varphi : Z^k(\Delta(t^0,\rho)\times \mathfrak{U}_2, \operatorname{Im}\alpha) \longrightarrow C^{k-1}(\Delta(t^0,\rho)\times\mathfrak{U}_1, \operatorname{Im}\alpha)$$

linear over $\mathbb{C}[t]$ *such that* $\delta\varphi$ *agrees with the restriction map and* $\|\varphi(\xi)\|_{\mathfrak{U}_1,t^0,\rho} \leqq C\|\xi\|_{\mathfrak{U}_2,t^0,\rho}$;

where C *is a constant independent of* ξ, t^0, *and* ρ .

Proof. Consider first the special case where X is a polydisc. We can assume without loss of generality that there exists an exact sequence

$$0 \longrightarrow \mathcal{O}^{p_\ell}_{\Delta\times X} \longrightarrow \cdots \longrightarrow \mathcal{O}^{p_1}_{\Delta\times X} \xrightarrow{\alpha} \mathcal{O}^{p_0}_{\Delta\times X} .$$

Let $(a)_m$ (respectively $(b)_m$) denote (a) (respectively (b)) for the case $\ell \leqq m$. By using (1.6) and (1.3) to obtain local solutions of (a) and by piecing together these local solutions by Čech cohomology, we conclude that $(a)_1$ holds and that $(b)_{m-1}$ implies $(a)_m$. From (3.3), we conclude that $(a)_m$ and $(b)_{m-1}$ imply $(b)_m$. Hence the special case follows by induction on m .

For the general case, we prove (a) first. We can assume without loss of generality that

i) X is a complex subspace of an open polydisc P , and

ii) there exists a commutative diagram of sheaf-homomorphisms

$$\begin{array}{ccc} \mathcal{O}^{\tilde{p}_1}_{\Delta\times P} & \xrightarrow{\tilde{\alpha}} & \mathcal{O}^{p_0}_{\Delta\times P} \\ \text{quot.}\downarrow & & \downarrow\text{quot.} \\ \mathcal{O}^{p_1}_{\Delta\times X} & \xrightarrow{\alpha} & \mathcal{O}^{p_0}_{\Delta\times X} \end{array}$$

such that Coker $\tilde{\alpha}$ is isomorphic to Coker α under the quotient map.

Then (a) follows from (3.1) and the special case. For (b) let m be a positive integer such that no more than m members of $\mathfrak{N}_1$ can intersect. We can assume without loss of generality that we have an exact sequence

$$\mathcal{O}^{p_m}_{\Delta\times X} \longrightarrow \cdots \longrightarrow \mathcal{O}^{p_1}_{\Delta\times X} \xrightarrow{\alpha} \mathcal{O}^{p_0}_{\Delta\times X} .$$

Then (b) follows from (a) and (3.3). Q.E.D.

(3.5) When the flatness condition on Coker α is dropped in (3.4), the conclusions of (3.4) remain valid with the following modifications.

i) ψ and φ are linear over $\mathbb{C}$, but may not be linear over $\mathbb{C}[t]$.

ii) For a fixed t^0, ρ has to be sufficiently strictly small.

iii) C may depend on t^0 and ρ.

However, if $t_n - t_n^0$ is not a zero-divisor of any stalk of

Coker α , then C can be chosen to be independent of ρ_n .

(3.6) Suppose X is a Stein complex space, $\mathcal{F}$ is a coherent analytic sheaf on $\Delta \times X$, and $\varphi : \mathcal{O}^p_{\Delta\times X} \longrightarrow \mathcal{F}$ is a sheaf-epimorphism. Let $\Delta(t^0,\rho) \subset\subset \Delta$ and U be a relatively compact open subset of X . For $f \in \Gamma(\Delta(t^0,\rho)\times U, \mathcal{F})$ define the <u>Grauert norm</u> $\|f\|_{U,t^0,\rho}$ as the infimum of $\|\tilde{f}\|_{U,t^0,\rho}$ where

$$\tilde{f} \in \Gamma(\Delta(t^0,\rho)\times U, \mathcal{O}^p_{\Delta\times X})$$

$$\varphi(\tilde{f}) = f .$$

A different choice of φ would give an equivalent semi-norm. When we use such a Grauert norm. we assume that a fixed φ is chosen.

Suppose $\mathfrak{U} = \{U_i\}$ is a collection of relatively compact open subsets of X . For

$$\xi = \{\xi_{i_0\cdots i_k}\} \in C^k(\Delta(t^0,\rho)\times\mathfrak{U}, \mathcal{F})$$

define

$$\|\xi\|_{\mathfrak{U},t^0,\rho} = \sup_{i_0,\ldots,i_k} \|\xi_{i_0\cdots i_k}\|_{U_{i_0}\cap\cdots\cap U_{i_k},t^0,\rho} .$$

When $t^0 = 0$, $\|\xi\|_{\mathfrak{U},t^0,\rho}$ is simply denoted by $\|\xi\|_{\mathfrak{U},\rho}$.

(3.7) <u>Proposition</u> (<u>Theorem B with Bounds</u>). <u>Suppose</u> X, G_ν

$\mathfrak{U}_\nu$ ($\nu = 1,2$) are as in (3.4). Suppose $\mathcal{F}$ is a coherent analytic sheaf on $\Delta \times X$ flat with respect to $\pi: \Delta \times X \longrightarrow \Delta$ and suppose $\alpha: \mathcal{O}^p_{\Delta \times X} \longrightarrow \mathcal{F}$ is a sheaf-epimorphism. Then there exists an open neighborhood Ω of 0 in Δ satisfying the following:

(a) for $\Delta(t^0,\rho) \subset \Omega$ there exists

$$\psi: \Gamma(\Delta(t^0,\rho) \times G_2, \mathcal{F}) \longrightarrow \Gamma(\Delta(t^0,\rho) \times G_1, \mathcal{O}^p_{\Delta \times X})$$

linear over $\mathbb{C}[t]$ such that $\alpha\psi$ equals the restriction map and $\|\psi(\eta)\|_{G_1,t^0,\rho} \leq C\|\eta\|_{G_2,t^0,\rho}$;

(b) for $\Delta(t^0,\rho) \subset \Omega$ there exists

$$\varphi: Z^k(\Delta(t^0,\rho) \times \mathfrak{U}_2, \mathcal{F}) \longrightarrow C^{k-1}(\Delta(t^0,\rho) \times \mathfrak{U}_1, \mathcal{F})$$

linear over $\mathbb{C}[t]$ such that $\delta\varphi$ equals the restriction map and $\|\varphi(\xi)\|_{\mathfrak{U}_1,t^0,\rho} \leq C\|\xi\|_{\mathfrak{U}_2,t^0,\rho}$;

where C is a constant independent of ξ, t^0, and ρ. (When $\mathcal{F}$ is not π-flat, for a fixed t^0, $\mathbb{C}$-linear ψ and φ exist for ρ sufficiently strictly small but may not be linear over $\mathbb{C}[t]$ and C may depend on t^0 and ρ. If $t_n - t_n^0$ is not a zero-divisor of any stalk of $\mathcal{F}$, C can be chosen to be independent of ρ_n.)

Proof. Follows from (1.6), (1.3), and (3.4). Q.E.D.

§4. Leray's Theorem with Bounds

(4.1) First we examine the diagram-chasing proof of the usual Leray's theorem without bounds. Suppose $\mathcal{F}$ is a coherent analytic sheaf on a complex space X and

$$\mathfrak{U} = \{U_\alpha\}_{\alpha \in A}$$

$$\mathfrak{V} = \{V_i\}_{i \in I}$$

are Stein open coverings of X such that $\mathfrak{V}$ refines $\mathfrak{U}$ with an index map $\tau : I \longrightarrow A$. Leray's theorem states that the restriction map

$$H^\ell(\mathfrak{U}, \mathcal{F}) \longrightarrow H^\ell(\mathfrak{V}, \mathcal{F})$$

is an isomorphism. Define $C^{\mu,\nu}(\mathfrak{U}, \mathfrak{V})$ as the set of all

$$\xi = \{ \xi^{\alpha_0 \ldots \alpha_\mu}_{i_0 \ldots i_\nu} \}$$

such that

$$\xi^{\alpha_0 \ldots \alpha_\mu}_{i_0 \ldots i} \in \Gamma(U_{\alpha_0} \cap \ldots \cap U_{\alpha_\mu} \cap V_{i_0} \cap \ldots \cap V_i, \mathcal{F})$$

is skew-symmetric in $\alpha_0, \ldots, \alpha_\mu$ and skew-symmetric in $i_0, \ldots, i_\nu$. Define

$$\delta_1 : C^{\mu,\nu}(\mathfrak{U}, \mathfrak{V}) \longrightarrow C^{\mu+1,\nu}(\mathfrak{U}, \mathfrak{V})$$

$$\delta_2 : C^{\mu,\nu}(\mathfrak{U}, \mathfrak{V}) \longrightarrow C^{\mu,\nu+1}(\mathfrak{U}, \mathfrak{V})$$

$$\theta_1 : C^{\nu}(\mathfrak{V}, F) \longrightarrow C^{0,\nu}(\mathfrak{U}, \mathfrak{V})$$

$$\theta_2 : C^{\mu}(\mathfrak{U}, F) \longrightarrow C^{\mu,0}(\mathfrak{U}, \mathfrak{V})$$

as follows:

$$(\delta_1 \xi)^{\alpha_0 \dots \alpha_{\mu+1}}_{i_0 \dots i_\nu} = \sum_{\lambda=0}^{\mu+1} (-1)^\lambda \xi^{\alpha_0 \dots \widehat{\alpha}_\lambda \dots \alpha_{\mu+1}}_{i_0 \dots i_\nu}$$

$$(\delta_2 \xi)^{\alpha_0 \dots \alpha_\mu}_{i_0 \dots i_{\nu+1}} = \sum_{\lambda=0}^{\nu+1} (-1)^\lambda \xi^{\alpha_0 \dots \alpha_\mu}_{i_0 \dots \widehat{i}_\lambda \dots i_{\nu+1}}$$

$$(\theta_1 \xi)^{\alpha_0}_{i_0 \dots i_\nu} = \xi_{i_0 \dots i_\nu} \Big|_{U_{\alpha_0} \cap V_{i_0} \cap \dots \cap V_{i_\nu}}$$

$$(\theta_2 \xi)^{\alpha_0 \dots \alpha_\mu}_{i_0} = \xi^{\alpha_0 \dots \alpha_\mu} \Big|_{U_{\alpha_0} \cap \dots \cap U_{\alpha_\mu} \cap V_{i_0}}$$

Consider the following commutative diagram:

$$\begin{array}{ccccccccc}
 & & 0 & & 0 & & 0 & & \\
 & & \downarrow & & \downarrow & & \downarrow & & \\
0 & \longrightarrow & \Gamma(X,\mathcal{F}) & \longrightarrow & C^0(\mathfrak{U},\mathcal{F}) & \xrightarrow{\delta} & C^1(\mathfrak{U},\mathcal{F}) & \xrightarrow{\delta} & \cdots \\
 & & \downarrow & & \downarrow \theta_2 & & \downarrow \theta_2 & & \\
0 & \longrightarrow & C^0(\mathcal{V},\mathcal{F}) & \xrightarrow{\theta_1} & C^{0,0}(\mathfrak{U},\mathcal{V}) & \xrightarrow{\delta_1} & C^{1,0}(\mathfrak{U},\mathcal{V}) & \xrightarrow{\delta_1} & \cdots \\
 & & \downarrow \delta & & \downarrow \delta_2 & & \downarrow \delta_2 & & \\
0 & \longrightarrow & C^1(\mathcal{V},\mathcal{F}) & \xrightarrow{\theta_1} & C^{0,1}(\mathfrak{U},\mathcal{V}) & \xrightarrow{\delta_1} & C^{1,1}(\mathfrak{U},\mathcal{V}) & \xrightarrow{\delta_1} & \cdots \\
 & & \downarrow \delta & & \downarrow \delta_2 & & \downarrow \delta_2 & & \\
 & & \vdots & & \vdots & & \vdots & &
\end{array}$$

A sequence

$$f_{*,\ell},\ f_{0,\ell-1},\ f_{1,\ell-2},\ \ldots,\ f_{\ell-2,1},\ f_{\ell-1,0},\ f_{\ell,*}$$

is called a <u>zigzag sequence</u> if

$$f_{*,\ell} \in Z^\ell(\mathcal{V},\mathcal{F})$$

$$f_{\nu,\ell-\nu-1} \in C^{\nu,\ell-\nu-1}(\mathfrak{U},\mathcal{V})$$

$$f_{\ell,*} \in Z^\ell(\mathfrak{U},\mathcal{F})$$

$$\theta_1 f_{*,\ell} + \delta_2 f_{0,\ell-1} = 0$$

$$\delta_1 f_{\nu-1,\ell-\nu} + \delta_2 f_{\nu,\ell-\nu-1} = 0$$

$$\theta_2 f_{\ell,*} + \delta_1 f_{\ell-1,0} = 0 .$$

The proof of Leray's theorem consists in showing that the

correspondence

(cohomology class of $f_{*,\ell}$) $\longleftrightarrow$ (cohomology class of $f_{\ell,*}$)

(where $f_{*,\ell}$ and $f_{\ell,*}$ are the end-terms of a zigzag sequence) defines an isomorphism between $H^{\ell}(\mathcal{V}, \mathcal{F})$ and $H^{\ell}(\mathfrak{U}, \mathcal{F})$ which agrees with the map $H^{\ell}(\mathfrak{U}, \mathcal{F}) \longrightarrow H^{\ell}(\mathcal{V}, \mathcal{F})$ defined by restriction. This is shown in the following three steps.

a) For every $f_{*,\ell} \in Z^{1}(\mathcal{V}, \mathcal{F})$ one can construct by the Theorem B of Cartan-Oka a zigzag sequence with $f_{*,\ell}$ as the first term. Likewise, for every $f_{\ell,*} \in Z^{1}(\mathfrak{U}, \mathcal{F})$ there is a zigzag sequence with $f_{\ell,*}$ as the last term.

b) If $f_{*,\ell}$, $f_{\nu,\ell-\nu-1}$ $(0 \leqq \nu < \ell-1)$, $f_{\ell,*}$ is a zigzag sequence, then $f_{*,\ell} \in B^{\ell-1}(\mathcal{V}, \mathcal{F})$ if and only if $f_{\ell,*} \in B^{\ell-1}(\mathfrak{U}, \mathcal{F})$. For the "if" part, let

$$f_{*,\ell} = \delta g_{*,\ell-1}$$

with

$$g_{*,\ell-1} \in C^{\ell-1}(\mathcal{V}, \mathcal{F}) .$$

Construct inductively

$$g_{\nu,\ell-\nu-2} \in C^{\nu,\ell-\nu-2}(\mathfrak{U}, \mathcal{V}) \qquad (0 \leqq \nu \leqq \ell-2)$$

$$g_{\ell-1,*} \in C^{\ell-1}(\mathfrak{U}, \mathcal{F})$$

such that

$$\delta_2 g_{0,\ell-2} = \theta_1 g_{*,\ell-1} + f_{0,\ell-1}$$

$$\delta_2 g_{\nu,\ell-\nu-2} = \delta_1 g_{\nu-1,\ell-\nu-1} + f_{\nu,\ell-\nu-1} \quad (1 \leqq \nu \leqq \ell - 2)$$

$$\theta_2 g_{\ell-1,*} = \delta_1 g_{\ell-2,0} + f_{\ell-1,0} .$$

The construction is possible, because of the Theorem B of Cartan-Oka and the following equations:

$$\delta_2(\theta_1 g_{*,\ell-1} + f_{0,\ell-1}) = \theta_1 \delta g_{*,\ell-1} + \delta_2 f_{0,\ell-1}$$

$$= \theta_1 f_{*,\ell} + \delta_2 f_{0,\ell-1} = 0$$

$$\delta_2(\delta_1 g_{\nu-1,\ell-\nu-1} + f_{\nu,\ell-\nu-1}) = \delta_1\delta_2 g_{\nu-1,\ell-\nu-1} + \delta_2 f_{\nu,\ell-\nu-1}$$

$$= \delta_1(\delta_1 g_{\nu-2,\ell-\nu-2} + f_{\nu-1,\ell-\nu-2}) + \delta_2 f_{\nu,\ell-\nu-1}$$

$$= \delta_1 f_{\nu-1,\ell-\nu-2} + \delta_2 f_{\nu,\ell-\nu-1} = 0 .$$

Finally, since

$$\theta_2 \delta g_{\ell-1,*} = \delta_1 \theta_2 g_{\ell-1,*} = \delta_1(\delta_1 g_{\ell-2,0} + f_{\ell-1,0})$$

$$= \delta_1 f_{\ell-1,0} = -\theta_2 f_{\ell,*} ,$$

it follows that $f_{\ell,*} = \delta(-g_{\ell-1,*})$. The proof of the "only if" part is analogous.

c) The correspondence agrees with the restriction map, because, if

$$\xi = \{\xi_{\alpha_0 \ldots \alpha_\ell}\} \in Z^\ell(\mathfrak{U},\mathcal{F}) ,$$

the sequence $f_{*,\ell}, f_{\nu,\ell-\nu-1}$ $(0 \leq \nu < \ell)$, $f_{\ell,*}$ defined by

$$(f_{*,\ell})_{i_0 \dots i_\ell} = \xi_{\tau(i_0)\dots\tau(i_\ell)} \Big| V_{i_0} \cap \dots \cap V_{i_\ell}$$

$$(f_{\nu,\ell-\nu-1})^{\alpha_0 \dots \alpha_\nu}_{i_0 \dots i_{\ell-\nu-1}}$$

$$= \xi_{\alpha_0 \dots \alpha_\nu \tau(i_0)\dots\tau(i_{\ell-\nu-1})} \Big| U_{\alpha_0} \cap \dots \cap U_{\alpha_\nu} \cap V_{i_0} \cap \dots V_{i_{\ell-\nu-1}}$$

$$f_{\ell,*} = \xi$$

is a zigzag sequence.

(4.2) Proposition (Leray's Theorem with Bounds). Suppose X is a complex space and $\mathcal{F}$ is a coherent analytic sheaf on $\Delta \times X$ flat with respect to $\pi: \Delta \times X \longrightarrow \Delta$. Suppose $\mathfrak{U}, \mathfrak{V}, \mathfrak{U}', \mathfrak{V}'$ are finite collections of Stein open subsets of X, each of which is relatively compact in some (but in general not the same) Stein open subset of X.

(a) (Surjectivity). Suppose $\mathfrak{V}' << \mathfrak{U}, \mathfrak{V}' << \mathfrak{V}$, and $|\mathfrak{U}| \subset\subset |\mathfrak{V}|$. Then there exists an open neighborhood Ω of 0 in Δ satisfying the following. For $\Delta(t^0,\rho) \subset \Omega$ and $\ell \geq 1$ there exists

$$(\varphi,\psi): Z^\ell(\Delta(t^0,\rho) \times \mathfrak{V}, \mathcal{F})$$

$$\longrightarrow Z^\ell(\Delta(t^0,\rho) \times \mathfrak{U}, \mathcal{F}) \oplus C^{\ell-1}(\Delta(t^0,\rho) \times \mathfrak{V}', \mathcal{F})$$

linear over $\mathbb{C}[t]$ such that

$$\xi = \varphi(\xi) + \delta\psi(\xi) \quad \underline{\text{on}} \quad \Delta(t^0,\rho)\times\mathcal{V}'$$

$$\mathrm{Max}\Big(\|\varphi(\xi)\|_{\mathfrak{U},t^0,\rho}, \|\psi(\xi)\|_{\mathcal{V}',t^0,\rho}\Big) \leq C\,\|\xi\|_{\mathcal{V},t^0,\rho}$$

where C is a constant independent of ξ, t^0, and ρ.

(b) (Injectivity). Suppose $\mathfrak{U} << \mathfrak{U}'$, $\mathcal{V} << \mathfrak{U}'$; and $|\mathfrak{U}| \subset\subset |\mathcal{V}|$. Then there exists an open neighborhood Ω of 0 in Δ satisfying the following. For $\Delta(t^0,\rho) \subset \Omega$ and $\ell \geq 1$ there exists a map θ from

$$\{(\xi,\eta) \in Z^\ell(\Delta(t^0,\rho)\times\mathfrak{U}',\mathcal{F}) \oplus C^{\ell-1}(\Delta(t^0,\rho)\times\mathcal{V},\mathcal{F}) \;\Big|\; \xi = \delta\eta \text{ on } \Delta(t^0,\rho)\times\mathcal{V}\}$$

to $C^{\ell-1}(\Delta(t^0,\rho)\times\mathfrak{U},\mathcal{F})$ linear over $\mathbb{C}[t]$ such that

$$\delta\theta(\xi,\eta) = \xi \quad \underline{\text{on}} \quad \Delta(t^0,\rho)\times\mathfrak{U}$$

$$\|\theta(\xi,\eta)\|_{\mathfrak{U},t^0,\rho} \leq C\,\mathrm{Max}\left(\|\xi\|_{\mathfrak{U}',t^0,\rho}, \|\eta\|_{\mathcal{V},t^0,\rho}\right),$$

where C is a constant independent of ξ, η, t^0, and ρ.

(When $\mathcal{F}$ is not π-flat, for a fixed t^0, $\mathbb{C}$-linear φ, ψ, θ exist for ρ sufficiently strictly small but may not be linear over $\mathbb{C}[t]$ and C may depend on t^0 and ρ. If $t_n - t_n^0$ is not a zero-divisor of any stalk of $\mathcal{F}$, then can be chosen to be independent of ρ_n.)

Proof. Follows the same line as in (4.1) except that (3.7)(b) (Theorem B with bounds) is used instead of the Theorem B of Cartan-Oka. Q.E.D.

Y-T. Siu

§5. Extension of Cohomology Classes

(5.1) Andreotti-Grauert [1] proved that a k-dimensional cohomology class with coefficients in a coherent analytic sheaf $\mathcal{F}$ can be extended across a strongly k-pseudoconvex boundary and across a strongly (r-k-1)-pseudoconcave boundary if codh $\mathcal{F} \geq r$. We need the corresponding result with bounds. So we are going to examine one key point of Andreotti-Grauert's result in such a way as can be carried over to the situation with bounds. The precise statement of the situation with bounds is given in (5.2).

Suppose X is a complex space and $\mathcal{F}$ is a coherent analytic sheaf on $\Delta \times X$. Suppose $\ell \geq 0$ and X_1, $D \subset\subset X$ are open subsets. Assume the following.

$$(\dagger)_\ell \begin{cases} H^\ell(\Delta \times (X_1 \cap D), \mathcal{F}) = 0 & \text{in case } \ell \geq 1 \\ H^\ell(\Delta \times D, \mathcal{F}) \longrightarrow H^\ell(\Delta \times (X_1 \cap D), \mathcal{F}) & \text{is surjective in case } \ell = 0 \\ H^{\ell+1}(\Delta \times D, \mathcal{F}) \longrightarrow H^{\ell+1}(\Delta \times (X_1 \cap D), \mathcal{F}) & \text{is injective} \end{cases}$$

Let $X_2 = X_1 \cup D$. From the Mayer-Vietoris sequence of $\mathcal{F}$ on $(\Delta \times X_1) \cup (\Delta \times D)$ it follows that the restriction map

$$H^\nu(\Delta \times X_2, \mathcal{F}) \longrightarrow H^\nu(\Delta \times X_1, \mathcal{F})$$

is surjective for $\nu = \ell$ and is injective for $\nu = \ell + 1$. Such an argument cannot be carried over to the case with bounds. So we look at the conclusion in another way. Introduce the following additional assumption

$$(X_1 - D)^- \cap (D - X_1)^- = \emptyset .$$

With this additional assumption, it is easy to see that one can choose a Stein open covering $\mathfrak{U}_1$ (respectively $\mathfrak{U}_2$) of X_1 (respecively D) such that $\mathfrak{U}_{12} := \mathfrak{U}_1 \cap \mathfrak{U}_2$ covers $X_1 \cap D$ and

(*) $U_1 \cap U_2 = \emptyset$ whenever $U_i \in \mathfrak{U}_1 \cup \mathfrak{U}_2 - \mathfrak{U}_i$ $(i = 1, 2)$.

Let $\mathfrak{U} = \mathfrak{U}_1 \cap \mathfrak{U}_2$. We are going to show, in a way that can be carried over to the case with bounds, that

$$H^\nu(\Delta \times \mathfrak{U}, \mathcal{F}) \longrightarrow H^\nu(\Delta \times \mathfrak{U}_1, \mathcal{F})$$

is surjective for $\nu = \ell$ and is injective for $\nu = \ell + 1$.

Let us consider surjectivity first. The case $\ell = 0$ is clear. Assume $\ell \geq 1$. Suppose $\xi \in Z^\ell(\Delta \times \mathfrak{U}_1, \mathcal{F})$. Since $H^\ell(\Delta \times \mathfrak{U}_{12}, \mathcal{F}) = 0$, there exists $\eta \in C^{\ell-1}(\Delta \times \mathfrak{U}_{12}, \mathcal{F})$ such that

$$\xi | \Delta \times \mathfrak{U}_{12} = \delta \eta .$$

Extend η trivially to $\tilde{\eta} \in C^{\ell-1}(\Delta \times \mathfrak{U}_1, \mathcal{F})$. Since $\xi - \delta\tilde{\eta}$ is 0 on $\Delta \times \mathfrak{U}_{12}$, by (*), $\xi - \delta\tilde{\eta}$ can be trivially extended to some element of $Z^\ell(\Delta \times \mathfrak{U}, \mathcal{F})$.

Now let us consider injectivity. Suppose $\xi \in Z^{\ell+1}(\Delta \times \mathfrak{U}, \mathcal{F})$ such that

$$\xi | \Delta \times \mathfrak{U}_1 = \delta \eta'$$

for some $\eta' \in C^{\ell}(\Delta \times \mathfrak{U}_1, \mathcal{F})$. Then there exists $\eta'' \in C^{\ell}(\Delta \times \mathfrak{U}_2, \mathcal{F})$ such that

$$\xi = \delta\eta'' \text{ on } \Delta \times \mathfrak{U}_{12} .$$

Then

$$\delta(\eta' - \eta'') = 0 \text{ on } \Delta \times \mathfrak{U}_{12} .$$

In case $\ell \geqq 1$, there exists $\zeta \in C^{\ell-1}(\Delta \times \mathfrak{U}_{12}, \mathcal{F})$ such that

$$\delta\zeta = \eta' - \eta'' \text{ on } \Delta \times \mathfrak{U}_{12} .$$

Extend ζ trivially to $\tilde{\zeta} \in C^{\ell-1}(\Delta \times \mathfrak{U}_2, \mathcal{F})$. Define $\eta \in C^{\ell}(\Delta \times \mathfrak{U}, \mathcal{F})$ by

$$\begin{cases} \eta = \eta' & \text{on } \Delta \times \mathfrak{U}_1 \\ \eta = \eta'' + \delta\tilde{\zeta} & \text{on } \Delta \times \mathfrak{U}_2 . \end{cases}$$

Then $\xi = \delta\eta$ on $\Delta \times \mathfrak{U}$. In case $\ell = 0$, $(\eta' - \eta'')|\Delta \times \mathfrak{U}_{12}$ can be extended to $\theta \in Z^{\ell}(\Delta \times \mathfrak{U}_2, \mathcal{F})$. Define $\eta \in C^{\ell}(\Delta \times \mathfrak{U}, \mathcal{F})$ by

$$\begin{cases} \eta = \eta' & \text{on } \Delta \times \mathfrak{U}_1 \\ \eta = \eta'' + \theta & \text{on } \Delta \times \mathfrak{U}_2 . \end{cases}$$

Then $\xi = \delta\eta$ on $\Delta \times \mathfrak{U}$.

Now we consider the strongly p-pseudoconvex case. Assume that there exist

i) a biholomorphic map γ embedding an open neighborhood

$\tilde{D} \subset\subset X$ of D onto a complex subspace of an open poly-disc P of $\mathbb{C}^N$

ii) real-valued C^2 functions $\varphi_1 \geq \varphi_2$ on X such that

a) $X_1 = \{\varphi_1 < 0\}$

b) $D = \tilde{D} \cap \{\varphi_2 < 0\}$

c) $\operatorname{Supp}(\varphi_1 - \varphi_2) \subset\subset \tilde{D}$

d) $\varphi_i \circ \gamma^{-1}$ is the restriction of a C^2 function $\tilde{\varphi}_i$ on P $(i = 1,2)$ whose restriction to $P \cap (\{x\} \times \mathbb{C}^{N-p+1})$ is strongly plurisubharmonic for every $x \in \mathbb{C}^{p-1}$

e) there exists an exact sequence

(#)

$$0 \longrightarrow {}_{n+N}\mathcal{O}^{p_m} \longrightarrow \cdots \longrightarrow {}_{n+N}\mathcal{O}^{p_0} \longrightarrow R^0\tilde{\gamma}_*(\mathcal{F}|\Delta \times D) \longrightarrow 0$$

on $\Delta \times P$, where $\tilde{\gamma} : \Delta \times D \longrightarrow \Delta \times G$ is defined by the identity map of Δ and γ.

Then, by an intermediate result of Andreotti-Grauert [1, p.217, Prop. 11],

$$H^{\ell}(P \cap \{\tilde{\varphi}_i < 0\}, {}_N\mathcal{O}) = 0 \quad \text{for} \quad \ell \geq p .$$

It follows that $(\dagger)_{\ell}$ is satisfied for $\ell \geq p$. In considering the situation with bounds, one has to apply the open mapping theorem to

$$C^{\ell-1}(\mathcal{V}, {}_N\mathcal{O}) \longrightarrow Z^{\ell}(\mathcal{V}, {}_N\mathcal{O}) \quad (\ell \geq p)$$

for a Stein open covering $\mathcal{V}$ of $P \cap \{\tilde{\varphi}_1 < 0\}$ in a way analogous to the proof of (3.3) and one has to apply (3.4)(a) to the sheaf-homomorphisms of (#).

For the strongly p-pseudoconcave case, we can assume that there exist γ, and φ_1, φ_2 as in the p-pseudoconvex case such that

$$X_1 = \{\varphi_2 > 0\}$$

$$X_1 \cup D = \{\varphi_1 > 0\}$$

and conditions c), d), e) of the p-pseudoconvex case are satisfied, where $m \leqq n + N - \operatorname{codh} \mathcal{F}$ on $\Delta \times D$, and, moreover, the Hartogs' figure

$$Q_1 := \left(P' \times (P'' - \overline{H''})\right) \cup (H_1' \times P'')$$

is contained in $P \cap \{\tilde{\varphi}_2 > 0\}$, where $P = P' \times P''$ with $P' \subset \mathbb{C}^p$ and $P'' \subset \mathbb{C}^{N-p}$, $\emptyset \neq H_1' \subset\subset P'$, $H'' \subset\subset P''$ are open polydiscs. Again, by the proof of an intermediate result of Andreotti-Grauert [1, p.222, Prop. 12],

$$H^{\ell}(P, {}_N\mathcal{O}) \longrightarrow H^{\ell}(P \cap \{\tilde{\varphi}_1 > 0\}, {}_N\mathcal{O})$$

is surjective for $\ell < N - p$. Hence the first two conditions of $(\dagger)_\ell$ are satisfied for $\ell < n + N - m - p$. To obtain the third condition of $(\dagger)_\ell$ for $\ell = n + N - m - p - 1$, we argue as follows. For every given $\varepsilon > 0$, we can assume that γ, $\tilde{\varphi}_1$, H_1', H'' are so chosen that, for some open polydisc $H_1' \subset H_2' \subset\subset P'$, the Hartogs' figure

$$Q_2 := \left(P' \times (P'' - \overline{H''})\right) \cup (H_2' \times P'')$$

contains $P \cap \{\tilde{\varphi}_1 > \varepsilon\}$ (cf. [1, pp.219-220]). It suffices to show that the restriction map

$$\tau : H^{N-p}(Q_2, {}_N\mathcal{O}) \longrightarrow H^{N-p}(Q_1, {}_N\mathcal{O})$$

is injective. Suppose $P'' = \prod_{j=1}^{N-p} P''_j$ and $H'' = \prod_{j=1}^{N-p} H''_j$ with $H''_j, P''_j \subset \mathbb{C}$. Let

$$U_0^{(1)} = H'_1 \times P''$$

$$U_0^{(2)} = H'_2 \times P''$$

$$U_j = P' \times \prod_{\mu=1}^{j-1} P''_\mu \times (P''_j - H''_j) \times \prod_{\nu=j+1}^{N-p} P''_\nu \quad (1 \leq j \leq N-p)$$

$$\mathfrak{U}_1 = \{U_0^{(1)}, U_1, \ldots, U_{N-p}\}$$

$$\mathfrak{U}_2 = \{U_0^{(2)}, U_1, \ldots, U_{N-p}\}$$

An element $\xi_i \in H^{N-p}(\mathfrak{U}_i, {}_N\mathcal{O})$ $(i = 1,2)$ is represented by a holomorphic function f_i on $U_0^{(i)} \cap U_1 \cap \ldots \cap U_{N-p}$, and $\xi_i = 0$ if and only if the holomorphic function

$$\sum_{\alpha_{p+1},\ldots,\alpha_N \leq -1} f^{(i)}_{\alpha_{p+1}\ldots\alpha_N} z_{p+1}^{\alpha_{p+1}} \ldots z_N^{\alpha_N}$$

on $U_0^{(i)} \cap U_1 \cap \ldots \cap U_{N-p}$ can be extended to a holomorphic function g_i on $U_1 \cap \ldots \cap U_{N-p}$ where

$$f_i = \sum_{\alpha_{p+1},\ldots,\alpha_N = -\infty}^{\infty} f^{(i)}_{\alpha_{p+1}\ldots\alpha_N} z_{p+1}^{\alpha_{p+1}} \ldots z_N^{\alpha_N}$$

is the Laurent series expansion in the last $N - p$ coordinates $z_{p+1}, \ldots, z_N$ of $\mathbb{C}^N$. Hence τ is injective. When

$\xi_i = 0$, the function f_i when regarded as an element of $Z^{N-p}(\mathfrak{U}_i, {}_N\mathcal{O})$ is the coboundary of

$$\{h^{(i)}_{j_0 \dots j_{N-p-1}}\} \in C^{N-p-1}(\mathfrak{U}_i, {}_N\mathcal{O})$$

where

$$h^{(i)}_{1 \dots (N-p)} = g_i$$

$$h^{(i)}_{01 \dots \hat{\nu} \dots (N-p)} = \sum_{\substack{\alpha_{p+1} \geq 0 \\ \vdots \\ \alpha_{p+\nu-1} \geq 0 \\ \alpha_{p+\nu} \leq -1}} f^{(i)}_{\alpha_{p+1} \dots \alpha_N} z_{p+1}^{\alpha_{p+1}} \dots z_N^{\alpha_N} .$$

It follows that, if f_1 is the restriction of f_2 and f_1 is the coboundary of some

$$\{s_{j_0 \dots j_{N-p-1}}\} \in C^{N-p-1}(\mathfrak{U}_1, {}_N\mathcal{O})$$

and if

$$s_{1 \dots (N-p-1)} = \sum_{\alpha_{p+1}, \dots, \alpha_N = -\infty}^{\infty} \sigma_{\alpha_{p+1} \dots \alpha_N} z_{p+1}^{\alpha_{p+1}} \dots z_N^{\alpha_N} ,$$

then

$$g_2 = \sum_{\alpha_{p+1}, \dots, \alpha_N \leq -1} \sigma_{\alpha_{p+1} \dots \alpha_N} z_{p+1}^{\alpha_{p+1}} \dots z_N^{\alpha_N}$$

and f_2 is the coboundary of $\{h^{(2)}_{j_0 \dots j_{N-p-1}}\}$. Hence, if the sup norms of $s_{j_0 \dots j_{N-p-1}}$ and f_2 are all $\leq e$, then the sup norms of $h^{(2)}_{j_0 \cdots j_{N-p-1}}$ are $\leqq$ Ce , where C is a constant depending only on P'' and H''. Now, to get the situation with bounds, one need only apply the open mapping theorem to

$$C^{\ell-1}(\mathcal{V}, {}_N\mathcal{O}) \longrightarrow Z^{\ell}(\mathcal{V}, {}_N\mathcal{O}) \quad (\ell < N-p)$$

for a Stein open covering $\mathcal{V}$ of $P \cap \{\tilde{\varphi}_2 > 0\}$ in a way analogous to the proof of (3.3) and apply (3.4)(a) to the sheaf-homomorphisms of (#).

(5.2) Proposition (Extension of Cohomology Classes with Bounds). Suppose X is a complex space, $\varphi: X \longrightarrow (a_*, b_*) \subset (-\infty, \infty)$ is a C^2 function, $a_* < a_\# < b_\# < b_*$, $p, q \geqq 1$, $r \geqq 0$, and $\mathcal{F}$ is a coherent analytic sheaf on $\Delta \times X$ such that

i) $\{\varphi < b\}^- = \{\varphi \leqq b\}$ for $b_\# < b < b_*$

ii) φ is strongly p-pseudoconvex on $\{\varphi > b_\#\}$ and strongly q-pseudoconvex on $\{\varphi < a_\#\}$

iii) the natural projection $(\operatorname{Supp}\mathcal{F}) \cap (\Delta \times \{a \leqq \varphi \leqq b\}) \longrightarrow$ is proper for $a_* < a < b < b_*$

iv) $\mathcal{F}$ is flat with respect to the projection $\pi: \Delta \times X \longrightarrow \Delta$

v) $\operatorname{codh} \mathcal{F} \geqq r$ on $\Delta \times \{\varphi < a_\#\}$

For $a_* < a < b < b_*$ let $X^b_a = \{a < \varphi < b\}$. Suppose

$\tilde{\Omega} \subset\subset \Delta$ is an open neighborhood of 0 and $\mathfrak{U}, \mathfrak{V}, \mathfrak{U}', \mathfrak{V}'$ are finite collections of Stein open subsets of X, each of which is relatively compact in some (but in general not the same) Stein open subset of X. Then, for $a_* < a_1 < a_\#$ and $b_\# < b_1 < b_*$, there exist $a_* < a_2 < a_1$ and $b_1 < b_2 < b_*$ (independent of $\tilde{\Omega}$) which satisfy the following.

(a) (Surjectivity). Suppose $a_2 \leq \tilde{a} < a \leq a_1$, $b_1 \leq b < \tilde{b} \leq b_2$, $\mathfrak{V}' << \mathfrak{U}$, $\mathfrak{V}' << \mathfrak{V}$, $|\mathfrak{V}'| \subset\subset X_a^b$, $|\mathfrak{U}| \subset\subset X_{\tilde{a}}^{\tilde{b}}$, and $(\tilde{\Omega} \times X_a^b) \cap \operatorname{Supp} \mathcal{F} \subset\subset \tilde{\Omega} \times |\mathfrak{V}|$. Then there exists an open neighborhood Ω of 0 in $\tilde{\Omega}$ with the following property. For $\Delta(t^0, \rho) \subset \Omega$ and $p \leq \ell < r - q - n$ there exists a map

$$(\varphi, \psi) : Z^\ell(\Delta(t^0,\rho) \times \mathfrak{V}, \mathcal{F})$$

$$\longrightarrow Z^\ell(\Delta(t^0,\rho) \times \mathfrak{U}, \mathcal{F}) \oplus C^{\ell-1}(\Delta(t^0,\rho) \times \mathfrak{V}', \mathcal{F})$$

linear over $\mathbb{C}[t]$ such that

$$\xi = \varphi(\xi) + \delta\psi(\xi) \quad \text{on} \quad \Delta(t^0,\rho) \times \mathfrak{V}'$$

$$\operatorname{Max}\left(\|\varphi(\xi)\|_{\mathfrak{U}, t^0, \rho}, \|\psi(\xi)\|_{\mathfrak{V}', t^0, \rho}\right) \leq C \|\xi\|_{\mathfrak{V}, t^0, \rho}$$

where C is a constant independent of ξ, t^0, and ρ.

(b) (Injectivity). Suppose $\mathfrak{U} << \mathfrak{U}'$, $\mathfrak{V} << \mathfrak{U}'$, $|\mathfrak{U}| \subset\subset X_{\tilde{a}}^{\tilde{b}}$, $(\tilde{\Omega} \times X_a^b) \cap \operatorname{Supp} \mathcal{F} \subset\subset \tilde{\Omega} \times |\mathfrak{V}|$, and $(\tilde{\Omega} \times X_{\tilde{a}}^{\tilde{b}}) \cap \operatorname{Supp} \mathcal{F} \subset\subset \tilde{\Omega} \times |\mathfrak{U}'|$. Then there exists an open neighborhood Ω of 0 in $\tilde{\Omega}$ with the following property. For $\Delta(t^0, \rho) \subset \Omega$ and $p < \ell \leq r - q - n$ there exists a map θ

from

$$\{(\xi,\eta) \in Z^{\ell}(\Delta(t^0,\rho)\times \mathfrak{U}',\mathcal{F}) \oplus C^{\ell-1}(\Delta(t^0,\rho)\times \mathcal{V},\mathcal{F}) \mid \xi = \delta\eta \text{ on } \Delta(t^0,\rho)\times\mathcal{V}\}$$

to $C^{\ell-1}(\Delta(t^0,\rho)\times\mathfrak{U},\mathcal{F})$ linear over $\mathbb{C}[t]$ such that

$$\delta\theta(\xi,\eta) = \xi \quad \text{on} \quad \Delta(t^0,\rho)\times\mathfrak{U}$$

$$\|\theta(\xi,\eta)\|_{\mathfrak{U},t^0,\rho} \leq C \operatorname{Max}\left(\|\xi\|_{\mathfrak{U}',t^0,\rho}, \|\eta\|_{\mathcal{V},t^0,\rho}\right),$$

where C is a constant independent of ξ, η, t^0, and ρ.

(When $\mathcal{F}$ is not π-flat, for a fixed t^0, $\mathbb{C}$-linear φ, ψ, θ exist for ρ sufficiently strictly small but may not be linear over $\mathbb{C}[t]$ and C may depend on t^0 and ρ. If $t_n - t_n^0$ is not a zero-divisor of any stalk of F, then C can be chosen to be independent of ρ_n.)

Y.-T. Siu

§6. Sheaf Systems

The treatment given here is a simplified version of the sheaf systems introduced by Forster-Knorr [3].

(6.1) Suppose $\mathfrak{U} = \{U_i\}_{i=1}^k$ is a collection of open subsets of a complex space X. Let A_ν be the set of all $(i_0, \ldots, i_\nu)$ with $1 \leq i_0 < \ldots < i_\nu \leq k$ and let $A = \bigcup_{\nu \geq 0} A_\nu$. For $\alpha = (i_0, \ldots, i_\nu) \in A$, denote $U_{i_0} \cap \ldots \cap U_{i_\nu}$ by U_α. For $\beta = (j_0, \ldots, j_\mu) \in A$, $\alpha \subset \beta$ means $\{i_0, \ldots, i_\nu\} \subset \{j_0, \ldots, j_\mu\}$.

A <u>sheaf system</u> on $\mathfrak{U}$ is defined as consisting of

i) a coherent analytic sheaf $\mathcal{G}_\alpha$ on U_α for $\alpha \in A$

ii) a sheaf-homomorphism $\psi_{\beta\alpha}: \mathcal{G}_\alpha | U_\beta \longrightarrow \mathcal{G}_\beta$ for $\alpha \subset \beta$

such that $\psi_{\alpha\alpha}$ is the identity map of $\mathcal{G}_\alpha$ and

$$\psi_{\gamma\beta} \circ (\psi_{\beta\alpha} | U_\gamma) = \psi_{\gamma\alpha}$$

for $\alpha \subset \beta \subset \gamma$.

A sheaf system $(\mathcal{G}_\alpha, \psi_{\beta\alpha})$ is said to be <u>free</u> if each $\mathcal{G}_\alpha$ is isomorphic to $\mathcal{O}_X^{p_\alpha} | U_\alpha$ for some p_α.

A <u>morphism</u> from a sheaf system $(\mathcal{G}_\alpha, \psi_{\beta\alpha})$ to a sheaf system $(\mathcal{H}_\alpha, \chi_{\beta\alpha})$ is defined as a collection (θ_α) of sheaf-homomorphisms $\theta_\alpha: \mathcal{G}_\alpha \longrightarrow \mathcal{H}_\alpha$ such that $\theta_\beta \circ \psi_{\beta\alpha} = \chi_{\beta\alpha} \circ (\theta_\alpha | U_\beta)$ for $\alpha \subset \beta$. The kernel of a morphism, an exact sequence of sheaf systems, and an epimorphism of sheaf systems are understood in the most obvious sense.

Suppose $\mathfrak{U}' = \{U_i'\}_{i=1}^{k'}$ is a collection of open subsets of X which refines $\mathfrak{U}$ by an index map $\tau : \{1,\dots,k'\} \longrightarrow \{1,\dots,k\}$, i.e. $U_i' \subset U_{\tau(i)}$. For a sheaf system $\mathcal{G} = (\mathcal{G}_\alpha, \psi_{\beta\alpha})$ on $\mathfrak{U}$, define a sheaf system $\tau^*\mathcal{G} = (\mathcal{H}_{\alpha'}, \chi_{\beta'\alpha'})$ on $\mathfrak{U}'$ as follows.

i) $\mathcal{H}_{\alpha'} = \mathcal{G}_{\tau(\alpha')} | U_{\alpha'}'$

ii) $\chi_{\beta'\alpha'} : \mathcal{H}_{\alpha'} \longrightarrow \mathcal{H}_{\beta'}$ equals $\psi_{\tau(\beta')\tau(\alpha')} | U_{\beta'}'$

where $\tau(i_0, \dots, i_\nu)$ means $(\tau(i_0), \dots, \tau(i_\nu))$. $\tau^*\mathcal{G}$ is also denoted by $\mathcal{G}|\mathfrak{U}'$.

Suppose $\mathcal{F}$ is a coherent analytic sheaf on $\Omega := |\mathfrak{U}|$. We can regard $\mathfrak{U}$ as a refinement of Ω with the unique index map σ and obtain $\sigma^*\mathcal{F}$. In such a case, we identify $\mathcal{F}$ with the sheaf system $\sigma^*\mathcal{F}$.

(6.2) <u>Lemma</u>. <u>Suppose U_i $(1 \leq i \leq k)$ is a relatively compact open subset of a Stein open subset $\tilde{U}_i$ of a complex space X and $\mathcal{F}$ is a coherent analytic sheaf on $\bigcup_{i=1}^{k} \tilde{U}_i$. Then there exists an exact sequence</u>

$$\dots \longrightarrow \mathcal{R}^m \longrightarrow \dots \longrightarrow \mathcal{R}^1 \longrightarrow \mathcal{R}^0 \longrightarrow \mathcal{F} \longrightarrow 0$$

<u>of sheaf systems on $\mathfrak{U} := \{U_i\}_{i=1}^{k}$ such that each $\mathcal{R}^m$ is free.</u>

<u>Proof</u>. Take Stein open subsets $U_i \subset\subset U_i' \subset\subset U_i'' \subset\subset \tilde{U}_i$ $(1 \leq i \leq k)$ and let $\mathfrak{U}' = \{U_i'\}_{i=1}^k$ and $\mathfrak{U}'' = \{U_i''\}_{i=1}^k$. It suffices to show that, for any sheaf system $\mathcal{G} = (\mathcal{G}_\alpha, \psi_{\beta\alpha})$ on $\mathfrak{U}''$, there exist a free sheaf system $\mathcal{A} = (\mathcal{A}_\alpha, \sigma_{\beta\alpha})$ on $\mathfrak{U}'$ and an epimorphism $\theta : \mathcal{A} \longrightarrow \mathcal{G}|\mathfrak{U}'$. We construct them as follows. Fix a multi-index α_0. It suffices to construct a free sheaf system $\mathcal{A}^{(\alpha_0)} = \left(\mathcal{A}_\alpha^{(\alpha_0)}, \sigma_{\beta\alpha}^{(\alpha_0)}\right)$ on $\mathfrak{U}'$ and a morphism $\theta^{(\alpha_0)} : \mathcal{A}^{(\alpha_0)} \longrightarrow \mathcal{G}|\mathfrak{U}'$ such that $\theta_{\alpha_0}^{(\alpha_0)}$ is surjective, because we can set $\mathcal{A} = \bigoplus_{\alpha_0} \mathcal{A}^{(\alpha_0)}$ and obtain θ from $\theta^{(\alpha_0)}$. There exists a sheaf-epimorphism $\eta : \mathcal{O}_X^p \longrightarrow \mathcal{G}_{\alpha_0}$ on U_{α_0}'. Define

$$\mathcal{A}_\alpha^{(\alpha_0)} = \mathcal{O}_X^p | U_\alpha' \quad \text{for } \alpha_0 \subset \alpha$$

$$\mathcal{A}_\alpha^{(\alpha_0)} = 0 \quad \text{for } \alpha_0 \not\subset \alpha$$

$$\sigma_{\beta\alpha}^{(\alpha_0)} = \text{the identity map of } \mathcal{O}_X^p | U_\beta' \quad \text{for } \alpha_0 \subset \alpha \subset \beta$$

$$\sigma_{\beta\alpha}^{(\alpha_0)} = 0 \quad \text{for } \alpha_0 \not\subset \alpha$$

$$\theta_\alpha^{(\alpha_0)} = \psi_{\alpha\alpha_0} \circ (\eta | U_\alpha) \quad \text{for } \alpha_0 \subset \alpha$$

$$\theta_\alpha^{(\alpha_0)} = 0 \quad \text{for } \alpha_0 \not\subset \alpha .$$

The construction is complete. Q.E.D.

(6.3) Suppose $\mathfrak{U} = \{U_i\}_{i=1}^k$ is a collection of open subsets of a complex space X and $\mathcal{G} = (\mathcal{G}_\alpha, \psi_{\beta\alpha})$ is a sheaf system on $\mathfrak{U}$. Introduce the cochain group

$$C^\ell(\mathfrak{U}, \mathcal{G}) = \prod_{\alpha \in A_\ell} \Gamma(U_\alpha, \mathcal{G}_\alpha)$$

(where A_ℓ is as in (6.1)) and define the coboundary map

$$\delta: C^\ell(\mathfrak{U}, \mathcal{G}) \longrightarrow C^{\ell+1}(\mathfrak{U}, \mathcal{G})$$

by

$$(\delta\xi)_\beta = \sum_{i=0}^{\ell+1} (-1)^i \psi_{\beta\beta_i}(\xi_{\beta_i})$$

where

$$\xi = (\xi_\alpha)_{\alpha \in A_\ell} \in C^\ell(\mathfrak{U}, \mathcal{G}),$$

$$\beta = (j_0, \ldots, j_{\ell+1}) \in A_{\ell+1}$$

$$\beta_i = (j_0, \ldots, \hat{j_i}, \ldots, j_{\ell+1}).$$

Suppose $\mathcal{F}$ is a coherent analytic sheaf on X and

$$(*) \qquad \cdots \longrightarrow \mathcal{R}^m \longrightarrow \cdots \longrightarrow \mathcal{R}^1 \longrightarrow \mathcal{R}^0 \longrightarrow \mathcal{F} \longrightarrow 0$$

is an exact sequence of sheaf systems on $\mathfrak{U}$, where each $\mathcal{R}^m$ is free. Consider the following commutative diagram.

$$
\begin{array}{ccccccccccc}
 & & 0 & & 0 & & 0 & & 0 & & \\
 & & \uparrow & & \uparrow & & \uparrow & & \uparrow & & \\
0 & \longrightarrow & C^0(\mathfrak{U},\mathcal{F}) & \xrightarrow{\delta_0} & C^1(\mathfrak{U},\mathcal{F}) & \xrightarrow{\delta_1} & C^2(\mathfrak{U},\mathcal{F}) & \xrightarrow{\delta_2} & C^3(\mathfrak{U},\mathcal{F}) & \longrightarrow & \cdots \\
 & & \uparrow d_{0,0} & & \uparrow d_{1,0} & & \uparrow d_{2,0} & & \uparrow d_{3,0} & & \\
0 & \longrightarrow & C^0(\mathfrak{U},\mathcal{R}^0) & \xrightarrow{\delta_{0,0}} & C^1(\mathfrak{U},\mathcal{R}^0) & \xrightarrow{\delta_{1,0}} & C^2(\mathfrak{U},\mathcal{R}^0) & \xrightarrow{\delta_{2,0}} & C^3(\mathfrak{U},\mathcal{R}^0) & \longrightarrow & \cdots \\
 & & \uparrow d_{0,1} & & \uparrow d_{1,1} & & \uparrow d_{2,1} & & \uparrow d_{3,1} & & \\
0 & \longrightarrow & C^0(\mathfrak{U},\mathcal{R}^1) & \xrightarrow{\delta_{0,1}} & C^1(\mathfrak{U},\mathcal{R}^1) & \xrightarrow{\delta_{1,1}} & C^2(\mathfrak{U},\mathcal{R}^1) & \xrightarrow{\delta_{2,1}} & C^3(\mathfrak{U},\mathcal{R}^1) & \longrightarrow & \cdots \\
 & & \uparrow d_{0,2} & & \uparrow d_{1,2} & & \uparrow d_{2,2} & & \uparrow d_{3,2} & & \\
0 & \longrightarrow & C^0(\mathfrak{U},\mathcal{R}^2) & \xrightarrow{\delta_{0,2}} & C^1(\mathfrak{U},\mathcal{R}^2) & \xrightarrow{\delta_{1,2}} & C^2(\mathfrak{U},\mathcal{R}^2) & \xrightarrow{\delta_{2,2}} & C^3(\mathfrak{U},\mathcal{R}^2) & \longrightarrow & \cdots \\
 & & \uparrow & & \uparrow & & \uparrow & & \uparrow & & \\
 & & \vdots & & \vdots & & \vdots & & \vdots & &
\end{array}
$$

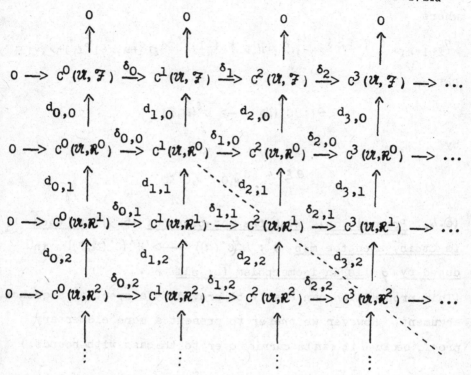

where the horizontal maps are coboundary maps and the vertical maps are defined by the morphisms of (*). Let

$$C^{\ell}(\mathfrak{U}) = \prod_{\nu=0}^{\infty} C^{\ell+\nu}(\mathfrak{U},\mathcal{R}^{\nu}) .$$

An element ξ of $C^{\ell}(\mathfrak{U})$ is given by $(\xi_{\ell+\nu,\nu})_{\nu=0}^{\infty}$ with $\xi_{\ell+\nu,\nu} \in C^{\ell+\nu}(\mathfrak{U},\mathcal{R}^{\nu})$. Define

$$\partial : C^{\ell}(\mathfrak{U}) \longrightarrow C^{\ell+1}(\mathfrak{U})$$

by

$$\partial\xi = \eta = (\eta_{\ell+1+\nu,\nu})_{\nu=0}^{\infty}$$

where

$$\eta_{\ell+1+\nu,\nu} = (-1)^{\nu} \delta_{\ell+\nu,} \xi_{\ell+\nu,\nu} + (-1)^{\nu+1} d_{\ell+1+\nu,\nu+1} \xi_{\ell+1+\nu,\nu+1} .$$

Define

$$\theta : C^{\ell}(\mathfrak{U}) \longrightarrow C^{\ell}(\mathfrak{U},\mathcal{F})$$

by

$$\theta \xi = d_{\nu,0} \xi_{\nu,0} .$$

(6.4) Proposition. With the notations of (6.3), if each U_i is Stein, then the map $\theta^*: H^{\ell}(C^{\cdot}(\mathfrak{U})) \longrightarrow H^{\ell}(C^{\cdot}(\mathfrak{U},\mathcal{F}))$ induced by θ is an isomorphism for all ℓ.

(This proposition can easily be proved by a spectral sequence argument. However we prefer to present a more elementary proof, because it can be carried over to the case with bounds.)

Proof. (a) (Surjectivity). Take $\tilde{\xi} \in Z^{\ell}(\mathfrak{U},\mathcal{F})$. By Theorem B of Cartan-Oka, one can construct, by induction on ν,

$$\xi_{\ell+\nu,\nu} \in C^{\ell+\nu}(\mathfrak{U},\mathcal{R}^{\nu}) \qquad (0 \leqq \nu < \infty)$$

such that

$$d_{\ell,0}\xi_{\ell,0} = \tilde{\xi}$$

$$d_{\ell+\nu,} \xi_{\ell+\nu,\nu} = \delta_{\ell+\nu-1,\nu-1}\xi_{\ell+\nu-1,\nu-1} \qquad (\geqq 1) .$$

Then

$$\xi : = (\xi_{\ell+\nu,\nu})_{\nu=0}^{\infty} \in Z^{\ell}(\mathfrak{U})$$

and $\theta\xi = \tilde{\xi}$.

(b) (Injectivity). Suppose

$$\xi = (\xi_{\ell+\nu,\nu})_{\nu=0}^{\infty} \in Z^{\ell}(\mathfrak{U})$$

is mapped to 0 in $H^{\ell}(C^{\cdot}(\mathfrak{U},\mathcal{F}))$. Then there exists $\zeta \in C^{\ell-1}(\mathfrak{U},\mathcal{F})$ such that

$$\delta_{\ell-1}\zeta = d_{\ell,0}\xi_{\ell,0}.$$

By Theorem B of Cartan-Oka, one can construct, by induction on ν,

$$\eta_{\ell-1+\nu,\nu} \in C^{\ell-1+\nu}(\mathfrak{U},\mathcal{R}^{\nu}) \quad (0 \leqq \nu < \infty)$$

such that

$$d_{\ell-1,0}\eta_{\ell-1,0} = \zeta$$

$$d_{\ell-1+\nu,\nu}\eta_{\ell-1+\nu,\nu} = (-1)^{\nu}\xi_{\ell-1+\nu,\nu-1} + \delta_{\ell-2+\nu,\nu-1}\eta_{\ell-2+\nu,\nu-1} \quad (\nu \geqq 1),$$

because

$$d_{\ell-1+\nu,\nu-1}\left((-1)^{\nu}\xi_{\ell-1+\nu,\nu-1} + \delta_{\ell-2+\nu,\nu-1}\eta_{\ell-2+\nu,\nu-1}\right)$$
$$= \delta_{\ell-2+\nu,\nu-2}\left((-1)^{\nu}\xi_{\ell-2+\nu,\nu-2} + d_{\ell-2+\nu,\nu-1}\eta_{\ell-2+\nu,\nu-1}\right)$$
$$= \delta_{\ell-2+\nu,\nu-2}\delta_{\ell-3+\nu,\nu-2}\eta_{\ell-3+\nu,\nu-2} = 0.$$

It follows that

$$\eta := (\eta_{\ell-1+\nu,\nu})_{\nu=0}^{\infty} \in C^{\ell-1}(\mathfrak{U})$$

satisfies $\partial\eta = \xi$. Q.E.D.

(6.5) For a given strongly (p,q)-pseudoconvex-pseudoconcave holomorphic map and for a given coherent analytic sheaf on the domain space, we are going to use the results and §5 and §6 to construct a sequence of complexes of Banach bundles which can be used to calculate certain direct images of the sheaf.

Suppose $X, \varphi, \mathcal{F}, p, q, r, X_a^b, \tilde{\Omega}, a_i, b_i$ $(i = 1,2)$ are as in (5.2). We arrive at such a situation by considering the graph of the strongly (p,q)-pseudoconvex-pseudoconcave holomorphic map and the trivial extension of the coherent analytic sheaf to the product after its transplantation to the graph.

Let $m \geq 1$. Choose

$$a_2 < c_m < \ldots < c_0 < a_1$$

$$b_1 < d_0 < \ldots < d_m < b_2 .$$

Choose finite collections

$$\mathfrak{U}_j \ll \mathfrak{U}_{j+1} \ll \tilde{\mathfrak{U}} \quad (1 \leq j < m)$$

of Stein open subsets of X such that

i) $(\tilde{\Omega} \times X_{c_j}^{d_j}) \cap \operatorname{Supp} \mathcal{F} \subset\subset \tilde{\Omega} \times |\mathfrak{U}_j| \subset\subset \tilde{\Omega} \times X_{c_{j+1}}^{d_{j+1}}$ $(0 \leq j < m)$

ii) there exists an exact sequence

$$\ldots \longrightarrow \mathcal{R}^\nu \longrightarrow \ldots \longrightarrow \mathcal{R}^1 \longrightarrow \mathcal{R}^0 \longrightarrow \mathcal{F} \longrightarrow 0$$

of sheaf systems on $\tilde{\Omega} \times \tilde{\mathfrak{U}}$ where each $\mathcal{R}^\nu$ is free.

By (6.3) and (6.4), for $1 \leq j \leq m$ and for any Stein open subset Q of $\tilde{\Omega}$, we have a complex

$$0 \longrightarrow C^0(Q \times \mathfrak{U}_j) \longrightarrow C^1(Q \times \mathfrak{U}_j) \longrightarrow \cdots \longrightarrow C^\ell(Q \times \mathfrak{U}_j) \longrightarrow \cdots$$

(constructed from the cochain groups of $\mathcal{R}^\nu | Q \times \mathfrak{U}_j$, $0 \leq \nu < \infty$) whose ℓ^{th} cohomology group is isomorphic to $H^\ell(Q \times \mathfrak{U}_j, \mathcal{F})$ $(0 \leq \ell < \infty)$. By letting Q vary, we obtain a complex of sheaves

$$0 \longrightarrow \mathcal{C}^0(\mathfrak{U}_j) \longrightarrow \mathcal{C}^1(\mathfrak{U}_j) \longrightarrow \cdots \longrightarrow \mathcal{C}^\ell(\mathfrak{U}_j) \longrightarrow \cdots$$

on $\tilde{\Omega}$ whose cohomology sheaves are isomorphic to the direct image of $\mathcal{F} | \tilde{\Omega} \times |\mathfrak{U}_j|$ under the projection $\tilde{\Omega} \times |\mathfrak{U}_j| \longrightarrow \tilde{\Omega}$. Now we turn to the situation with bounds. Fix $1 \leq j \leq m$. Let $\mathfrak{U}_j = \{U_i\}_{i=1}^k$. Let A_ℓ $(0 \leq \ell < \infty)$ and U_α $(\alpha \in A_\ell)$ be as in (6.1). Define the subgroup $C_b^\ell(Q \times \mathfrak{U}_j, \mathcal{R}^\nu)$ of $C^\ell(Q \times \mathfrak{U}_j, \mathcal{R}^\nu)$ as follows. We say that an element

$$f = (f_\alpha)_{\alpha \in A_\ell} \in C^\ell(Q \times \mathfrak{U}_j, \mathcal{R}^\nu)$$

belongs to $C_b^\ell(Q \times \mathfrak{U}_j, \mathcal{R}^\nu)$ if and only if $f_\alpha \in \Gamma(Q \times U_\alpha, (\mathcal{R}^\nu | \mathfrak{U}_j)_\alpha)$ satisfies

$$\|f_\alpha\|_{Q \times U_\alpha} < \infty$$

for all $\alpha \in A_\ell$. Let

$$C_b^\ell(Q \times \mathfrak{U}_j) = \prod_{\nu=0}^{\infty} C_b^{\ell+\nu}(Q \times \mathfrak{U}_j, \mathcal{R}^\nu) .$$

As in the case without bounds, we have a complex

$$0 \longrightarrow C_b^0(Q \times \mathfrak{U}_j) \longrightarrow C_b^1(Q \times \mathfrak{U}_j) \longrightarrow \cdots \longrightarrow C_b^\ell(Q \times \mathfrak{U}_j) \longrightarrow \cdots .$$

As Q varies, we obtain a complex of sheaves

$$0 \longrightarrow \mathcal{C}_b^0(\mathfrak{U}_j) \longrightarrow \mathcal{C}_b^1(\mathfrak{U}_j) \longrightarrow \cdots \longrightarrow \mathcal{C}_b^\ell(\mathfrak{U}_j) \longrightarrow \cdots$$

on $\tilde{\Omega}$. Since each $\mathcal{R}^\nu$ is free, this complex of sheaves is naturally isomorphic to

$$(\#)_j \quad 0 \longrightarrow \mathcal{O}(E_j^0) \longrightarrow \mathcal{O}(E_j^1) \longrightarrow \cdots \longrightarrow \mathcal{O}(E_j^\ell) \longrightarrow \cdots$$

where $\mathcal{O}(E_j^\ell)$ is the sheaf of germs of holomorphic sections of a trivial Banach bundle E_j^ℓ on $\tilde{\Omega}$. Let $\mathcal{H}_j^\ell$ be the ℓth cohomology sheaf of $(\#)_j$. Let $\pi_j: \tilde{\Omega} \times X_{c_j}^{d_j} \longrightarrow \tilde{\Omega}$ be the natural projection map. We have natural maps defined by restrictions:

$$\mathcal{H}_j^\ell \longrightarrow R^\ell(\pi_j)_*\mathcal{F} \longrightarrow \mathcal{H}_{j-1}^\ell \longrightarrow R^\ell(\pi_{j-1})_*\mathcal{F}.$$

For $p \leq \ell < r - q - n$, by the bumping techniques of Andreotti-Grauert [1] and by (5.2), one concludes that the maps

$$R^\ell(\pi_j)_*\mathcal{F} \longrightarrow R^\ell(\pi_{j-1})_*\mathcal{F}$$

$$\mathcal{H}_j^\ell \longrightarrow \mathcal{H}_{j-1}^\ell$$

satisfy certain conditions of surjectivity and injectivity. These conditions can be translated into relations between $\mathcal{H}_j^\ell$ and $R^\ell(\pi_j)_*\mathcal{F}$. The coherence of $R^\ell(\pi_j)_*\mathcal{F}$ will be investigated by working with $\mathcal{H}_j^\ell$. These statements will be made precise and presented in detail in Part II.

PART II THE POWER SERIES METHOD

§7 Finite Generation with Bounds

(7.1) In (6.5) we constructed a sequence of complexes of trivial Banach bundles and stated that the direct images, whose coherence we are interested in, are related to the cohomology sheaves of the complexes. Now we are going to consider abstractly a sequence of complexes of holomorphic Banach bundles satisfying certain conditions and derive conclusions concerning the finite generation of the cohomology sheaves of the complexes.

First we introduce some notations. Suppose B is a trivial bundle with a Banach space $(F, \|\cdot\|_F)$ as fiber. Let $\mathcal{O}(B)$ be the sheaf of germs of holomorphic sections of B. For $t^0 \in \Delta$ and $d \in \mathbb{N}_*^n$ let

$$\mathcal{O}(B)(t^0,d) = \mathcal{O}(B)\Big/\sum_{d_i \neq \infty} (t_i - t_i)^{d_i}\,\mathcal{O}(B)\,.$$

When $t^0 = 0$, $\mathcal{O}(B)(t^0,d)$ is simply denoted by $\mathcal{O}(B)(d)$. By identifying $\mathcal{O}(B)(t^0,d)$ with its zeroth direct image under the natural projection from Δ to

$$\Delta^* := \Delta \cap \{t_i = t_i^0 \ \text{ for } \ d_i \neq \infty\}\,,$$

we can regard $\mathcal{O}(B)(t^0,d)$ as an analytic sheaf over the reduced space Δ^*. $\mathcal{O}(B)(t^0,d)$ is the sheaf of germs of holomorphic sections of a trivial bundle over Δ^* whose fiber is the direct sum of $\prod_{d_i \neq \infty} d_i$ copies of F. For

$$f \in \Gamma(\Delta(t^0,\rho), \mathcal{O}(B)(t^0,d))$$

with power series expansion

$$\sum_{\substack{\nu \in \mathbb{N}_0^n \\ \nu_i < d_i}} f_\nu \left(\frac{t - t^0}{\rho}\right)^\nu$$

define

$$\|f\|_{B,t^0,d,\rho} = \sup_{\nu_i < d_i} \|f_\nu\|_F .$$

Denote by $B(t^0,d,\rho)$ the set of all

$$f \in \Gamma(\Delta(t^0,\rho), \mathcal{O}(B)(t^0,d))$$

with $\|f\|_{B,t^0,d,\rho} < \infty$. When $d = (\infty, \ldots, \infty)$, $B(t^0,d,\rho)$ is simply denoted by $B(\rho)$.

Now we introduce a formulation in terms of abstract complexes of holomorphic Banach bundles. Let

$$\begin{array}{ccccccccc}
\cdots & \longrightarrow & E_\alpha^{\nu-1} & \xrightarrow{\delta_\alpha^{\nu-1}} & E_\alpha^\nu & \xrightarrow{\delta_\alpha^\nu} & E_\alpha^{\nu+1} & \longrightarrow & \cdots \\
 & & \downarrow & & \downarrow r_{\alpha+1,\alpha}^\nu & & \downarrow & & \\
\cdots & \longrightarrow & E_{\alpha+1}^{\nu-1} & \longrightarrow & E_{\alpha+1}^\nu & \longrightarrow & E_{\alpha+1}^{\nu+1} & \longrightarrow & \cdots
\end{array}$$

be a commutative diagram of trivial Banach bundles and bundle-homomorphisms on Δ with $\delta_\alpha^\nu \delta_\alpha^{\nu-1} = 0$, where $\alpha_1 \leqq \alpha \leqq \alpha_2$ for some integers α_1, α_2; that is, we have a sequence of complexes. Some statements concerning these complexes which we will consider later are true only for $\alpha_1 + c \leqq \alpha \leqq \alpha_2 - c$ where c is a positive integer depending only on n (and other given numbers). Such a restriction will be clear from the proofs and will not be explicitly stated. We will be interested in the behavior of these complexes in a neighborhood of 0. So we will sometimes replace Δ by some suitable $\Delta(\rho^0)$.

For $t^0 \in \Delta$ and $d \in \mathbb{N}_*^n$ let $\mathcal{H}_\alpha^\nu[t^0,d]$ denote the νth cohomology sheaf of the complex

$$\cdots \longrightarrow \mathcal{O}(E_\alpha^{\nu-1})(t^0,d) \longrightarrow \mathcal{O}(E_\alpha^\nu)(t^0,d) \longrightarrow \mathcal{O}(E_\alpha^{\nu+1})(t^0,d) \longrightarrow \cdots .$$

When $t^0 = 0$, $\mathcal{H}_\alpha^\nu[t^0,d]$ is simply denoted by $\mathcal{H}_\alpha^\nu[d]$, and, when $d = (\infty, \ldots, \infty)$, it is simply denoted by $\mathcal{H}_\alpha^\nu$.

For $\alpha < \beta$ let

$$r_{\beta\alpha}^\nu = r_{\beta,\beta-1}^\nu r_{\beta-1,\beta-2}^\nu \cdots r_{\alpha+1,\alpha}^\nu .$$

For $t^0 \in \Delta$ and $d \in \mathbb{N}_*^n$ let

$$\|\cdot\|_{\alpha,t^0,d,\rho} = \|\cdot\|_{E_\alpha,t^0,d,\rho} .$$

When $d = (\infty, \ldots, \infty)$, $\|\cdot\|_{\alpha,t^0,d,\rho}$ is simply denoted by $\|\cdot\|_{\alpha,t^0,\rho}$

Consider the following conditions $(E)^\nu$, $(M)^\nu$, $(F)^\nu$, $(B)^\nu_n$.

$(E)^\nu$ (Quasi-epimorphism with Bounds). There exists a constant C with the following property. For $\alpha < \beta$, $\xi \in E^\nu_\beta(t^0,d,\rho)$ with $\delta^\nu_\beta \xi = 0$ there exist

$$\tilde{\xi} := \varphi_{\alpha,\beta,t^0,d,\rho}(\xi) \in E^\nu_\alpha(t^0,d,\rho) \quad \text{with} \quad \delta^\nu_\alpha \xi = 0$$

$$\eta := \psi_{\alpha,\beta,t^0,d,\rho}(\xi) \in E^{\nu-1}_{\beta+1}(t^0,d,\rho)$$

such that

i) $r^\nu_{\beta+1,\beta}\xi = r^\nu_{\beta+1,\alpha}\tilde{\xi} + \delta^{\nu-1}_{\beta+1}\eta$

ii) $\operatorname{Max}\left(\|\tilde{\xi}\|^\nu_{\alpha,t^0,d,\rho}, \|\eta\|^{\nu-1}_{\beta+1,t^0,d,\rho}\right) \leqq C\|\xi\|^\nu_{\beta,t^0,d,\rho}$

iii) $\varphi_{\alpha,\beta,t^0,d,\rho}$ and $\psi_{\alpha,\beta,t^0,d,\rho}$ are linear over $\mathbb{C}[t]$.

$(M)^\nu$ (Quasi-monomorphism with Bounds). There exists a constant C with the following property. For $\alpha < \beta$, $\xi \in E^\nu_\alpha(t^0,d,\rho)$ and $\eta \in E^{\nu-1}_\beta(t^0,d,\rho)$ satisfying $\delta^\nu_\alpha\xi = 0$ and $r^\nu_{\beta\alpha}\xi = \delta^{\nu-1}_\beta\eta$ there exists

$$\tilde{\eta} := \varphi_{\alpha,\beta,t^0,d,\rho}(\xi,\eta) \in E^{\nu-1}_{\alpha+1}(t^0,d,\rho)$$

such that

i) $r^\nu_{\alpha+1,\alpha}\xi = \delta^{\nu-1}_{\alpha+1}\tilde{\eta}$

ii) $\|\tilde{\eta}\|^{\nu-1}_{\alpha+1,t^0,d,\rho} \leq C \operatorname{Max}\left(\|\xi\|^{\nu}_{\alpha,t^0,d,\rho}, \|\eta\|^{\nu-1}_{\beta,t^0,d,\rho}\right)$

iii) $\varphi_{\alpha,\beta,t^0,d,\rho}$ is bilinear over $\mathbb{C}[t]$.

$(F)^{\nu}$ (<u>Finite-dimensionality along the Fibers</u>). For $t^0 \in \Delta$ and $d \in \mathbb{N}^n$ there exists a commutative diagram of continuous linear maps

$$\begin{array}{ccccccccc}
\cdots \longrightarrow & E^{\nu-1}_{\alpha}(t^0,d,\rho) & \longrightarrow & E^{\nu}_{\alpha}(t^0,d,\rho) & \longrightarrow & E^{\nu+1}_{\alpha}(t^0,d,\rho) & \longrightarrow \cdots \\
& \downarrow & & \downarrow & & \downarrow & \\
\cdots \longrightarrow & F^{\nu-1}_{\alpha,t^0,d,\rho} & \longrightarrow & F^{\nu}_{\alpha,t^0,d,\rho} & \longrightarrow & F^{\nu+1}_{\alpha,t^0,d,\rho} & \longrightarrow \cdots \\
& \downarrow & & \downarrow & & \downarrow & \\
\cdots \longrightarrow & E^{\nu-1}_{\alpha+1}(t^0,d,\rho) & \longrightarrow & E^{\nu}_{\alpha+1}(t^0,d,\rho) & \longrightarrow & E^{\nu+1}_{\alpha+1}(t^0,d,\rho) & \longrightarrow \cdots
\end{array}$$

where

i) the composite vertical maps are $r^{\mu}_{\alpha+1,\alpha}$

ii) $F^{\mu}_{\alpha,t^0,d,\rho} \longrightarrow E^{\mu}_{\alpha+1}(t^0,d,\rho)$ factors through a Hilbert space

iii) the middle row is a complex of Fréchet spaces

iv) $\dim_{\mathbb{C}} H^{\nu}(F^{\bullet}_{\alpha,t^0,d,\rho}) < \infty$

v) $H^{\nu}(F^{\bullet}_{\alpha,t^0,d,\rho}) \longrightarrow H^{\nu}(F^{\bullet}_{\alpha+1,t^0,d,\rho})$ is bijective

vi) $H^{\nu+1}(F^{\bullet}_{\alpha,t^0,d,\rho}) \longrightarrow H^{\nu+1}(F^{\bullet}_{\alpha+1,t^0,d,\rho})$ is injective

$(B)^{\nu}_{n}$ (<u>Finite Generation with Bounds</u>).

a) $\mathrm{Im}(\mathcal{H}^{\nu}_{\alpha,t^0} \longrightarrow \mathcal{H}^{\nu}_{\alpha+1,t^0})$ is finitely generated over ${}_n\mathcal{O}_{t^0}$ for $t^0 \in \Delta$.

b) Let $\xi^{(1)}, \ldots, \xi^{(k)} \in E^{\nu}_{\alpha}(t^0,\rho^0)$ with $\delta^{\nu}_{1}\xi^{(i)} = 0$ $(1 \leq i \leq k)$ and let A be the ${}_n\mathcal{O}_{t^0}$-submodule of $\mathcal{H}^{\nu}_{\alpha+1,t^0}$. Then, for ρ sufficiently strictly small, there exists a constant C_{ρ} satisfying the following. If $\xi \in E^{\nu}_{\alpha}(t^0,\rho)$ with $\delta^{\nu}_{\alpha}\xi = 0$ such that the image of ξ in $\mathcal{H}^{\nu}_{\alpha+1,t^0}$ belongs to A, then there exist

$$a^{(1)}, \ldots, a^{(k)} \in \Gamma(\Delta(t^0,\rho), {}_n\mathcal{O})$$

$$\eta \in E^{\nu-1}_{\alpha+1}(t^0,\rho)$$

such that

$$r^{\nu}_{\alpha+1,\alpha}\xi = \sum_{i=1}^{k} a^{(i)} r^{\nu}_{\alpha+1,\alpha}\xi^{(i)} + \delta^{\nu-1}_{\alpha+1}\eta$$

$$|a^{(i)}|_{t^0,\rho} \leq C_{\rho}\|\xi\|^{\nu}_{\alpha,t^0,\rho}$$

$$\|\eta\|^{\nu-1}_{\alpha+1,t^0,\rho} \leq C_{\rho}\|\xi\|^{\nu}_{\alpha,t^0,\rho} .$$

(7.2) **Proposition.** $(E)^{\nu}$, $(M)^{\nu+1}$, $(B)^{\nu}_{n-1} \Longrightarrow (B)^{\nu}_{n}$.

Proof. It suffices to prove the case $t^0 = 0$. Let

$$\bar{d} = (\infty, \ldots, \infty, 1) \in \mathbb{N}^n_*$$
$$\overline{\mathcal{H}^{\nu}_{\alpha,0}} = \mathcal{H}^{\nu}_{\alpha}[\bar{d}]_0$$
$$H^{\nu}_{\alpha+1} = \operatorname{Im}(\mathcal{H}^{\nu}_{\alpha,0} \longrightarrow \mathcal{H}^{\nu}_{\alpha+1,0})$$
$$\overline{H^{\nu}_{\alpha+1}} = \operatorname{Im}(\overline{\mathcal{H}^{\nu}_{\alpha,0}} \longrightarrow \overline{\mathcal{H}^{\nu}_{\alpha+1,0}}) .$$

Since $\overline{H^{\nu}_{\alpha+1}}$ is finitely generated over ${}_{n-1}\mathcal{O}_0$, there exist

$$\xi^{(1)}, \ldots, \xi^{(k)} \in E^{\nu}_{\alpha}(\rho^0) \quad \text{(for some } \rho^0\text{)}$$

with $\delta^{\nu}_1 \xi^{(i)} = 0$ $(1 \leq i \leq k)$ such that they generate $\operatorname{Im}(H^{\nu}_{\alpha+1} \longrightarrow \overline{H^{\nu}_{\alpha+1}})$ over ${}_{n-1}\mathcal{O}_0$. Assume $\rho_n < \frac{\rho^0_n}{2}$. For $\rho = (\rho_1, \ldots, \rho_n)$, let

$$\bar{\rho} = (\rho_1, \ldots, \rho_{n-1})$$
$$\rho^* = (\rho_1, \ldots, \rho_{n-1}, \rho^0_n) .$$

In this proof clearly we can assume without loss of generality that δ^{μ}_{γ} and $r^{\mu}_{\gamma+1,\gamma}$ are norm-nonincreasing. In other proofs where we can also make such an assumption without loss of generality, it will not be explicitly mentioned. Since $t^0 = 0$, we will drop t^0 from all notations if no confusion can arise. We break up the proof into four parts.

(I) First we prove the following.

(*) For $\bar{\rho}$ sufficiently strictly small, there exists $C_{\bar{\rho}}$ such that, if $\xi \in E^{\nu}_{\alpha}(\rho)$ and $\delta^{\nu}_{\alpha}\xi = 0$, then there exists $\theta \in E^{\nu}_{\alpha+1}(\rho^*)$ satisfying

$$\delta^{\nu}_{\alpha+1}\left(\theta + r^{\nu}_{\alpha+1,\alpha} \sum_{\lambda=1}^{\infty} \xi_{\lambda}\left(\frac{t_n}{\rho_n}\right)^{\lambda-1}\right) = 0$$

$$\|\theta\|^{\nu}_{\alpha+1,\rho^*} \leqq \rho_n \; C_{\bar{\rho}} \; \|\xi\|^{\nu}_{\alpha,\rho}$$

where $\xi = \sum_{\lambda=0}^{\infty} \xi_{\lambda}\left(\frac{t_n}{\rho_n}\right)^{\lambda}$ is the power series expansion of ξ in t_n.

By $(B)^{\nu}_{n-1}$ for $\bar{\rho}$ sufficiently strictly small there exist

$a^{(i)} \in \Gamma(\Delta(\rho^*), {}_n\mathcal{O})$ independent of t_n

$\eta \in E^{\nu-1}_{\alpha+1}(\rho^*)$ independent of t_n

$\zeta \in E^{\nu}_{\alpha+1}(\rho^*)$

such that

$$r^{\nu}_{\alpha+1,\alpha}\xi = \sum_{i=1}^{k} a^{(i)} r^{\nu}_{\alpha+1,\beta}\xi^{(i)} + \delta^{\nu-1}_{\alpha+1}\eta + \left(\frac{t_n}{\rho_n}\right)\zeta$$

$$|a^{(i)}|_{\rho^*} \leqq C'_{\bar{\rho}} \|\xi\|^{\nu}_{\alpha,\rho}$$

$$\|\eta\|^{\nu-1}_{\alpha+1,\rho^*} \leqq C'_{\bar{\rho}} \|\xi\|^{\nu}_{\alpha,\rho} \; .$$

Let $\xi_* = \sum_{\lambda=1}^{\infty} \xi_\lambda \left(\frac{t_n}{\rho_n}\right)^{\lambda-1}$. Then

$$r^{\nu}_{\alpha+1,\alpha}\xi_0 - \sum_{i=1}^{k} a^{(i)} r^{\nu}_{\alpha+1,\alpha}\xi^{(i)} - \delta^{\nu-1}_{\alpha+1}\eta = \left(\frac{t_n}{\rho_n}\right)(\zeta - r^{\nu}_{\alpha+1,\alpha}\xi_*)$$

Define

$$\theta = \zeta - r^{\nu}_{\alpha+1,\alpha}\xi_* .$$

Then

$$\theta + r^{\nu}_{\alpha+1,\alpha}\xi_* = \zeta .$$

Hence

$$\delta^{\nu}_{\alpha+1}(\theta + r^{\nu}_{\alpha+1,\alpha}\xi_*) = 0 .$$

Moreover,

$$\left\|\left(\frac{t_n}{\rho_n}\right)\theta\right\|^{\nu}_{\alpha+1,\rho*} = \left\|r^{\nu}_{\alpha+1,\alpha}\left(\xi_0 - \sum_{i=1}^{k} a^{(i)}\xi^{(i)}\right) - \delta^{\nu-1}_{\alpha+1}\eta\right\|^{\nu}_{\alpha+1,\rho*}$$

$$\leqq C''_{\bar\rho}\|\xi\|^{\nu}_{\alpha,\rho}$$

where $C''_{\bar\rho}$ is a constant depending on $\bar\rho$. It follows that

$$\|\theta\|^{\nu}_{\alpha+1,\rho*} = \left\|\left(\frac{t_n}{\rho^0_n}\right)\theta\right\|^{\nu}_{\alpha+1,\rho*} = \left(\frac{\rho_n}{\rho^0_n}\right)\left\|\left(\frac{t_n}{\rho_n}\right)\theta\right\|^{\nu}_{\alpha+1,\rho*} \leqq \rho_n\left(\frac{C''_{\bar\rho}}{\rho^0_n}\right)\|\xi\|^{\nu}_{\alpha,\rho} .$$

Therefore (*) is proved.

(II) Next we prove the following.

(†) For $\bar\rho$ sufficiently strictly small, there exists $D_{\bar\rho}$ such that, if $\xi \in E^\nu_\alpha(\rho)$ and $\delta^\nu_\alpha \xi = 0$ and $m \geqq 1$, then there exists $\theta_m \in E^\nu_{\alpha+1}(\rho^*)$ such that

$$\delta^\nu_{\alpha+1}\left(\theta_m + r^\nu_{\alpha+1,\alpha} \sum_{\lambda=m}^\infty \xi_\lambda \left(\frac{t_n}{\rho_n}\right)^{\lambda-m}\right) = 0$$

$$\|\theta_m\|^\nu_{\alpha+1,\rho^*} \leqq \rho_n D_{\bar\rho} \|\xi\|^\nu_{\alpha,\rho}$$

where $\xi = \sum_{\lambda=0}^\infty \xi_\lambda \left(\frac{t_n}{\rho_n}\right)^\lambda$ is the power series expansion of ξ in t_n.

We are going to construct θ_m by induction on m. The case $m = 1$ has been done in (*) of (I). Suppose we have θ_m. Let

$$\tau = r^\nu_{\alpha+1,\alpha}\left(\sum_{\lambda=m}^\infty \xi_\lambda \left(\frac{t_n}{\rho_n}\right)^{\lambda-m}\right) + \theta_m .$$

Then $\delta^\nu_{\alpha+1}\tau = 0$. Let $e = \|\xi\|^\nu_{\alpha,\rho}$. Then

$$\|\tau\|^\nu_{\alpha+1,\rho} \leqq e + \|\theta_m\|^\nu_{\alpha+1,\rho} \leqq e + \rho_n D_{\bar\rho}\, e .$$

We assume that ρ_n is so small that $1 + \rho_n D_{\bar\rho} \leqq 2$. So

$$\|\tau\|^\nu_{\alpha+1,\rho} \leqq 2e .$$

Let $\tau = \tau_0 + \left(\frac{t_n}{\rho_n}\right)\tau_*$, where τ_0 is independent of t_n. By applying (*) of (I) to τ, we obtain $\chi \in E^{\nu}_{\alpha+2}(\rho^*)$ such that

$$\delta^{\nu}_{\alpha+2,\alpha+1}(\chi + r^{\nu}_{\alpha+2,\alpha+1}\tau_*) = 0$$

$$\|\chi\|^{\nu}_{\alpha+2,\rho*} \leq \rho_n C_{\bar{\rho}} 2e .$$

Let

$$\xi' = \sum_{\lambda=0}^{m} \xi_\lambda \left(\frac{t_n}{\rho_n}\right)^{\lambda}$$

$$\xi_{\#} = \sum_{\lambda=m+1}^{\infty} \xi_\lambda \left(\frac{t_n}{\rho_n}\right)^{\lambda-m-1} .$$

We are going to apply $(M)^{\nu+1}$ to

$$r^{\nu+1}_{\alpha+2,\alpha}(\delta^{\nu}_{\alpha}\xi_{\#}) = \delta^{\nu}_{\alpha+2}(r^{\nu}_{\alpha+2,\alpha}\xi_{\#} - r^{\nu}_{\alpha+2,\alpha+1}\tau_* - \chi) .$$

For this purpose we need estimates on the norms of the two expressions in parentheses.

First we are going to obtain an estimate on the norm of $\delta^{\nu}_{\alpha}\xi_{\#}$. Since

$$\left(\frac{t_n}{\rho_n}\right)^{m+1} \delta^{\nu}_{\alpha}\xi_{\#} = -\delta^{\nu}_{\alpha}\xi' ,$$

by defining

$$\delta_\alpha^\nu \xi_\# = -\left(\frac{t_n}{\rho_n}\right)^{-(m+1)} \delta_\alpha^\nu \xi'$$

for $t_n \neq 0$, we can regard $\delta_\alpha^\nu \xi_\#$ as an element of $E_\alpha^{\nu+1}(\rho^*)$. Because

$$\left\| \left(\frac{t_n}{\rho_n}\right)^{m+1} \delta_\alpha^\nu \xi_\# \right\|_{\alpha,\rho^*}^{\nu+1} = \left\| \delta_\alpha^\nu \xi' \right\|_{\alpha,\rho^*}^{\nu+1} \leq \left\| \xi' \right\|_{\alpha,\rho^*}^{\nu+1} \leq \left(\frac{\rho_n^0}{\rho_n}\right)^m e ,$$

it follows that

$$\left\| \delta_\alpha^\nu \xi_\# \right\|_{\alpha,\rho^*}^{\nu+1} = \left\| \left(\frac{t_n}{\rho_n^0}\right)^{m+1} \delta_\alpha^\nu \xi_\# \right\|_{\alpha,\rho^*}^{\nu+1}$$

$$= \left(\frac{\rho_n}{\rho_n^0}\right)^{m+1} \left\| \left(\frac{t_n}{\rho_n}\right)^{m+1} \delta_\alpha^\nu \xi_\# \right\|_{\alpha,\rho^*}^{\nu+1} \leq \frac{\rho_n}{\rho_n^0} e .$$

Now we are going to obtain an estimate on the norm of $(r_{\alpha+2,\alpha}^\nu \xi_\# - r_{\alpha+2,\alpha+1}^\nu \tau_* - \chi)$. From the definition of τ,

$$\tau_0 + \left(\frac{t_n}{\rho_n}\right)\tau_* = r_{\alpha+2,\alpha}^\nu \xi_m + \left(\frac{t_n}{\rho_n}\right) r_{\alpha+1,\alpha}^\nu \xi_\# + \theta_m .$$

$$\left\| \left(\frac{t_n}{\rho_n}\right) (r_{\alpha+1,\alpha}^\nu \xi_\# - \tau_*) \right\|_{\alpha+1,\,*}^\nu$$

$$= \left\| \tau_0 - r_{\alpha+1,\alpha}^\nu \xi_m - \theta_m \right\|_{\alpha+1,\,*}^\nu \leq 2e + e + \rho_n D_{\bar\rho} e \leq 4e$$

(where we have defined

$$r^{\nu}_{\alpha+1,\alpha}\xi_{\#} - \tau_{*} = \left(\frac{t_n}{\rho_n}\right)^{-1}(\tau_0 - r^{\nu}_{\alpha+1,\alpha}\xi_m - \theta_m)$$

for $t_n \neq 0$). It follows that

$$\left\| r^{\nu}_{\alpha+1,\alpha}\xi_{\#} - \tau_{*}\right\|^{\nu}_{\alpha+1,\rho^*} = \frac{\rho_n}{\rho_n^0}\left\| \left(\frac{t_n}{\rho_n}\right)(r^{\nu}_{\alpha+1,\alpha}\xi_{\#} - \tau_{*})\right\|^{\nu}_{\alpha+1,\rho^*}$$

$$\leq \frac{\rho_n}{\rho_n^0} 4e .$$

Hence

$$\left\| r^{\nu}_{\alpha+2,\alpha}\xi_{\#} - r^{\nu}_{\alpha+2,\alpha+1}\tau_{*} - \chi\right\|^{\nu}_{\alpha+2,\rho^*} \leq \frac{\rho_n}{\rho_n^0} 4e + \rho_n C_{\bar{\rho}} 2e .$$

By $(M)^{\nu+1}$, there exists $\theta_{m+1} \in E^{\nu}_{\alpha+1}(\rho^*)$ such that

$$r^{\nu+1}_{\alpha+1,\alpha}\delta^{\nu}_{\alpha}\xi_{\#} = \delta^{\nu}_{\alpha+1}\theta_{m+1}$$

$$\left\|\theta_{m+1}\right\|^{\nu}_{\alpha+1,\rho^*} \leq C' \operatorname{Max}\left(\frac{\rho_n}{\rho_n^0} e, \frac{\rho_n}{\rho_n^0} 4e + \rho_n C_{\bar{\rho}} 2e\right),$$

where C' is the constant coming from $(M)^{\nu+1}$. This finishes the proof of (†) if we put

$$D_{\bar{\rho}} \geq C'\left(\frac{4}{\rho_n^0} + 2C_{\bar{\rho}}\right).$$

(III) Now we apply $(B)^{\nu}_{n-1}$ to get the finite generation of $H^{\nu}_{\alpha+1}$ over ${}_n\mathcal{O}_0$. Let

$$\theta_m = \sum_{\lambda=0}^{\infty} \theta_{m\lambda}\left(\frac{t_n}{\rho_n}\right)^{\lambda}$$

be the power series expansion of θ_m in t_n. Since

$$\delta_{\alpha+1}\left(\theta_m + r^{\nu}_{\alpha+1,\alpha} \sum_{\lambda=m}^{\infty} \xi_\lambda\left(\frac{t_n}{\rho_n}\right)^{\lambda-m}\right) = 0$$

and

$$\left\|\theta_{m0} + r^{\nu}_{\alpha+1,\alpha}\xi_m\right\|^{\nu}_{\alpha+1,\rho*} \leqq (\rho_n D_{\overline{\rho}} + 1)e ,$$

by applying $(B)^{\nu}_{n-1}$ to the ${}_{n-1}\mathcal{O}_0$-submodule $\mathrm{Im}(H^{\nu}_{\alpha+2} \longrightarrow \overline{H}^{\nu}_{\alpha+2})$ of $\overline{\mathcal{H}}^{\nu}_{\alpha+2,0}$ generated by the images of $r^{\nu}_{\alpha+1,\alpha}\xi^{(i)}$ $(1 \leqq i \leqq k)$, we can find, for $\overline{\rho}$ sufficiently strictly small,

$a^{(i)}_m \in \Gamma(\Delta(\rho^*), {}_n\mathcal{O})$ independent of t_n

$\eta_m \in E^{\nu-1}_{\alpha+2}(\rho^*)$ independent of t_n

$\zeta_m \in E^{\nu}_{\alpha+2}(\rho^*)$

such that

$$(\#) \quad r^{\nu}_{\alpha+2,\alpha+1}\theta_{m0} + r^{\nu}_{\alpha+2,\alpha}\xi_m = \sum_{i=1}^{k} a^{(i)}_m r^{\nu}_{\alpha+2,\alpha}\xi^{(i)} + \delta^{\nu-1}_{\alpha+2}\eta + \left(\frac{t_n}{\rho^0_n}\right)\zeta_m$$

$$|a_m^{(i)}|_{\rho*} \leq C''_{\bar{\rho}}(\rho_n D_{\bar{\rho}} + 1)e$$

$$\|\eta_m\|^{\nu-1}_{\alpha+2,\rho*} \leq C''_{\bar{\rho}}(\rho_n D_{\bar{\rho}} + 1)e$$

$$\|\zeta_m\|^{\nu}_{\alpha+2,\rho*} \leq C''_{\bar{\rho}}(\rho_n D_{\bar{\rho}} + 1)e ,$$

where $C''_{\bar{\rho}}$ is the constant from $(B)^{\nu}_{n-1}$.

From (#) it follows that

$$\|\zeta_m\|^{\nu}_{\alpha+2,\rho*} \leq C''' C''_{\bar{\rho}}(\rho_n D_{\bar{\rho}} + 1)e$$

where C''' is a constant.

By multiplying (#) by $\left(\frac{t_n}{\rho_n}\right)^m$ and summing over m, we obtain

$$r^{\nu}_{\alpha+2,\alpha}\xi + r^{\nu}_{\alpha+2,\alpha+1}\tilde{\theta} = \sum_{i=1}^{k} a^{(i)} r^{\nu}_{\alpha+2,\alpha}\xi^{(i)} + \delta^{\nu-1}_{\alpha+2}\eta + \left(\frac{t_n}{\rho_n^0}\right)\zeta$$

where

$$\tilde{\theta} = \sum_{m=0}^{\infty} \theta_{m0}\left(\frac{t_n}{\rho_n}\right)^m$$

$$a^{(i)} = \sum_{m=0}^{\infty} a_m^{(i)}\left(\frac{t_n}{\rho_n}\right)^m$$

$$\eta = \sum_{m=0}^{\infty} \eta_m\left(\frac{t_n}{\rho_n}\right)^m$$

$$\zeta = \sum_{m=0}^{\infty} \zeta_m\left(\frac{t_n}{\rho_n}\right)^m$$

It follows that

$$\|\tilde{\theta}\|^{\nu}_{\alpha+1,\rho} \leq \rho_n D_{\bar{\rho}} e$$

$$|a^{(i)}|_{\rho} \leq C''_{\bar{\rho}}(\rho_n D_{\bar{\rho}} + 1)e$$

$$\|\eta\|^{\nu}_{\alpha+2,\rho} \leq C''_{\bar{\rho}}(\rho_n D_{\bar{\rho}} + 1)e$$

$$\|\zeta\|^{\nu}_{\alpha+2,\rho} \leq \tilde{C}C''' C''_{\bar{\rho}}(\rho_n D_{\bar{\rho}} + 1)e$$

where $\tilde{C}$ is a constant.

$$\left\| r^{\nu}_{\alpha+2,\alpha}\left(\xi - \sum_{i=1}^{k} a^{(i)}\xi^{(i)}\right) - \delta^{\nu-1}_{\alpha+2}\eta \right\|^{\nu}_{\alpha+2},$$

$$= \left\| r^{\nu}_{\alpha+2,\alpha+1}\tilde{\theta} - \left(\frac{t_n}{\rho_n^0}\right)\zeta \right\|^{\nu}_{\alpha+2,\rho} \leq \rho_n D_{\bar{\rho}} e + \frac{\rho_n}{\rho_n^0}\tilde{C}C''' C''_{\bar{\rho}}\,(\rho_n D_{\bar{\rho}} + 1)e\,.$$

By $(E)^{\nu}$, we can find

$$\tilde{\xi} \in E^{\nu}_{\alpha}(\rho) \quad \text{with} \quad \delta^{\nu}_{\alpha}\tilde{\xi} = 0$$

$$\tilde{\eta} \in E^{\nu-1}_{\alpha+3}(\rho)$$

such that

$$r^{\nu}_{\alpha+3,\alpha+2}\left(r^{\nu}_{\alpha+2,\alpha}\left(\xi - \sum_{i=1}^{k} a^{(i)}\xi^{(i)}\right) - \delta^{\nu-1}_{\alpha+2}\eta\right) = r^{\nu}_{\alpha+3,\alpha}\tilde{\xi} + \delta^{\nu-1}_{\alpha+3}\tilde{\eta}$$

$$\mathrm{Max}\left(\|\tilde{\xi}\|^{\nu}_{\alpha,\rho}, \|\tilde{\eta}\|^{\nu}_{\alpha+3,\rho}\right) \leq \hat{C}\left(\rho_n D_{\bar{\rho}} e + \frac{\rho_n}{\rho_n^0}\tilde{C}C''' C''_{\bar{\rho}}\,(\rho_n D_{\bar{\rho}} + 1)e\right)$$

where $\hat{C}$ is the constant from $(E)^{\nu}$. Choose ρ_n so small that

$$\hat{C}\left(\rho_n D_{\bar{\rho}} + \frac{\rho_n}{\rho_n^0} CC''' C''_{\bar{\rho}}(\rho_n D_{\bar{\rho}} + 1)\right) \leq \frac{1}{2} .$$

Define

$$F_i(\xi) = a^{(i)}$$

$$\Phi(\xi) = r^{\nu-1}_{\alpha+3,\alpha+2}\eta + \tilde{\eta}$$

$$\Psi(\xi) = \tilde{\xi} .$$

Let Ψ^{λ} denote $\Psi \circ \cdots \circ \Psi$ (λ times). Then

$$r^{\nu}_{\alpha+3,\alpha}\xi = \sum_{i=1}^{k}\left(\sum_{\lambda=1}^{\infty} F_i(\Psi^{\lambda}(\xi))\right) r^{\nu}_{\alpha+3,\alpha}\xi^{(i)} + \delta^{\nu-1}_{\alpha+3}\sum_{\lambda=1}^{\infty}\Phi(\Psi^{\lambda}(\xi))$$

$$\left|\sum_{\lambda=1}^{\infty} F_i(\Psi^{\lambda}(\xi))\right| \leq 2C''_{\bar{\rho}}(\rho_n D_{\bar{\rho}} + 1)e$$

$$\left\| \sum_{\lambda=1}^{\infty}\Phi(\Psi^{\lambda}(\xi))\right\|^{\nu-1}_{\alpha+3,\rho} \leq 2C''_{\bar{\rho}}(\rho_n D_{\bar{\rho}} + 1)e + e .$$

This is almost but not exactly the result we want concerning finite generation, because the equation is in $E_{\alpha+3}$ instead of $E_{\alpha+1}$. We are going to remedy it by using $(E)^{\nu}$ first before using the argument to get the equation.

By $(E)^{\nu}$ there exist $\hat{\xi}^{(i)} \in E^{\nu}_{\alpha-2}(\rho^0)$ $(1 \leq i \leq k)$ with $\delta^{\nu}_{\alpha-2}\hat{\xi}^{(i)} = 0$ such that they generate $\mathrm{Im}(H^{\nu}_{\alpha-1} \longrightarrow \overline{H}^{\nu}_{\alpha-1})$ over ${}_{n-1}\mathcal{O}_0$. (We retain ρ^0 and k

simply to avoid the introduction of more symbols and such a retention clearly does not result in any loss of generality.) By applying $(E)^\nu$ to ξ, we can find

$$\hat{\xi} \in E^\nu_{\alpha-2}(\rho) \quad \text{with} \quad \delta^\nu_{\alpha-2}\hat{\xi} = 0$$

$$\hat{\eta} \in E^\nu_{\alpha+1}(\rho)$$

such that

$$r^\nu_{\alpha+1,\alpha}\xi = r^\nu_{\alpha+1,\alpha-2}\hat{\xi} + \delta^{\nu-1}_{\alpha+1}\hat{\eta}$$

$$\|\hat{\xi}\|^\nu_{\alpha-2,\rho} \leqq \hat{C}e$$

$$\|\hat{\eta}\|^\nu_{\alpha+1,\rho} \leqq \hat{C}e .$$

By repeating the preceding argument with $\hat{\xi}$, $\hat{\xi}(i)$ (instead of $\xi, \xi^{(i)}$), we obtain (for ρ sufficiently strictly small)

$$a^*_i \in \Gamma(\Delta(\rho), {}_n\mathcal{O})$$

$$\eta^* \in E^\nu_{\alpha+1}(\rho)$$

such that

$$r^\nu_{\alpha+1,\alpha-2}\hat{\xi} = \sum_{i=1}^{k} a^*_1 r^\nu_{\alpha+1,\alpha-2}\hat{\xi}^{(i)} + \delta^{\nu-1}_{\alpha+1}\eta^*$$

$$|a^*_i|_\rho \leqq 2C''_{\bar{\rho}}(\rho_n D_{\bar{\rho}} + 1)\hat{C}e$$

$$\|\eta^*\|^{\nu-1}_{\alpha+1} \leqq 2C''_{\bar{\rho}}(\rho_n D_{\bar{\rho}} + 1)\hat{C}e + \hat{C}e$$

(where again $C''_{\bar{\rho}}$, $D_{\bar{\rho}}$ are retained simply to avoid more symbols and such a retention does not result in any loss of generality). It follows that

$$r^{\nu}_{\alpha+1,\alpha}\xi = \sum_{i=1}^{k} a_i^* r^{\nu}_{\alpha+1,\alpha-2}\xi^{(i)} + \delta^{\nu-1}_{\alpha+1}(\hat{\eta} + \eta^*) .$$

Hence $H^{\nu}_{\alpha+1}$ is finitely generated over ${}_n\mathcal{O}_0$. We have actually proved more than this, namely, we have shown that the finite generation is with bounds when generators are chosen in a certain way.

(IV) We are going to prove the full strength of $(B)^{\nu}_n$ b) by invoking the existence of privileged polydiscs (in the sense of Grauert). Suppose $\check{\xi}^{(i)} \in E^{\nu}_{\alpha}(\rho^0)$ $(1 \leq i \leq k)$ with $\delta\check{\xi}^{(i)} = 0$. Let A be the ${}_n\mathcal{O}_0$-submodule of $\mathcal{H}^{\nu}_{\alpha+1,0}$ generated by $\check{\xi}^{(1)}, \ldots, \check{\xi}^{(k)}$.

By $(E)^{\nu}$ and the finite generation of $\overline{H^{\nu}_{\alpha-1}}$ over ${}_{n-1}\mathcal{O}_0$, (after shrinking ρ^0) there exist $\zeta^{(i)} \in E^{\nu}_{\alpha-2}(\rho^0)$ $(1 \leq i \leq \ell)$ with $\delta^{\nu}_{\alpha-2}\zeta^{(i)} = 0$ such that $\zeta^{(1)}, \ldots, \zeta^{(\ell)}$ generate $\operatorname{Im}(H^{\nu}_{\alpha-1} \longrightarrow \overline{H^{\nu}_{\alpha-1}})$. Let

$$f: {}_n\mathcal{O}_0^{\ell} \longrightarrow \mathcal{H}^{\nu}_{\alpha+1,0}/A$$

be induced by the $1 \times \ell$ matrix

$$\left(r^{\nu}_{\alpha+1,\alpha-2}\zeta^{(1)}, \ldots, r^{\nu}_{\alpha+1,\alpha-2}\zeta^{(\ell)}\right)$$

(an element of ${}_n\mathcal{O}_0^{\ell}$ being represented by a column ℓ-vector). There exist (after shrinking ρ^0)

$$v^{(i)} = \left(v_1^{(i)}, \ldots, v_{\ell}^{(i)}\right) \in \Gamma(\Delta(\rho^0), {}_n\mathcal{O}^{\ell}) \quad (1 \leq i \leq m)$$

such that the germs of $v^{(1)}, \ldots, v^{(m)}$ at 0 generate $\operatorname{Ker} f$. There exist (after shrinking ρ^0)

$$u_j^{(i)} \in \Gamma(\Delta(\rho^0), {}_n\mathcal{O})$$
$$\varkappa^{(i)} \in E_{\alpha+1}^{\nu-1}(\rho^0) \qquad (1 \leq i \leq m,\ 1 \leq j \leq \ell)$$

such that

$$\sum_{j=1}^{\ell} v_j^{(i)} r_{\alpha+1,\alpha-2}^{\nu} \zeta^{(j)} = \sum_{j=1}^{k} u_j^{(i)} r_{\alpha+1,\alpha}^{\nu} \check{\xi}^{(j)} + \delta_{\alpha+1}^{\nu-1} \varkappa^{(i)}$$

$$(1 \leq i \leq m).$$

Let $\varphi: {}_n\mathcal{O}^m \longrightarrow {}_n\mathcal{O}^\ell$ on $\Delta(\rho^0)$ be defined by the matrix $\left(v_j^{(i)}\right)_{1 \leq i \leq m,\ 1 \leq j \leq \ell}$.

Now, take $\xi \in E_\alpha^\nu(\rho)$ with $\delta_\alpha^\nu \xi = 0$ such that the image of ξ in $\mathcal{H}_{\alpha+1,0}^\nu$ belongs to A. Let $e = \|\xi\|_{\alpha,\rho}^\nu$. By (III), for ρ sufficiently strictly small, there exist

$$b_i \in \Gamma(\Delta(\rho), {}_n\mathcal{O})$$
$$\pi \in E_{\alpha+1}^{\nu-1}(\rho)$$

such that

$$r_{\alpha+1,\alpha}^\nu \xi = \sum_{i=1}^{\ell} b_i r_{\alpha+1,\alpha-2}^\nu \zeta^{(i)} + \delta_{\alpha+1}^{\nu-1} \pi$$

$$|b_i|_\rho \leq C_\rho^* e$$

$$\|\pi\|_{\alpha+1,\rho}^{\nu-1} \leq C_\rho^* e$$

where $C^*_{\bar\rho}$ is a constant depending on $\bar\rho$. Since the image of ξ in $\mathcal{H}^\nu_{\alpha+1,0}$ belongs to A, it follows that

$$(b_1, \ldots, b_\ell)_0 \in \operatorname{Ker} f = (\operatorname{Im}\varphi)_0 .$$

By (1.6), for ρ sufficiently strictly small, $\Delta(\rho)$ is a $(\operatorname{Coker}\varphi)$-privileged neighborhood of 0 in the sense of Grauert. Hence, for such ρ, there exist

$$c_i \in \Gamma(\Delta(\rho), {}_n\mathcal{O}) \quad (1 \leqq i \leqq m)$$

such that

$$(b_1, \ldots, b_\ell) = \sum_{i=1}^m c_i(v_1^{(i)}, \ldots, v_\ell^{(i)})$$

$$|c_i|_\rho \leqq C^\#_\rho C^*_{\bar\rho} e ,$$

where $C^\#_\rho$ is a constant depending on ρ. It follows that

$$r^\nu_{\alpha+1,\alpha}\xi = \sum_{i=1}^k \Big(\sum_{j=1}^m c_j u_i^{(j)}\Big) r^\nu_{\alpha+1,\alpha}\check\xi^{(i)} + \delta^{\nu-1}_{\alpha+1}\Big(\sum_{j=1}^m c_j \varkappa^{(j)} + \pi\Big).$$

This concludes the proof of $(B)^\nu_n$. Q.E.D.

Notice that, if t_n is not a zero-divisor of $(\operatorname{Coker}\varphi)_0$, then all the constants involved in the proof are independent of ρ_n.

(7.3) Observe that, by the open mapping theorem, $(F)^\nu \Longrightarrow (B)^\nu_0$. Moreover, by the condition of factorization through a Hilbert space given in $(F)^\nu$, in the statement of

$(B)_0^\nu$ we can choose $a^{(i)}$ and η so that the map

$$\xi \longmapsto (a^{(1)}, \ldots, a^{(k)}, \eta)$$

is linear over $\mathbb{C}[t]$.

By induction on n, it follows from (7.2) that $(E)^\nu$, $(M)^{\nu+1}$, $(F)^\nu \Longrightarrow (B)_n^\nu$. Under the assumption of $(E)^\nu$, $(M)^{\nu+1}$, and $(F)^\nu$, we are going to show, by induction on n that the natural map θ from

$$\operatorname{Im}(\mathcal{H}^\nu_{\alpha,t^0} \longrightarrow \mathcal{H}^\nu_{\alpha+1,t^0})$$

to

$$\operatorname{Im}(\mathcal{H}^\nu_{\beta,t^0} \longrightarrow \mathcal{H}^\nu_{\beta+1,t^0})$$

is an isomorphism. We can assume without loss of generality that $t^0 = 0$. The case $n = 0$ follows immediately from $(F)^\nu$. The surjectivity of θ follows from $(E)^\nu$. Suppose $\theta(\xi^*) = 0$ for some

$$\xi^* \in \operatorname{Im}(\mathcal{H}^\nu_{\alpha,0} \longrightarrow \mathcal{H}^\nu_{\alpha+1,0}) .$$

By $(E)^\nu$, ξ^* is the image of some $\xi \in \mathcal{H}^\nu_{\alpha-1,0}$. Take arbitrarily $m \geqq 1$. Let $d^m = (\infty, \ldots, \infty, m) \in \mathbb{N}^n_*$. Let $\bar{\xi} \in \mathcal{H}^\nu_{\alpha-1}[d^m]_0$ be the image of ξ. Since the image of $\bar{\xi}$ in $\mathcal{H}^\nu_{\beta+1}[d^m]_0$ is 0, by induction hypothesis the image of $\bar{\xi}$ in $\mathcal{H}^\nu_\alpha[d^m]_0$ is 0. From the cohomology sequence

$$\cdots \longrightarrow \mathcal{H}^{\nu}_{\alpha,0} \longrightarrow \mathcal{H}^{\nu}_{\alpha,0} \longrightarrow \mathcal{H}_{\alpha}[d^m]_0 \longrightarrow \mathcal{H}^{\nu+1}_{\alpha,0} \longrightarrow \cdots$$

of the short exact sequence

$$0 \longrightarrow \mathcal{O}(E^{\cdot}_{\alpha})_0 \overset{\sigma}{\longrightarrow} \mathcal{O}(E^{\cdot}_{\alpha})_0 \longrightarrow \mathcal{O}(E^{\cdot}_{\alpha})_0(d^m) \longrightarrow 0$$

(where σ is defined by multiplication by t_n^m), it follows that the image of ξ in $\mathcal{H}^{\nu}_{\alpha,0}$ belongs to $t_n^m \mathcal{H}^{\nu}_{\alpha,0}$. Hence

$$\xi^* \in t_n^m \operatorname{Im}(\mathcal{H}^{\nu}_{\alpha,0} \longrightarrow \mathcal{H}^{\nu}_{\alpha+1,0}) .$$

Since $\operatorname{Im}(\mathcal{H}^{\nu}_{\alpha,0} \longrightarrow \mathcal{H}^{\nu}_{\alpha+1,0})$ is finitely generated over ${}_n\mathcal{O}_0$, it follows from the arbitrariness of m, that $\xi^* = 0$.

From now on, whenever $(E)^{\nu}$, $(M)^{\nu+1}$, and $(F)^{\nu}$ are satisfied, we denote $\operatorname{Im}(\mathcal{H}^{\nu}_{\alpha} \longrightarrow \mathcal{H}^{\nu}_{\alpha+1})$ by $\mathcal{H}^{\nu}$. $\mathcal{H}^{\nu}$ is independent of the choice of α.

(7.4) Let us investigate under which circumstances, in the statement of $(B)^{\nu}_n$, we can choose $a^{(i)}$ and η so that the map

$$\xi \longmapsto (a^{(1)}, \ldots, a^{(k)}, \eta)$$

is linear over $\mathbb{C}[t]$. Looking at the proof of (7.2), one easily sees that this is the case if

i) $(B)^{\nu}_{n-1}$ has the corresponding property of $\mathbb{C}[t_1, \ldots, t_{n-1}]$-linearity for the ${}_{n-1}\mathcal{O}_0$-submodule $\operatorname{Im}(H^{\nu}_{\alpha-1} \longrightarrow \overline{H^{\nu}_{\alpha-1}})$ of $\overline{\mathcal{H}^{\nu}_{\alpha-1,0}}$.

ii) Coker φ is locally free at 0.

For $1 \leqq \ell \leqq n$ let

$$d^{(\ell)} = (\underbrace{\infty, \ldots, \infty}_{\ell} \underbrace{1, \ldots, 1}_{n-\ell}) \in \mathbb{N}_*^n$$

Denote

$$\operatorname{Im}\left(\mathcal{H}_\beta^\nu[d^{(\ell)}] \longrightarrow \mathcal{H}_{\beta+1}^\nu[d^{(\ell)}]\right)$$

by $\mathcal{H}^\nu[d^{(\ell)}]$ which by (7.3) is independent of the choice of β.

The above condition ii) is satisfied if $A = H_{\alpha+1}^\nu$. By induction on n, condition ii) is satisfied if the natural maps from $\mathcal{H}^\nu[d^{(\ell)}]_0$ to $\mathcal{H}^\nu[d^{(\ell-1)}]_0$ are surjective for $1 < \ell \leqq n$. (A by-product of this surjectivity condition is that all the constants in the proof of (7.2) are independent of ρ, because of the last sentence of (7.2)). From the exact sequence

$$0 \longrightarrow \mathcal{O}(E_\beta^\bullet)(d^{(\ell)}) \xrightarrow{\sigma} \mathcal{O}(E_\beta^\bullet)(d^{(\ell)}) \longrightarrow \mathcal{O}(E_\beta^\bullet)(d^{(\ell-1)}) \longrightarrow 0$$

(where σ is defined by multiplication by t_ℓ), we obtain the exact sequence

$$\begin{aligned}\cdots \longrightarrow \mathcal{H}_\beta^\mu(d^{(\ell)}) &\longrightarrow \mathcal{H}_\beta^\mu(d^{(\ell)}) \\ &\longrightarrow \mathcal{H}_\beta^\mu(d^{(\ell-1)}) \longrightarrow \mathcal{H}_\beta^{\mu+1}(d^{(\ell)}) \longrightarrow \cdots.\end{aligned}$$

Hence the surjectivity condition just mentioned is satisfied if $\mathcal{H}_0^\mu = 0$ for $\nu < \mu < \nu + n$. Of course, in general, this last condition is not satisfied. Under the assumption of $(E)^{\mu}$, $(M)^{\mu}$, $(F)^{\mu}$ for certain μ's, we are going to modify the complexes so that the new complexes satisfy this last condition.

(7.5) Suppose $L^{\nu} = \Delta \times \mathbb{C}^{p_\nu}$ and

$$\cdots \longrightarrow L^{\nu-1} \longrightarrow L^{\nu} \xrightarrow{\partial^{\nu}} L^{\nu+1} \longrightarrow \cdots$$

is a complex in which the maps of (holomorphic) bundle-homomorphisms. Let

$$\begin{array}{ccccccccc} \cdots \longrightarrow & L^{\nu-1} & \longrightarrow & L^{\nu} & \xrightarrow{\partial^{\nu}} & L^{\nu+1} & \longrightarrow \cdots \\ & \downarrow & & \downarrow \sigma_{\alpha}^{\nu} & & \downarrow & \\ \cdots \longrightarrow & E_{\alpha}^{\nu-1} & \longrightarrow & E_{\alpha}^{\nu} & \longrightarrow & E_{\alpha}^{\nu+1} & \longrightarrow \cdots \end{array}$$

be a commutative diagram of bundle-homomorphisms such that

$$\sigma_{\beta}^{\nu} = r_{\beta\alpha}^{\nu} \sigma_{\alpha}^{\nu} \quad \text{for } \alpha < \beta .$$

Note that, to have σ_{α}^{ν}, it suffices to have $\sigma_{\alpha_1}^{\nu}$ and then define σ_{α}^{ν} as $r_{\alpha\alpha_1}^{\nu}\sigma_{\alpha_1}^{\nu}$. Let the complex

$$\cdots \longrightarrow \tilde{E}_{\alpha}^{\nu-1} \longrightarrow \tilde{E}_{\alpha} \xrightarrow{\tilde{\delta}_{\alpha}^{\nu}} \tilde{E}_{\alpha}^{\nu+1} \longrightarrow \cdots$$

be the mapping cone of the above commutative diagram; that is,

$$\tilde{E}_{\alpha}^{\nu} = E_{\alpha}^{\nu} \oplus L^{\nu+1}$$

and

$$\tilde{\delta}_\alpha^\nu : E_\alpha^\nu \oplus L^{\nu+1} \longrightarrow E_\alpha^{\nu+1} \oplus L^{\nu+2}$$

is given by

$$\tilde{\delta}_\alpha^\nu(\xi \oplus f) = (\delta_\alpha^\nu \xi + (-1)^{\nu+1}\sigma^{\nu+1} f) \oplus \partial^{\nu+1} f .$$

Define

$$\tilde{r}_{\beta\alpha}^\nu : \tilde{E}_\alpha^\nu \longrightarrow \tilde{E}_\beta^\nu$$

by

$$\tilde{r}_{\beta\alpha}^\nu(\xi \oplus f) = (r_{\beta\alpha}^\nu \xi) \oplus f .$$

We are going to prove the following three statements (after a possible replacement of Δ by any open polydisc relatively compact in Δ).

a) The complexes $E_\alpha^\bullet$ satisfy $(E)^\nu$ and $(M)^{\nu+1} \Longrightarrow$ the complexes $\tilde{E}_\alpha^\bullet$ satisfy $(E)^\nu$.

b) The complexes $E_\alpha^\bullet$ satisfy $(M)^\nu \Longrightarrow$ the complexes $\tilde{E}_\alpha^\bullet$ satisfy $(M)^\nu$.

c) The complexes $E_\alpha^\bullet$ satisfy $(F)^\nu \Longrightarrow$ the complexes $\tilde{E}_\alpha^\bullet$ satisfy $(F)^\nu$.

Statement c) is clear. Let us prove b): Suppose

$$\xi \oplus f \in \tilde{E}_\alpha^\nu(t^0,d,\rho) \text{ with } \tilde{\delta}_\alpha^\nu(\xi \oplus f) = 0$$

and

$$\tilde{r}_{\beta\alpha}^\nu(\xi \oplus f) = \tilde{\delta}_\beta^{\nu-1}(\xi' \oplus f')$$

for some $\xi' \oplus f' \in \tilde{E}_\beta^{\nu-1}(t^0,d,\rho)$ and $\alpha < \beta$. From the definition of $\tilde{\delta}_\alpha^\nu$, $\tilde{\delta}_\beta^{\nu-1}$ we have

i) $$\delta_\alpha^\nu \xi + (-1)^{\nu+1}\sigma_\alpha^{\nu+1} f = 0$$

ii) $$r_{\beta\alpha}^\nu \xi = \delta_\beta^{\nu-1}\xi' + (-1)^\nu \sigma_\beta^\nu f'$$

iii) $$f = \partial^\nu f' .$$

It follows from $\delta_\alpha^\nu \sigma_\alpha^\nu = \sigma_\alpha^{\nu+1}\partial^\nu$ and i), iii) that

$$\delta_\alpha^\nu(\xi + (-1)^{\nu+1}\sigma_\alpha^\nu f') = 0 .$$

By applying $(M)^\nu$ of the complexes $E_\gamma^\bullet$ to the equation (obtained from ii))

$$r_{\beta\alpha}^\nu(\xi + (-1)^{\nu+1}\sigma_\alpha^\nu f') = \delta_\beta^{\nu-1}\xi'$$

we obtain

$$\xi'' \in E_{\alpha+1}^{\nu-1}(t^0,d,\rho)$$

such that

$$r_{\alpha+1,\alpha}^\nu(\xi + (-1)^{\nu+1}\sigma_\alpha^\nu f') = \delta_{\alpha+1}^{\nu-1}\xi'' .$$

$$\|\xi''\|_{\alpha+1,t^0,d,\rho}^{\nu-1} \leq C \operatorname{Max}\left(\|\xi + (-1)^{\nu+1}\sigma_\alpha^\nu f'\|_{\alpha,t^0,d,\rho}^\nu , \|\xi'\|_{\beta,t^0,d,\rho}^{\nu-1}\right).$$

So we have

$$\tilde{r}_{\alpha+1,\alpha}^\nu(\xi \oplus f) = \tilde{\delta}_{\alpha+1}^{\nu-1}(\xi'' \oplus f') .$$

The requirement on the estimation of the norm of $\xi'' \oplus f'$ by the norms of $\xi \oplus f$ and $\xi' \oplus f'$ is clearly satisfied.

Now we prove a), Suppose $\alpha < \beta$

$$\xi \oplus f \in \tilde{E}^{\nu}_{\beta}(t^0, d, \rho) \quad \text{with} \quad \tilde{\delta}^{\nu}_{\beta}(\xi \oplus f) = 0, \quad \text{i.e.}$$

$$\begin{cases} \delta^{\nu}_{\beta}\xi + (-1)^{\nu+1}\sigma^{\nu+1}_{\beta} f = 0 \\ \partial^{\nu+1} f = 0 . \end{cases}$$

From the first equation, we obtain

$$r^{\nu}_{\beta,\alpha-1}\sigma^{\nu+1}_{\alpha-1} f = \delta^{\nu}_{\beta}(-1)^{\nu}\xi$$

By applying property $(M)^{\nu+1}$ of the complexes $E^{\bullet}_{\gamma}$ to this equation, (since $\delta^{\nu+1}_{\alpha-1}\sigma^{\nu+1}_{\alpha-1} f = \sigma^{\nu+1}_{\alpha-1}\partial^{\nu+1} f = 0$), we obtain

$$\xi' \in E^{\nu}_{\alpha}(t^0, d, \rho)$$

such that

$$r^{\nu}_{\alpha,\alpha-1}\sigma^{\nu+1}_{\alpha-1} f = \delta^{\nu}_{\alpha}\xi'$$

$$\|\xi'\|^{\nu}_{\alpha, t^0, d, \rho} \leq C \operatorname{Max}\left(\|\sigma^{\nu+1}_{\alpha-1} f\|^{\nu+1}_{\alpha-1, t^0, d, \rho}, \|\xi\|^{\nu}_{\beta, t^0, d, \rho}\right).$$

It follows that

$$\delta^{\nu}_{\beta}(\xi + (-1)^{\nu+1} r^{\nu}_{\beta\alpha}\xi') = 0 .$$

By applying property $(E)^{\nu}$ of the complexes $E^{\bullet}_{\gamma}$ to $\xi + (-1)^{\nu+1} r^{\nu}_{\beta\alpha}\xi'$, we obtain

$$\tilde{\xi} \in E_\alpha^\nu(t^0,d,\rho) \quad \text{with} \quad \delta_\alpha^\nu \tilde{\xi} = 0$$

$$\eta \in E_{\beta+1}^{\nu-1}(t^0,d,\rho)$$

such that

$$r_{\beta+1,\beta}^\nu(\xi + (-1)^{\nu+1} r_{\beta\alpha}^\nu \xi') = r_{\beta+1,\alpha}^\nu \tilde{\xi} + \delta_{\beta+1}^{\nu-1}\eta$$

$$\operatorname{Max}\left(\|\tilde{\xi}\|_{\alpha,t^0,d,\rho}^\nu , \|\eta\|_{\beta+1,t^0,d,\rho}^{\nu-1}\right) \leq C' \|\xi + (-1)^{\nu+1} r_{\beta\alpha}^\nu \xi'\|_{\beta,t^0,d,\rho}^\nu .$$

Then

$$(\xi' + \xi) \oplus f \in \tilde{E}_\alpha^\nu(t^0,d,\rho) \quad \text{with} \quad \tilde{\delta}_\alpha^\nu((\xi' + \xi) \oplus f) = 0$$

$$\eta \oplus 0 \in \tilde{E}_{\beta+1}^{\nu-1}$$

$$\tilde{r}_{\beta+1,\beta}^\nu(\xi \oplus f) = \tilde{r}_{\beta+1,\alpha}^\nu((\xi' + \tilde{\xi}) \oplus f) + \tilde{\delta}_{\beta+1}^{\nu-1}(\eta \oplus 0) .$$

The requirement on the estimation of the norms of $(\xi' + \xi) \oplus f$ and $\eta \oplus 0$ by the norm of $\xi \oplus f$ is clearly satisfied.

(7.6) <u>Proposition</u>. <u>Suppose, for</u> $p \leq \nu \leq s$, <u>the properties</u> $(E)^\nu$, $(M)^{\nu+1}$, $(F)^\nu$ <u>hold for the complexes</u> $E_\alpha^\bullet$. <u>Then there exists a complex</u>

$$0 \longrightarrow L^p \longrightarrow L^{p+1} \longrightarrow \cdots \longrightarrow L^s \longrightarrow 0$$

<u>of trivial vector bundles of finite rank on</u> $\Delta(\rho^0)$ (<u>where</u> $\rho^0 \in \mathbb{R}_+^n$ <u>and</u> $\Delta(\rho^0) \subset \Delta$) <u>in which the maps are holomorphic bundle-homomorphisms and there exists a commutative diagram</u>

$$\begin{array}{ccc} & L^{\bullet} & \\ {\scriptstyle \sigma_\alpha^{\bullet}} \swarrow & & \searrow {\scriptstyle \sigma_{\alpha+1}^{\bullet}} \\ E_\alpha^{\bullet} & \xrightarrow[r_{\alpha+1,\alpha}^{\bullet}]{} & E_{\alpha+1}^{\bullet} \end{array}$$

<u>of complex-homomorphisms on</u> $\Delta(\rho^0)$ <u>such that the mapping cones</u> $\tilde{E}_\alpha^{\bullet}$ <u>of</u> $\sigma_\alpha^{\bullet}$ <u>satisfy</u>

$$\operatorname{Im}\left(H^\nu(\mathcal{O}(\tilde{E}_\alpha^{\bullet})_0) \longrightarrow H^\nu(\mathcal{O}(\tilde{E}_{\alpha+1}^{\bullet})_0)\right) = 0$$

<u>for</u> $p \leqq \nu \leqq s$.

<u>Proof</u>. We are going to prove by descending induction on μ that, for $p \leqq \mu \leqq s+1$, there exist a complex

$$0 \longrightarrow {}_\mu L^\mu \xrightarrow{{}_\mu\partial^\mu} {}_\mu L^{\mu+1} \xrightarrow{{}_\mu\partial^{\mu+1}} \cdots \longrightarrow {}_\mu L^s \longrightarrow 0$$

of trivial vector bundles of finite rank on some $\Delta(\rho^0)$ and a commutative diagram

$$\begin{array}{ccc} & {}_\mu L^{\bullet} & \\ {\scriptstyle {}_\mu\sigma_\alpha^{\bullet}} \swarrow & & \searrow {\scriptstyle {}_\mu\sigma_{\alpha+1}^{\bullet}} \\ E_\alpha^{\bullet} & \xrightarrow[r_{\alpha+1,\alpha}^{\bullet}]{} & E_{\alpha+1}^{\bullet} \end{array}$$

such that the mapping cones ${}_\mu\tilde{E}_\alpha^{\bullet}$ of ${}_\mu\sigma_\alpha^{\bullet}$ satisfy

$$\operatorname{Im}\left(H^\nu(\mathcal{O}({}_\mu\tilde{E}_\alpha^{\bullet})_0) \longrightarrow H^\nu(\mathcal{O}({}_\mu\tilde{E}_{\alpha+1}^{\bullet})_0)\right) = 0$$

for $\mu \leqq \nu \leqq s$.

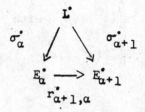

The case $\mu = s+1$ is trivial, because one can set ${}_{\mu}L^{\bullet} = 0$. To go from the step μ to the step $\mu-1$, we observe that, by (7.5), ${}_{\mu}\tilde{E}^{\bullet}_{\alpha}$ has properties $(E)^{\mu-1}, (M)^{\mu}$, and $(F)^{\mu-1}$. By (7.2), ${}_{\mu}\tilde{E}^{\bullet}_{\alpha}$ has property $(B)^{\mu-1}_{n}$.

$$A: = \operatorname{Im}\left(H^{\nu}\left(\mathcal{O}({}_{\mu}\tilde{E}^{\bullet}_{\alpha})_0\right) \longrightarrow H^{\nu}\left(\mathcal{O}({}_{\mu}\tilde{E}^{\bullet}_{\alpha+1})_0\right)\right)$$

is finitely generated over ${}_{n}\mathcal{O}_0$. One can find (after shrinking ρ^0)

$$\xi^{(1)}, \ldots, \xi^{(k)} \in {}_{\mu}\tilde{E}^{\mu-1}_{\alpha_1}(\rho^0) \quad \text{with} \quad {}_{\mu}\delta^{\mu-1}_{\alpha_1}\xi^{(i)} = 0$$

whose images in A generate A over ${}_{n}\mathcal{O}_0$. Define

$$\begin{aligned} {}_{\mu-1}L^{\mu-1} &= \Delta(\rho^0) \times \mathbb{C}^k \\ {}_{\mu-1}L^{\nu} &= {}_{\mu}L^{\nu} \\ {}_{\mu-1}\partial^{\nu} &= {}_{\mu}\partial^{\nu} \qquad\qquad (\mu \leqq \nu \leqq s) \\ {}_{\mu-1}\sigma^{\nu}_{\alpha} &= {}_{\mu}\sigma^{\nu}_{\alpha} \end{aligned}$$

Let

$$(\tau_1, \tau_2): {}_{\mu-1}L^{\mu-1} \longrightarrow {}_{\mu}\tilde{E}^{\mu-1}_{\alpha_1} = E^{\mu-1}_{\alpha_1} \oplus {}_{q}L^{\mu}$$

be defined by $\xi^{(1)}, \ldots, \xi^{(k)}$. Define

$$\begin{aligned} {}_{\mu-1}\sigma^{\mu-1}_{\alpha} &= (-1)^{\mu-1} r^{\mu-1}_{\alpha\alpha_1}\tau_1 \, . \\ {}_{\mu-1}\partial^{\mu-1} &= \tau_2 \, . \end{aligned}$$

Then the complex ${}_{\mu-1}L^{\bullet}$ and the map ${}_{\mu-1}\sigma^{\bullet}_{\alpha}$ satisfy the requirement. Q.E.D.

§8. Right Inverses of Coboundary Maps

(8.1) As in §7, suppose

$$\begin{array}{ccccccccc}
\cdots & \longrightarrow & E_\alpha^{\nu-1} & \longrightarrow & E_\alpha^{\nu} & \xrightarrow{\delta_\alpha^\nu} & E_\alpha^{\nu+1} & \longrightarrow & \cdots \\
 & & \downarrow & & \downarrow r_{\alpha+1,\alpha}^{\nu} & & \downarrow & & \\
\cdots & \longrightarrow & E_\alpha^{\nu-1} & \longrightarrow & E_{\alpha+1}^{\nu} & \longrightarrow & E_{\alpha+1}^{\nu+1} & \longrightarrow & \cdots
\end{array}$$

$(\alpha_1 \leq \alpha \leq \alpha_2)$ is a commutative diagram of trivial Banach bundles and bundle-homomorphisms on Δ with $\delta_\alpha^\nu \delta_\alpha^{\nu-1} = 0$. We use the same notations as in §7. Fix p. We assume that

i) the complexes $E_\alpha^\bullet$ satisfy $(E)^\nu$, $(M)^{\nu+1}$, and $(F)^{\nu-}$ for $p \leq \nu \leq p+n$.

ii) $\operatorname{Im}(\mathcal{H}_{\alpha,0}^\nu \longrightarrow \mathcal{H}_{\alpha+1,0}^\nu) = 0$ for $p \leq \nu \leq p+n$ and all α.

By the results of §7, $(B)_n^p$ and $(B)_n^{p+1}$ hold with an additional statement of $\mathbb{C}[t]$-linearity. So, for ρ sufficiently strictly small and for $\nu = p, p+1$, we have maps Φ_α^ν from

$$\{\xi \in E_\alpha^\nu(\rho) \mid \delta_\alpha^\nu \xi = 0\}$$

to $E_{\alpha+1}^{\nu-1}(\rho)$ linear over $\mathbb{C}[t]$ such that

$$r_{\alpha+1,\alpha}^\nu \xi = \delta_{\alpha+1}^{\nu-1} \Phi_\alpha^\nu \xi$$

$$\|\Phi_\alpha^\nu \xi\|_{\alpha+1,\rho}^{\nu-1} \leq C \|\xi\|_{\alpha,\rho}^\nu$$

where C is a constant. It follows that, for $\xi \in E_\alpha^p(\rho)$, we have

$$\delta_{\alpha+2}^{p-1}\left(\Phi_{\alpha+1}^p(r_{\alpha+1,\alpha}^p\xi - \Phi_\alpha^{p+1}\delta_\alpha^p\xi)\right) = r_{\alpha+1,\alpha+2}^p(r_{\alpha+1,\alpha}^p\xi - \Phi_\alpha^{p+1}\delta_\alpha^p\xi).$$

By replacing Δ by $\Delta(\rho)$ for some sufficiently strictly small ρ, we assume that Φ_α^p and Φ_α^{p+1} are defined for $\rho = (1, \ldots, 1)$ and, from now on, Φ_α^p and Φ_α^{p+1} denote the maps for that particular ρ.

(8.2) We are going to define

$$\Psi_\alpha^p: \mathcal{O}(E_\alpha^p) \longrightarrow \mathcal{O}(E_{\alpha+2}^{p-1})$$

such that

$$(*) \qquad \delta_{\alpha+2}^{p-1}\Psi_\alpha^p\delta_\alpha^{p-1} = r_{\alpha+2,\alpha}^p\delta_\alpha^{p-1}$$

on $\mathcal{O}(E_\alpha^{p-1})$.

For $\Delta(t^0,\rho) \subset \Delta$ and $\xi \in E_\alpha^p(t^0,\rho)$ with power series expansion

$$\xi = \sum_{\lambda \in \mathbb{N}_0^n} \xi_\lambda \left(\frac{t - t^0}{\rho}\right)^\lambda$$

define

$$\Psi_\alpha^p\xi = \sum_{\lambda \in \mathbb{N}_0^n} \left(\frac{t - t^0}{\rho}\right)^\lambda \Phi_{\alpha+1}^p(r_{\alpha+1,\alpha}^p\xi_\lambda - \Phi_\alpha^{p+1}\delta_\alpha^p\xi_\lambda)$$

where ξ_λ is regarded as an element of $E_\alpha^p(1, \ldots, 1)$. Because of the norm estimates for $\Phi^p_{\alpha+1}, \Phi^{p+1}_\alpha$, we have $\Psi^p_\alpha \xi \in E^{p-1}_{\alpha+2}(t^0,\rho)$.

To verify that this definition can give us a sheaf-homomorphism, we have to prove its compatibility with restrictions. Suppose $\Delta(t',\rho') \subset\subset \Delta(t^0,\rho)$. Let $\xi' \in E^p_\alpha(t',\rho')$ be the restriction of ξ. Then we have the power series expansion

$$\xi' = \sum_{\lambda \in \mathbb{N}_0^n} \left(\sum_{\substack{\gamma \in \mathbb{N}_0^n \\ \gamma \geqq \lambda}} \binom{\gamma}{\lambda} \left(\frac{t'-t^0}{\rho}\right)^{\gamma-\lambda} \xi_\gamma \left(\frac{\rho'}{\rho}\right)^\lambda \right) \left(\frac{t-t'}{\rho'}\right)^\lambda .$$

Hence

$$\Psi^p_\alpha \xi' = \sum_\lambda \left(\frac{t-t'}{\rho'}\right)^\lambda \Phi^p_{\alpha+1}\left(r^p_{\alpha+1,\alpha} - \Phi^{p+1}_\alpha \delta^p_\alpha\right)\left(\sum_{\gamma \geqq \lambda} \binom{\gamma}{\lambda}\left(\frac{t'-t^0}{\rho}\right)^{\gamma-\lambda} \xi_\gamma \left(\frac{\rho'}{\rho}\right)^\lambda\right).$$

It follows from the $\mathbb{C}$-linearity of $\Phi^p_{\alpha+1}, \Phi^{p+1}_\alpha, \delta^p_\alpha$ and their norm estimates that

$$\Psi^p_\alpha \xi' = \sum_\lambda \left(\frac{t-t'}{\rho'}\right)^\lambda \sum_{\gamma \geqq \lambda} \binom{\gamma}{\lambda}\left(\frac{t'-t^0}{\rho}\right)^{\gamma-\lambda} \left(\frac{\rho'}{\rho}\right)^\lambda \Phi^p_{\alpha+1}\left(r^p_{\alpha+1,\alpha} - \Phi^{p+1}_\alpha \delta^p_\alpha\right)\xi_\gamma$$

equals the restriction of $\Psi^p_\alpha \xi$ to $\Delta(t',\rho')$.

It remains to verify the identity (*). For $\Delta(t^0,\rho) \subset\subset \Delta$, take

$$\eta = \sum_\lambda \eta_\lambda \left(\frac{t-t^0}{\rho}\right)^\lambda \in E^{p-1}_\alpha(t^0,\rho) .$$

Let

$$\delta_\alpha^{p-1}\eta_\lambda = \sum_\gamma \zeta_{\lambda\gamma}\left(\frac{t - t^0}{\rho}\right)^\gamma .$$

Then

$$\delta_\alpha^{p-1}\eta = \sum_\lambda \left(\sum_{\gamma \leqq \lambda} \zeta_{\lambda-\gamma,\gamma}\right)\left(\frac{t - t^0}{\rho}\right)^\lambda$$

and, by definition of Ψ_α^p,

$$\Psi_\alpha^p \delta_\alpha^{p-1}\eta = \sum_\lambda \left(\frac{t - t^0}{\rho}\right)^\lambda \Phi_{\alpha+1}^p (r_{\alpha+1,\alpha}^p - \Phi_\alpha^{p+1}\delta_\alpha^p)\left(\sum_{\gamma \leqq \lambda} \zeta_{\lambda-\gamma,\gamma}\right)$$

Since $\Phi_{\alpha+1}^p, \Phi_\alpha^{p+1}$ both are linear over $\mathbb{C}[t]$ and have norm estimates, it follows that

$$\Psi_\alpha^p \delta_\alpha^{p-1}\eta = \Phi_{\alpha+1}^p (r_{\alpha+1,\alpha}^p - \Phi_\alpha^{p+1}\delta_\alpha^p)\delta_\alpha^{p-1}\eta .$$

Hence

$$\delta_{\alpha+2}^{p-1}\Psi_\alpha^p \delta_\alpha^{p-1}\eta = r_{\alpha+2,\alpha}^p \delta_\alpha^{p-1}\eta$$

and (*) follows.

§9. Global Isomorphism

(9.1) Suppose $E_\alpha^\cdot$ is a sequence of complexes of trivial Banach bundles as in §7. Fix two integers $s \geqq p$. Assume that, for $p \leqq \nu \leqq s$ and $\alpha < \beta$, the natural map from $\mathrm{Im}(\mathcal{H}_\alpha^\nu \longrightarrow \mathcal{H}_{\alpha+1}^\nu)$ to $\mathrm{Im}(\mathcal{H}_\beta^\nu \longrightarrow \mathcal{H}_{\beta+1}^\nu)$ is an isomorphism. Let

$$\mathcal{H}^\nu = \mathrm{Im}(\mathcal{H}_\alpha^\nu \longrightarrow \mathcal{H}_{\alpha+1}^\nu) \qquad (p \leqq \nu \leqq s) .$$

Assume that $\mathcal{H}^p, \mathcal{H}^{p+1}, \dots, \mathcal{H}^{s-1}$ are coherent on Δ. Let

$$\mathcal{B}_\alpha^\nu = \mathrm{Im}\left(\mathcal{O}(E_\alpha^{\nu-1}) \xrightarrow{\delta_\alpha^{\nu-1}} \mathcal{O}(E_\alpha^\nu)\right)$$

$$\mathcal{Z}_\alpha^\nu = \mathrm{Ker}\left(\mathcal{O}(E_\alpha^\nu) \xrightarrow{\delta_\alpha^\nu} \mathcal{O}(E_\alpha^{\nu+1})\right)$$

and, for any open polydisc $\Omega \subset \Delta$, let

$$H_\alpha^\nu(\Omega) = H^\nu\left(\Gamma(\Omega, \mathcal{O}(E_\alpha^\cdot))\right) .$$

Suppose that there exists, for every α, a sheaf-homomorphism

$$\Theta_\alpha^p : \mathcal{B}_\alpha^p \longrightarrow \mathcal{O}(E_{\alpha+2}^{p-1})$$

such that

$$\delta_{\alpha+2}^{p-1}\Theta_\alpha^p = r_{\alpha+2,\alpha}^p .$$

We are going to prove the following two statements for any open polydisc $\Omega \subset \Delta$.

i) $H_\alpha^s(\Omega) \longrightarrow \Gamma(\Omega, \mathcal{H}^s)$ is surjective.

ii) $\operatorname{Ker}\left(H^s_\alpha(\Omega) \longrightarrow \Gamma(\Omega, \mathcal{H}^s)\right) \subset \operatorname{Ker}\left(H^s_\alpha(\Omega) \longrightarrow H^s_{\alpha+p-\nu+3}(\Omega)\right)$

Let

$$\tilde{\mathcal{Z}}^\nu_\alpha = \mathcal{B}^\nu_\alpha + r^\nu_{\alpha,\alpha-1}\mathcal{Z}^\nu_{\alpha-1} \subset \mathcal{O}(E^\nu_\alpha) .$$

Consider the following two statements.

$1)_\nu$ $H^k(\Omega, \mathcal{B}^\nu_\alpha) \longrightarrow H^k(\Omega, \mathcal{B}^\nu_{\alpha+\nu-p+2})$ has zero image for $k \geq 1$.

$2)_\nu$ $H^k(\Omega, \mathcal{Z}^\nu_\alpha) \longrightarrow H^k(\Omega, \mathcal{Z}^\nu_{\alpha+\nu-p+2})$ has zero image for $k \geq 1$.

First, let us show that $1)_\nu \Longrightarrow 2)_\nu$ for $p \leq \nu < s$ The commutative diagram with exact rows

$$(*)_\nu \qquad \begin{array}{ccccccc} 0 \longrightarrow & \mathcal{B}^\nu_\alpha & \longrightarrow & \tilde{\mathcal{Z}}^\nu_\alpha & \longrightarrow & \mathcal{H}^\nu & \longrightarrow 0 \\ & \downarrow & & \downarrow & & \| & \\ 0 \longrightarrow & \mathcal{B}^\nu_{\alpha+\nu-p+2} & \longrightarrow & \tilde{\mathcal{Z}}^\nu_{\alpha+\nu-p+2} & \longrightarrow & \mathcal{H}^\nu & \longrightarrow 0 \end{array}$$

yields the commutative diagram with exact rows

$$\begin{array}{ccccc} H^k(\Omega, \mathcal{B}^\nu_\alpha) & \longrightarrow & H^k(\Omega, \tilde{\mathcal{Z}}^\nu_\alpha) & \longrightarrow & H^k(\Omega, \mathcal{H}^\nu) \\ \downarrow & & \downarrow & & \| \\ H^k(\Omega, \mathcal{B}^\nu_{\alpha+\nu-p+2}) & \longrightarrow & H^k(\Omega, \tilde{\mathcal{Z}}^\nu_{\alpha+\nu-p+2}) & \longrightarrow & H^k(\Omega, \mathcal{H}^\nu) . \end{array}$$

Since $\mathcal{H}^\nu$ is coherent, $H^k(\Omega, \mathcal{H}^\nu) = 0$ for $k \geq 1$. The result follows.

Next, we want to show that $2)_\nu \Longrightarrow 1)_{\nu+1}$ for any ν. The commutative diagram with exact rows

$$(\dagger)^\nu_\alpha \qquad \begin{array}{ccccccccc} 0 & \longrightarrow & \mathcal{Z}^\nu_\alpha & \longrightarrow & \mathcal{O}(E^\nu_\alpha) & \longrightarrow & \mathcal{B}^{\nu+1}_\alpha & \longrightarrow & 0 \\ & & \downarrow & & \downarrow & & \downarrow & & \\ 0 & \longrightarrow & \mathcal{Z}^\nu_{\alpha+\nu-p+3} & \longrightarrow & \mathcal{O}(E^\nu_{\alpha+\nu-p+3}) & \longrightarrow & \mathcal{B}^{\nu+1}_{\alpha+\nu-p+3} & \longrightarrow & 0 \end{array}$$

yields the commutative diagram with exact rows

$$\begin{array}{ccccc} H^k(\Omega,\mathcal{O}(E^\nu_\alpha)) & \longrightarrow & H^k(\Omega,\mathcal{B}^{\nu+1}_\alpha) & \longrightarrow & H^{k+1}(\Omega,\mathcal{Z}^\nu_\alpha) \\ \downarrow & & \downarrow & & \downarrow \sigma \\ H^k(\Omega,\mathcal{O}(E^\nu_{\alpha+\nu-p+3})) & \longrightarrow & H^k(\Omega,\mathcal{B}^{\nu+1}_{\alpha+\nu-p+3}) & \longrightarrow & H^{k+1}(\Omega,\mathcal{Z}^\nu_{\alpha+\nu-p+3}) \, . \end{array}$$

The result follows from $H^k(\Omega, \mathcal{O}(E^\nu_{\alpha+\nu-p+3})) = 0 \quad (k \geq 1)$ and the fact that σ factors through the map

$$H^{k+1}(\Omega, \tilde{\mathcal{Z}}^\nu_{\alpha+1}) \longrightarrow H^{k+1}(\Omega, \tilde{\mathcal{Z}}^\nu_{\alpha+\nu-p+3}) \, .$$

Now we are ready to prove i) and ii). The existence of Θ, together with the vanishing of $H^k(\Omega,\mathcal{O}(E^p_{\alpha+2})) = 0$ for $k \geq 1$, implies that $1)_p$ holds. Since $1)_\nu \Longrightarrow 2)_\nu$ for $p \leq \nu < s$ and $2)_\nu \Longrightarrow 1)_\nu$ for any ν, we have $1)_\nu$ for $p \leq \nu \leq s$ and $2)_\nu$ for $p \leq \nu < s$. The diagram $(*)_s$ yields the commutative diagram with exact rows

$$\begin{array}{ccccc}
\Gamma(\Omega, \tilde{\mathfrak{Z}}^s_\alpha) & \longrightarrow & \Gamma(\Omega, \mathcal{H}^s) & \longrightarrow & H^1(\Omega, \mathcal{B}^s_\alpha) \\
\downarrow & & \| & & \downarrow \\
(\Omega, \tilde{\mathfrak{Z}}^s_{\alpha+s-p+2}) & \xrightarrow{\tau} & \Gamma(\Omega, \mathcal{H}^s) & \longrightarrow & H^1(\Omega, \mathcal{B}^s_{\alpha+s-p+2})
\end{array}$$

By $1)_s$, τ is surjective and i) follows. To prove ii), we have to distinguish between the case $s = p$ and the case $s > p$. The case $s = p$ follows immediately from the existence of $\Theta^p_{\alpha+1}$. Suppose $s > p$. The diagram $(\dagger)^{s-1}_{\alpha+1}$ yields the commutative diagram with exact rows

$$\begin{array}{ccccc}
\Gamma(\Omega, \mathcal{O}(E^{s-1}_{\alpha+1})) & \longrightarrow & \Gamma(\Omega, \mathcal{B}^s_{\alpha+1}) & \longrightarrow & H^1(\Omega, \mathfrak{Z}^{s-1}_{\alpha+1}) \\
\downarrow & & \downarrow & & \downarrow \chi \\
\Gamma(\Omega, \mathcal{O}(E^{s-1}_{\alpha+s-p+3})) & \longrightarrow & \Gamma(\Omega, \mathcal{B}^s_{\alpha+s-p+3}) & \longrightarrow & H^1(\Omega, \mathfrak{Z}^{s-1}_{\alpha+s-p+3})
\end{array}$$

Suppose an element ξ of $\Gamma(\Omega, \mathfrak{Z}^s_\alpha)$ is mapped to 0 in $\Gamma(\Omega, \mathcal{H}^s)$. Then

$$r^s_{\alpha+1,\alpha}\xi \in \Gamma(\Omega, \mathcal{B}^s_{\alpha+1}) .$$

Since χ factors through the map

$$H^1(\Omega, \tilde{\mathfrak{Z}}^{s-1}_{\alpha+2}) \longrightarrow H^1(\Omega, \tilde{\mathfrak{Z}}^{s-1}_{\alpha+s-p+3}) ,$$

it follows from $2)_{s-1}$ that there exists

$$\eta \in \Gamma(\Omega, \mathcal{O}(E^{s-1}_{\alpha+s-p+3}))$$

such that

$$\delta^{s-1}_{\alpha+s-p+3}\eta = r^{s}_{\alpha+s-p+3,\alpha}\xi .$$

Hence ii) is proved.

(9.2) Suppose $E^{\bullet}_{\alpha}$ is a sequence of complexes of trivial Banach bundles as in §7. Fix two integers $s \geq p$. Assume that the complexes $E^{\bullet}_{\alpha}$ satisfy $(E)^{\nu}$, $(M)^{\nu+1}$, and $(F)^{\nu}$ for $p \leq \nu \leq \mathrm{Max}(s,p+n)$. By (7.6) (after replacing Δ by some $\Delta(\rho^0)$) there exists a complex $L^{\bullet}$ of trivial vector bundles of finite rank on Δ and there exists a commutative diagram

$$\begin{array}{ccc} & L^{\bullet} & \\ \sigma^{\bullet}_{\alpha} \swarrow & & \searrow \sigma^{\bullet}_{\alpha+1} \\ E^{\bullet}_{\alpha} & \xrightarrow[r^{\bullet}_{\alpha+1,\alpha}]{} & E^{\bullet}_{\alpha+1} \end{array}$$

of complex-homomorphisms on Δ such that the mapping cones $\tilde{E}^{\bullet}_{\alpha}$ of $\sigma^{\bullet}_{\alpha}$ satisfy

$$\mathrm{Im}\left(H^{\nu}(\mathcal{O}(\tilde{E}^{\bullet}_{\alpha})_0) \longrightarrow H^{\nu}(\mathcal{O}(\tilde{E}^{\bullet}_{\alpha+1})_0)\right) = 0$$

for $p \leq \nu \leq s$. By the results of §8, (after replacing Δ by some $\Delta(\rho^0)$) there exists a sheaf-homomorphism

$$\Psi^{p}_{\alpha}: \mathcal{O}(\tilde{E}^{p}_{\alpha}) \longrightarrow \mathcal{O}(\tilde{E}^{p-1}_{\alpha+2})$$

such that

$$\tilde{\delta}^{p-1}_{\alpha+2}\Psi^{p}_{\alpha}\tilde{\delta}^{p-1}_{\alpha} = \tilde{r}^{p}_{\alpha+2,\alpha}\tilde{\delta}^{p-1}_{\alpha}$$

on $\mathcal{O}(\tilde{E}^{p-1}_{\alpha})$, where $\tilde{\delta}^{\nu}_{\beta}$, $\tilde{r}^{\nu}_{\beta}$ are as in (7.5).

For any object derived from $E^{\bullet}_{\alpha}$, we put a $\sim$ on top of its symbol to denote the corresponding object derived from $\tilde{E}^{\bullet}_{\alpha}$. For any open polydisc $\Omega \subset \Delta$ and for $t^0 \in \Delta$ and $d \in \mathbb{N}^n_*$ let

$$H^{\nu}_{\alpha}(\Omega, t^0, d) = H^{\nu}\left(\Gamma(\Omega, \mathcal{O}(E^{\bullet}_{\alpha})(t^0,d))\right) .$$

For $t^0 \in \Delta$ and $d \in \mathbb{N}^n_*$, Ψ^p_{α} induces a sheaf-homomorphism $\Theta^p_{\alpha}(t^0,d)$ from

$$\tilde{\mathcal{B}}^p_{\alpha}(t^0,d) := \operatorname{Im}\left(\mathcal{O}(\tilde{E}^{p-1}_{\alpha})(t^0,d) \longrightarrow \mathcal{O}(\tilde{E}^p_{\alpha})(t^0,d)\right)$$

to $\mathcal{O}(\tilde{E}^{p-1}_{\alpha+2})(t^0,d)$ such that

$$\tilde{\delta}^{p-1}_{\alpha+2}\Theta^p_{\alpha}(t^0,d) = \tilde{r}^p_{\alpha+2,\alpha}$$

on $\tilde{\mathcal{B}}^p_{\alpha}(t^0,d)$. By applying (9.1) to the complexes of bundles associated to $\mathcal{O}(\tilde{E}^{\bullet}_{\alpha})(t^0,d)$, we obtain the following. If $\tilde{\mathcal{H}}^p[t^0,d], \ldots, \tilde{\mathcal{H}}^{s-1}[t^0,d]$ are coherent on Δ for all $t^0 \in \Delta$ and all $d \in \mathbb{N}^n_*$ with $d_n \neq \infty$, then, for $p \leq \nu \leq s$,

i) $\tilde{H}^{\nu}_{\alpha}(\Omega, t^0, d) \longrightarrow \Gamma(\Omega, \tilde{\mathcal{H}}^{\nu}[t^0,d])$ is surjective

ii) $\operatorname{Ker}\left(\tilde{H}^{\nu}_{\alpha}(\Omega,t^0,d) \longrightarrow \Gamma(\Omega, \tilde{\mathcal{H}}^{\nu}[t^0,d])\right)$

$$\subset \operatorname{Ker}\left(\tilde{H}^{\nu}_{\alpha}(\Omega,t^0,d) \longrightarrow \tilde{H}^{\nu}_{\alpha+\nu-p+3}(\Omega,t^0,d)\right)$$

for any open polydisc $\Omega \subset \Delta$ and for all $t^0 \in \Delta$ and all $d \in \mathbb{N}^n_*$ with $d_n \neq \infty$.

Since $\mathcal{O}(\tilde{E}^{\bullet}_{\alpha})(t^0,d)$ is the mapping cone of

$$\mathcal{O}(L^{\bullet}_{\alpha})(t^0,d) \longrightarrow \mathcal{O}(E^{\bullet}_{\alpha})(t^0,d)$$

and $\Gamma(\Omega, \mathcal{O}(\tilde{E}^{\bullet}_{\alpha})(t^0,d))$ is the mapping cone of

$$\Gamma(\Omega, \mathcal{O}(L^{\bullet}_{\alpha})(t^0,d)) \longrightarrow \Gamma(\Omega, \mathcal{O}(E^{\bullet}_{\alpha})(t^0,d)) ,$$

we have the following two long exact sequences:

$$\cdots \longrightarrow \mathcal{H}^{\nu}_{\alpha}[t^0,d] \longrightarrow \tilde{\mathcal{H}}^{\nu}_{\alpha}[t^0,d]$$

$$\longrightarrow \mathcal{H}^{\nu+1}(\mathcal{O}(L^{\bullet})(t^0,d)) \longrightarrow \mathcal{H}^{\nu+1}_{\alpha}[t^0,d] \longrightarrow \cdots$$

$$\cdots \longrightarrow H^{\nu}_{\alpha}(\Omega,t^0,d) \longrightarrow \tilde{H}^{\nu}_{\alpha}(\Omega,t^0,d)$$

$$\longrightarrow H^{\nu+1}(\Gamma(\Omega, \mathcal{O}(L^{\bullet})(t^0,d))) \longrightarrow H^{\nu+1}_{\alpha}(\Omega,t^0,d) \longrightarrow \cdots$$

From the first long exact sequence and $(M)^{\nu+1}$ $(p \leqq \nu < s)$, it follows that

$$\mathcal{H}^{p}(\mathcal{O}(L^{\bullet})(t^0,d)) \longrightarrow \mathcal{H}^{p}[t^0,d] \longrightarrow \tilde{\mathcal{H}}^{p}[t^0,d] \longrightarrow$$
$$\mathcal{H}^{p+1}(\mathcal{O}(L^{\bullet})(t^0,d)) \longrightarrow \cdots \longrightarrow \mathcal{H}^{s-1}[t^0,d] \longrightarrow$$
$$\tilde{\mathcal{H}}^{s-1}[t^0,d] \longrightarrow \mathcal{H}^{s}(\mathcal{O}(L^{\bullet})(t^0,d)) \longrightarrow \mathcal{H}^{s}[t^0,d]$$

is exact. From the sharp form of the Five-Lemma, we conclude the following. <u>If</u> $\mathcal{H}^p[t^0,d], \ldots, \mathcal{H}^s[t^0,d]$ <u>are coherent on</u> Δ <u>for all</u> $t^0 \in \Delta$ <u>and all</u> $d \in \mathbb{N}^n_*$ <u>with</u> $d_n \neq \infty$, <u>then</u>

i) <u>for</u> $p \leqq \nu \leqq s$, $H^{\nu}_{\alpha}(\Omega,t^0,d) \longrightarrow \Gamma(\Omega,\mathcal{H}^{\nu}[t^0,d])$ <u>is surjective</u>

ii) <u>for</u> $p < \nu \leqq s$, $\operatorname{Ker}(H^{\nu}_{\alpha}(\Omega,t^0,d) \longrightarrow \Gamma(\Omega,\mathcal{H}^{\nu}[t^0,d]))$ $\subset \operatorname{Ker}(H^{\nu}_{\alpha}(\Omega,t^0,d) \longrightarrow H^{\nu}_{\alpha+\nu-p+3}(\Omega,t^0,d))$ <u>for any open</u>

<u>polydisc</u> $\Omega \subset \Delta$ <u>and for all</u> $t^0 \in \Delta$ <u>and all</u> $d \in \mathbb{N}_*^n$ <u>with</u> $d_n \neq \infty$.

§10. Proof of Coherence

(10.1) Suppose $E^{\bullet}_{\alpha}$ is a sequence of complexes of trivial Banach bundles as in §7. Fix two integers p, s such that $s \geqq p + n$. Assume that the complexes $E^{\bullet}_{\alpha}$ satisfy $(E)^{\nu}$, $(M)^{\nu+1}$, $(F)^{\nu}$ for $p \leqq \nu \leqq s$. We use the notations of §9. We assume the following two conditions for $p \leqq \nu \leqq s$, $\alpha < \beta$, $t^0 \in \Delta$, $d \in \mathbb{N}^n_*$ and for any open polydisc $\Omega \subset \Delta$.

a) $$\operatorname{Im}\left(H^{\nu}_{\alpha}(\Omega,t^0,d) \longrightarrow H_{\beta+1}(\Omega,t^0,d)\right) = \operatorname{Im}\left(H^{\nu}_{\beta}(\Omega,t^0,d) \longrightarrow H^{\nu}_{\beta+1}(\Omega,t^0,d)\right)$$

b) $$\operatorname{Ker}\left(H^{\nu+1}_{\alpha}(\Omega,t^0,d) \longrightarrow H^{\nu+1}_{\beta}(\Omega,t^0,d)\right) \subset \operatorname{Ker}\left(H^{\nu+1}_{\alpha}(\Omega,t^0,d) \longrightarrow H^{\nu+1}_{\alpha+1}(\Omega,t^0,d)\right).$$

We are going to prove by induction on n that $\mathcal{H}^{\nu}$ is coherent on Δ for $p \leqq \nu < s$. The case $n = 0$ is trivial. Assume $n \geqq 1$. The induction hypothesis states that $\mathcal{H}^{\nu}[t^0,d]$ is coherent for $p \leqq \nu < s$, $t^0 \in \Delta$, and $d \in \mathbb{N}^n_*$ with $d_n \neq \infty$. Since coherence is a local property, by (9.2) without loss of generality we can assume the following for any open polydisc $\Omega \subset \Delta$ and for all $t^0 \in \Delta$ and all $d \in \mathbb{N}^n_*$ with $d_n \neq \infty$.

i) For $p \leqq \nu < s$, $H^{\nu}_{\alpha}(\Omega,t^0,d) \longrightarrow \Gamma(\Omega,\mathcal{H}^{\nu}[t^0,d])$ is surjective.

ii) For $p < \nu < s$, $\operatorname{Ker}\left(H^{\nu}_{\alpha}(\Omega, t^0, d) \longrightarrow \Gamma(\Omega, \mathcal{H}^{\nu}[t^0, d])\right)$

$$\subset \operatorname{Ker}\left(H^{\nu}_{\alpha}(\Omega, t^0, d) \longrightarrow H^{\nu}_{\alpha+\nu-p+3}(\Omega, t^0, d)\right) .$$

As in (7.3) we use the following notation. For $m \geqq 1$, let

$$d^m = (\infty, \ldots, \infty, m) \in \mathbb{N}^n_* .$$

(10.2) Lemma. <u>Suppose</u> $p \leqq \nu < s$ <u>and</u> t_n <u>is not a zero-divisor for</u> $t_n^m \mathcal{H}_0^{\nu+1}$. <u>If</u> $\rho < \rho'$ <u>in</u> $\mathbb{R}^n_+$ <u>and</u> ρ <u>is sufficiently strictly small, then for</u> $t^0 \in \Delta(\rho)$ <u>and</u> $\lambda \in \mathbb{N}_0$

$$\operatorname{Im}\left(H^{\nu}_{\alpha}(\Delta(\rho'), t^0, d^{\lambda+m}) \longrightarrow H^{\nu}_{\alpha+1}(\Delta(\rho), t^0, d^{\lambda})\right)$$

<u>is contained in</u>

$$\operatorname{Im}\left(H^{\nu}_{\alpha+1}(\Delta(\rho)) \longrightarrow H^{\nu}_{\alpha+1}(\Delta(\rho), t^0, d^{\lambda})\right) .$$

<u>Proof</u>. Consider the following commutative diagram

$$\begin{array}{ccccccc}
H^{\nu}_{\alpha+1}(\Delta(\rho)) & \xrightarrow{\sigma} & H^{\nu}_{\alpha+1}(\Delta(\rho), t^0, d^{\lambda}) & \xrightarrow{\tau} & H^{\nu+1}_{\alpha+1}(\Delta(\rho)) & & \\
\uparrow & & \uparrow \theta & & \uparrow h & & \\
H^{\nu}_{\alpha}(\Delta(\rho')) & \longrightarrow & H^{\nu}_{\alpha}(\Delta(\rho'), t^0, d^{\lambda+m}) & \xrightarrow{\tau'} & H^{\nu+1}_{\alpha}(\Delta(\rho')) & \longrightarrow & H^{\nu+1}_{\alpha}(\Delta(\rho')) \\
 & & & & \chi \downarrow & & \chi \downarrow \\
 & & & & \mathcal{H}_0^{\nu+1} & \xrightarrow{\varphi} & \mathcal{H}_0^{\nu+1}
\end{array}$$

which comes from the commutative diagram

$$\begin{array}{ccccccccc}
0 & \longrightarrow & \mathcal{O}(E^{\bullet}_{\alpha+1}) & \xrightarrow{c} & \mathcal{O}(E^{\bullet}_{\alpha+1}) & \longrightarrow & \mathcal{O}(E^{\bullet}_{\alpha+1})(t^0, d^{\lambda}) & \longrightarrow & 0 \\
 & & b\uparrow & & r^{\bullet}_{\alpha+1,\alpha}\uparrow & & \uparrow & & \\
0 & \longrightarrow & \mathcal{O}(E^{\bullet}_{\alpha}) & \xrightarrow{a} & \mathcal{O}(E^{\bullet}_{\alpha}) & \longrightarrow & \mathcal{O}(E^{\bullet}_{\alpha})(t^0, d^{\lambda+m}) & \longrightarrow & 0
\end{array}$$

where

i) χ is the natural map

ii) a is defined by multiplication by $(t_n - t_n^0)^{\lambda+m}$

iii) b is defined by multiplication by $(t_n - t_n^0)^m$

iv) c is defined by multiplication by $(t_n - t_n^0)^{\lambda}$.

Let

$$f: H_{\alpha}^{\nu+1}(\Delta(\rho')) \longrightarrow H_{\alpha}^{\nu+1}(\Delta(\rho'))$$

$$\psi : \mathcal{H}_0^{\nu+1} \longrightarrow \mathcal{H}_0^{\nu+1}$$

be defined by multiplication by $(t_n - t_n^0)^m$. Let

$$g: H_{\alpha}^{\nu+1}(\Delta(\rho')) \longrightarrow H_{\alpha+1}^{\nu+1}(\Delta(\rho))$$

be induced by $r_{\alpha+1,\alpha}^{\nu+1}$. By $(B)_n^{\nu+1}$ (applied to $A = 0$), for ρ sufficiently strictly small, Ker $\chi \subset$ Ker g . Since t_n is not a zero-divisor of $t_n^m \mathcal{H}_0^{\nu+1}$, it follows that

(*) Ker $\varphi \subset$ Ker ψ

for $t_n^0 = 0$. When $t_n^0 = 0$, both φ and ψ are isomorphisms and, hence, (*) trivially holds.

One has

$$\varphi\chi\tau' = 0 \Rightarrow \psi\chi\tau' = 0 \Rightarrow \chi f\tau' = 0$$

$$\Rightarrow gf\tau' = 0 \Rightarrow h\tau' = 0 ,$$

because $h = gf$. It follows that $\tau\theta = 0$ and $\operatorname{Im}\theta \subset \operatorname{Im}\sigma$.

Q.E.D.

The following lemma is a strengthened form of (10.2) in codimension ≥ 1. Its proof is similar to that of (10.2). Its consequence (10.5) will be needed only for the proof of the coherence of $\mathcal{H}^p$.

(10.3) Lemma. *Suppose* $s \geq p+2$, $1 \leq \ell < n$, $d_{\ell+1}, \ldots, d_n \in \mathbb{N}$, $\Omega \subset\subset \Delta$ *is an open polydisc, and* $t^0 \in \Omega$. *For* $\lambda \in \mathbb{N}_*$ *let*

$$e^\lambda = (\infty, \ldots, \infty, \lambda, d_{\ell+1}, \ldots, d_n) \in \mathbb{N}_*^n .$$

Then there exists $m \in \mathbb{N}_0$ *such that, for* $\lambda \in \mathbb{N}$, $\operatorname{Im}\left(H^p_\alpha(\Omega, t^0, e^{\lambda+m}) \longrightarrow H^p_{\alpha+3}(\Omega, t^0, e^\lambda)\right)$ *is contained in* $\operatorname{Im}\left(H^p_{\alpha+3}(\Omega, t^0, e^\infty) \longrightarrow H^p_{\alpha+3}(\Omega, t^0, e^\lambda)\right)$.

Proof. Since $\mathcal{H}^{p+1}[t^0, e^\infty]$ is coherent on Δ, by considering the increasing sequence of subsheaves consisting of the kernels of the sheaf-homomorphisms $\mathcal{H}^{p+1}[t^0, e^\infty] \longrightarrow \mathcal{H}^{p+1}[t^0, e^\infty]$ defined by multiplication by $(t_\ell - t^0_\ell)^m$ as m varies, we conclude that there exists $m \in \mathbb{N}_0$ such that $t_\ell - t^0_\ell$ is not a zero-divisor of $(t_\ell - t^0_\ell)^m \mathcal{H}^{p+1}[t^0, e^\infty]_x$ for $x \in \Omega$.

Consider the following commutative diagram

$$\begin{array}{ccccccc}
H^p_{\alpha+3}(\Omega,t^0,e^\infty) & \xrightarrow{\sigma} & H^p_{\alpha+3}(\Omega,t^0,e^\lambda) & \xrightarrow{\tau} & H^{p+1}_{\alpha+3}(\Omega,t^0,e^\infty) & & \\
\uparrow & & \uparrow \theta & & \uparrow h & & \\
H^p_\alpha(\Omega,t^0,e^\infty) & \longrightarrow & H^p_\alpha(\Omega,t^0,e^{\lambda+m}) & \xrightarrow{\tau'} & H^{p+1}_\alpha(\Omega,t^0,e^\infty) & \longrightarrow & H^{p+1}_\alpha(\Omega,t^0,e^\infty) \\
 & & & & \chi \downarrow & & \chi \downarrow \\
 & & & & \Gamma(\Omega,\mathcal{H}^{p+1}[t^0,e^\infty]) & \xrightarrow{\varphi} & \Gamma(\Omega,\mathcal{H}^{p+1}[t^0,e^\infty])
\end{array}$$

which comes from the commutative diagram

$$\begin{array}{ccccccccc}
0 & \longrightarrow & \mathcal{O}(E^\cdot_{\alpha+3})(t^0,e^\infty) & \xrightarrow{c} & \mathcal{O}(E^\cdot_{\alpha+3})(t^0,e^\infty) & \longrightarrow & \mathcal{O}(E^\cdot_{\alpha+3})(t^0,e^\lambda) & \longrightarrow & 0 \\
 & & b \uparrow & & \uparrow r^\cdot_{\alpha+3,\alpha} & & \uparrow & & \\
0 & \longrightarrow & \mathcal{O}(E^\cdot_\alpha)(t^0,e^\infty) & \xrightarrow{a} & \mathcal{O}(E^\cdot_\alpha)(t^0,e^\infty) & \longrightarrow & \mathcal{O}(E^\cdot_\alpha)(t^0,e^{\lambda+m}) & \longrightarrow & 0
\end{array}$$

where

i) χ is the natural map

ii) a is defined by multiplication by $(t_\ell - t_\ell^0)^{\lambda+m}$

iii) b is defined by multiplication by $(t_\ell - t_\ell^0)^m$

iv) c is defined by multiplication by $(t_\ell - t_\ell^0)^\lambda$.

Let

$$f: H^{p+1}_\alpha(\Omega,t^0,e^\infty) \longrightarrow H^{p+1}_\alpha(\Omega,t^0,e^\infty)$$

$$\psi: \Gamma(\Omega,\mathcal{H}^{p+1}[t^0,e^\infty]) \longrightarrow \Gamma(\Omega,\mathcal{H}^{p+1}[t^0,\infty])$$

be defined by multiplication by $(t_\ell - t_\ell^0)^m$. Let

$$g: H_{\alpha}^{p+1}(\Omega, t^0, e^{\infty}) \longrightarrow H_{\alpha+3}^{p+1}(\Omega, t^0, e^{\infty})$$

be induced by $r_{\alpha+3,\alpha}^{p+1}$. By (10.1) ii), $\operatorname{Ker} \chi \subset \operatorname{Ker} g$. By the choice of m, $\operatorname{Ker} \varphi \subset \operatorname{Ker} \psi$. The conclusion follows from repeating verbatim the last paragraph of the proof of (10.2). Q.E.D.

(10.4) Lemma. *Suppose* $s \geqq p+2$, $0 \leqq \ell < n$, $d_{\ell+1}, \ldots, d_n \in \mathbb{N}$, $\Omega \subset\subset \Delta$ *is an open polydisc, and* $t^0 \in \Omega$. *For* $\lambda \in \mathbb{N}_*$ *let* $e^{\lambda} = (\infty, \ldots, \infty, \lambda, d_{\ell+1}, \ldots, d_n) \in \mathbb{N}_*^n$. *Then for* $(f_1, \ldots, f_\ell) \in \mathbb{N}^\ell$ *there exists* $(g_1, \ldots, g_\ell) \in \mathbb{N}^\ell$ *such that, if* $\xi \in H_{\alpha}^p(\Omega, t^0, e^{\infty})$ *and the image of* ξ *in* $\mathcal{H}^p[t^0, e^{\infty}]_{t^0}$ *belongs to* $\sum_{i=1}^{\ell} (t_i - t_i^0)^{g_i} \mathcal{H}^p[t^0, e^{\infty}]_{t^0}$, *then the image of* ξ *in* $H_{\alpha+3\ell}^p(\Omega, t^0, e^{\infty})$ *belongs to* $\sum_{i=1}^{\ell} (t_i - t_i^0)^{f_i} H_{\alpha+3\ell}^p(\Omega, t^0, e^{\infty})$.

Proof. We prove by induction on ℓ. The case $\ell = 0$ is trivial. By (10.3) there exists $m \in \mathbb{N}_0$ such that

$$\operatorname{Im}\left(H_{\alpha}^p(\Omega, t^0, e^{\lambda+m}) \longrightarrow H_{\alpha+3}^p(\Omega, t^0, e^{\lambda})\right)$$

is contained in

$$\operatorname{Im}\left(H_{\alpha+3}^p(\Omega, t^0, e^{\infty}) \longrightarrow H_{\alpha+3}^p(\Omega, t^0, e^{\lambda})\right).$$

Let $g_\ell = f_\ell + m$. By induction hypothesis there exists $(g_1, \ldots, g_{\ell-1}) \in \mathbb{N}^{\ell-1}$ such that, if the image of ξ in

$\mathcal{H}^p[t^0, e^{g_\ell}]_{t^0}$ belongs to

$$\sum_{i=1}^{\ell-1} (t_i - t_i^0)^{g_i} \mathcal{H}^p[t^0, e^{g_\ell}]_{t^0} ,$$

then the image $\bar{\xi}$ of ξ in $H^p_{\alpha+3(\ell-1)}(\Omega, t^0, e^{g_\ell})$ belongs to

$$\sum_{i=1}^{\ell-1} (t_i - t_i^0)^{f_i} H^p_{\alpha+3(\ell-1)}(\Omega, t^0, e^{g_\ell}) .$$

Observe that, if the image of ξ in $\mathcal{H}^p[t^0, e^\infty]_{t^0}$ belongs to

$$\sum_{i=1}^{\ell} (t_i - t_i^0)^{g_i} \mathcal{H}^p[t^0, e^\infty]_{t^0} ,$$

then the image of ξ in $\mathcal{H}^p[t^0, e^{g_\ell}]_{t^0}$ belongs to

$$\sum_{i=1}^{\ell-1} (t_i - t_i^0)^{g_i} \mathcal{H}^p[t^0, e^{g_\ell}]_{t^0} .$$

Hence

$$\bar{\xi} = \sum_{i=1}^{\ell-1} (t_i - t_i^0)^{f_i} \bar{\xi}_i$$

for some $\bar{\xi}_i \in H^p_{\alpha+3(\ell-1)}(\Omega, t^0, e^{g_\ell})$. By (10.3) the image of $\bar{\xi}_i$ in $H^p_{\alpha+3\ell}(\Omega, t^0, e^{f_\ell})$ equals the image in $H^p_{\alpha+3\ell}(\Omega, t^0, e^{f_\ell})$ of some $\xi_i \in H^p_{\alpha+3\ell}(\Omega, t^0, e^\infty)$ $(1 \leq i < \ell)$. It follows that the image ξ^* of ξ in $H^p_{\alpha+3\ell}(\Omega, t^0, e^\infty)$ satisfies

$$\xi^* - \sum_{i=1}^{\ell-1} (t_i - t_i^0)^{f_i} \xi_i \in (t_\ell - t_\ell^0)^{f_\ell} H^p_{\alpha+3\ell}(\Omega, t^0, e^\infty) .$$

Q.E.D.

(10.5) Lemma. If $s \geqq p+2$, then for $\rho \in \mathbb{R}^n_+$ sufficiently strictly small the following holds. For $t^0 \in \Delta(\rho)$ and $f \in \mathbb{N}^n$ there exists $g \in \mathbb{N}^n$ such that, if $\xi \in H^p_\alpha(\Delta(\rho))$ and the image of ξ in $\mathcal{H}^p_{t^0}$ belongs to $\sum_{i=1}^n (t_i - t_i^0)^{g_i} \mathcal{H}^p_{t^0}$, then the image of ξ in $H^p_{\alpha+3n-2}(\Delta(\rho))$ belongs to $\sum_{i=1}^n (t_i - t_i^0)^{f_i} H^p_{\alpha+3n-2}(\Delta(\rho))$.

Proof. Since $\mathcal{H}^{p+1}_0$ is finitely generated over ${}_n\mathcal{O}_0$, there exists $m \in \mathbb{N}_0$ such that t_n is not a zero-divisor for $t_n^m \mathcal{H}^{p+1}_0$. Let $g_n = f_n + m$. By (10.4) (applied to $\ell = n - 1$), there exists $(g_1, \ldots, g_{n-1}) \in \mathbb{N}^{n-1}$ such that if the image of ξ in $\mathcal{H}^p[t^0, d^{g_n}]_{t^0}$ belongs to

$$\sum_{i=1}^{n-1} (t_i - t_i^0)^{g_i} \mathcal{H}^p[t^0, d^{g_n}]_{t^0} ,$$

then the image $\bar{\xi}$ of ξ in $H^p_{\alpha+3(n-1)}(\Delta(\rho), t^0, d^{g_n})$ belongs to

$$\sum_{i=1}^{n-1} (t_i - t_i^0)^{f_i} H^p_{\alpha+3(n-1)}(\Delta(\rho), t^0, d^{g_n}) .$$

Observe that, if the image of ξ in $\mathcal{H}^p_{t^0}$ belongs to

$$\sum_{i=1}^{n} (t_i - t_i^0)^{g_i} \mathcal{H}^p_{t^0} ,$$

then the image of ξ in $\mathcal{H}^p[t^0, d^{g_n}]_{t^0}$ belongs to

$$\sum_{i=1}^{n} (t_i - t_i^0)^{g_i} \mathcal{H}^p_{t^0}[t^0, d^{g_n}]_{t^0} .$$

Hence,

$$\bar{\xi} = \sum_{i=1}^{n-1} (t_i - t_i^0)^{f_i} \bar{\xi}_i$$

for some $\bar{\xi}_i \in H^p_{\alpha+3(n-1)}(\Delta(\rho), t^0, d^{g_n})$. By (10.2), for ρ sufficiently strictly small, the image of $\bar{\xi}_i$ in $H^p_{\alpha+3n-2}(\Delta(\rho), t^0, d^{f_n})$ equals the image in $H^p_{\alpha+3n-2}(\Delta(\rho), t^0, d^{f_n})$ of some $\xi_i \in H^p_{\alpha+3n-2}(\Delta(\rho))$. It follows that the image ξ^* of ξ in $H^p_{\alpha+3n-2}(\Delta(\rho))$ satisfies

$$\xi^* - \sum_{i=1}^{n-1} (t_i - t_i^0)^{f_i} \xi_i \in (t_n - t_n^0)^{f_n} H^p_{\alpha+3n-2}(\Delta(\rho)) .$$

Q.E.D.

(10.6) <u>Theorem</u>. <u>$\mathcal{H}^\nu$ is coherent on</u> Δ <u>for</u> $p \leqq \nu < s$.

<u>Proof</u>. It suffices to prove that $\mathcal{H}^\nu$ is coherent at 0 . We break up the proof into three parts.

(I) We first show that $\mathcal{H}^\nu$ is of finite type at 0 .

Take $\rho' < \rho'' < \rho'''$ in $\mathbb{R}^n_+$ and $t^0 \in \Delta(\rho')$. By $(B)^\nu_n$, for ρ' sufficiently strictly small,

$$(*)\quad \begin{cases} \operatorname{Im}\left(H^\nu_\alpha(\Delta(\rho'')) \longrightarrow H^\nu_{\alpha+1}(\Delta(\rho'))\right) \text{ generates a finitely} \\ \text{generated } \Gamma(\Delta(\rho'), {}_n\mathcal{O})\text{-module.} \end{cases}$$

Since $\mathcal{H}^{\nu+1}_0$ is finitely generated over ${}_n\mathcal{O}_0$, there exists $m \in \mathbb{N}_0$ such that t_n is not a zero-divisor of $t^m_n \mathcal{H}^{\nu+1}_0$. By (10.2), for ρ'' sufficiently strictly small,

$$(\#)\quad \begin{cases} \operatorname{Im}\left(H^\nu_{\alpha-1}(\Delta(\rho'''), t^0, d^{m+1}) \longrightarrow H^\nu_\alpha(\Delta(\rho''), t^0, d^1)\right) \text{ is con-} \\ \text{tained in } \operatorname{Im}\left(H^\nu_\alpha(\Delta(\rho'')) \longrightarrow H^\nu_\alpha(\Delta(\rho''), t^0, d^1)\right). \end{cases}$$

By (10.1) i), we have the surjectivity of

$$(\dagger)\quad H^\nu_{\alpha-1}(\Delta(\rho'''), t^0, d^{m+1}) \longrightarrow \Gamma(\Delta(\rho'''), \mathcal{H}^\nu[t^0, d^{m+1}]).$$

It follows from the coherence of $\mathcal{H}^\nu[t^0, d^{m+1}]$ and the Theorem A of Cartan-Oka that

$${}_n\mathcal{O}_{t^0}\,\Gamma(\Delta(\rho'''), \mathcal{H}^\nu[t^0, d^{m+1}]) \longrightarrow \mathcal{H}^\nu_{t^0}$$

is surjective. From (#) and (†) we conclude that

$$\mathcal{H}^\nu_{t^0} \subset {}_n\mathcal{O}_{t^0} \operatorname{Im}\left(H^\nu_\alpha(\Delta(\rho'')) \longrightarrow \mathcal{H}^\nu_{t^0}\right) + (t_n - t^0_n)\mathcal{H}^\nu_{t^0}.$$

Since $\mathcal{H}^\nu_{t^0}$ is finitely generated over ${}_n\mathcal{O}_0$, by Nakayama's Lemma,

$$\mathcal{H}^\nu_{t^0} \subset {}_n\mathcal{O}_{t^0} \operatorname{Im}\left(H^\nu_\alpha(\Delta(\rho'')) \longrightarrow \mathcal{H}^\nu_{t^0}\right).$$

It follows from (*) that $\mathcal{H}^{\nu}|\Delta(\rho')$ is generated by a finite number of elements $\xi_1, \ldots, \xi_k \in H^{\nu}_{\alpha}(\rho'')$. Let $\bar{\xi}_i \in \Gamma(\Delta(\rho''), \mathcal{H}^{\nu})$ be induced by ξ_i $(1 \leq i < k)$.

Next we prove that the relation sheaf $\mathcal{R}$ of $\bar{\xi}_1|\Delta(\rho'), \ldots, \bar{\xi}_k|\Delta(\rho')$ is coherent at 0. We distinguish between the case $\nu > p$ and the case $\nu = p$. Take $\rho < \rho'$ in $\mathbb{R}^n_+$.

(II) The case $\nu < p$. Suppose

$$(a_1, \ldots, a_k) \in \mathcal{R}_{t^0}$$

for some $t^0 \in \Delta(\rho)$; that is $\sum_{i=1}^{k} a_i(\bar{\xi}_i)_{t^0} = 0$. Let $\mathfrak{m}_{t^0}$ denote the maximal ideal of ${}_n\mathcal{O}_{t^0}$ as well as the ideal sheaf on $\mathbb{C}^n$ for the reduced subspace $\{t^0\}$. By taking the λth partial sum of the power series of a_i at t^0, we obtain $b_i \in \Gamma(\Delta(\rho''), {}_n\mathcal{O})$ such that

$$(b_i)_{t^0} - a_i \in \mathfrak{m}^{\lambda}_{t^0} \quad (1 \leq i \leq k).$$

It follows that

$$\sum_{i=1}^{k} b_i \bar{\xi}_i \in \Gamma(\Delta(\rho''), \mathfrak{m}^{\lambda}_{t^0}\mathcal{H}^{\nu}).$$

Since $\mathcal{H}^{\nu}[t^0, d^{\lambda+m}]$ is coherent on Δ, by the Theorem B of Cartan-Oka

$$\Gamma(\Delta(\rho''), \mathfrak{m}^{\lambda}_{t^0}\mathcal{H}^{\nu}[t^0, d^{\lambda+m}])$$

equals

$$\Gamma(\Delta(\rho''), \mathfrak{m}^{\lambda}_{t^0})\Gamma(\Delta(\rho''), \mathcal{H}^{\nu}[t^0, d^{\lambda+m}]) .$$

Hence the image of $\sum_{i=1}^{k} b_i \bar{\xi}_i$ in $\Gamma(\Delta(\rho''), \mathcal{H}^{\nu}[t^0, d^{\lambda+m}])$ belongs to

$$\Gamma(\Delta(\rho''), \mathfrak{m}^{\lambda}_{t^0})\Gamma(\Delta(\rho''), \mathcal{H}^{\nu}[t^0, d^{\lambda+m}]) .$$

By (10.1) i) and ii), the image of $\sum_{i=1}^{k} b_i \xi_i$ in $H^{\nu}_{\alpha+\nu-p+3}(\Delta(\rho''), t^0, d^{\lambda+m})$ belongs to the image of

$$\Gamma(\Delta(\rho''), \mathfrak{m}^{\lambda}_{t^0}) H^{\nu}(\Delta(\rho''), t^0, d^{\lambda+m})$$

in $H^{\nu}_{\alpha+\nu-p+3}(\Delta(\rho''), t^0, d^{\lambda+m})$. By (10.2), for ρ' sufficiently strictly small, the image of $\sum_{i=1}^{k} b_i \xi_i$ in $H^{\nu}_{\alpha+\nu-p+3}(\Delta(\rho'), t^0, d^{\lambda})$ belongs to the image of

$$\Gamma(\Delta(\rho'), \mathfrak{m}^{\lambda}_{t^0}) H^{\nu}_{\alpha+1}(\Delta(\rho'))$$

in $H^{\nu}_{\alpha+\nu-p+3}(\Delta(\rho'), t^0, d^{\lambda})$. It follows that the image of $\sum_{i=1}^{k} b_i \xi_i$ in $H^{\nu}_{\alpha+\nu-p+3}(\Delta(\rho'))$ belongs to

$$\Gamma(\Delta(\rho'), \mathfrak{m}^{\lambda}_{t^0}) H^{\nu}_{\alpha+\nu-p+3}(\Delta(\rho')) + (t_n - t^0_n)^{\lambda} H^{\nu}_{\alpha+\nu-p+3}(\Delta(\rho')) .$$

By $(B)^{\nu}_{n}$, for ρ sufficiently strictly small, the image of $\sum_{i=1}^{k} b_i \xi_i$ in $H^{\nu}_{\alpha+\nu-p+4}(\Delta(\rho))$ equals the image of $\sum_{i=1}^{k} c_i \xi_i$ in

$H^{\nu}_{\alpha+\nu-p+4}(\Delta(\rho))$ for some

$$c_i \in \Gamma(\Delta(\rho), \mathfrak{m}^{\lambda}_{t^0}) \qquad (1 \leqq i \leqq k) .$$

Since

$$(a_1, \ldots, a_k) - (b_1 - c_1, \ldots, b_k - c_k)_{t^0} \in \mathfrak{m}^{\lambda}_{t^0}$$

and

$$(b_1 - c_1, \ldots, b_k - c_k) \in \Gamma(\Delta(\rho), \mathcal{R}) ,$$

it follows that

$$\mathcal{R}_{t^0} \subset \mathfrak{m}^{\lambda}_{t^0} \, {}_n\mathcal{O}^k_{t^0} + {}_n\mathcal{O}^k_{t^0}\Gamma(\Delta(\rho), \mathcal{R}) .$$

Since λ is arbitrary,

$$\mathcal{R}_{t^0} = {}_n\mathcal{O}^k_{t^0}\Gamma(\Delta(\rho), \mathcal{R}) .$$

Hence $\mathcal{R}$ is coherent on $\Delta(\rho)$.

(III) The case $\nu = p$. The only difference between this case and the previous case is that, when $\nu = p$, (10.1) ii) no longer implies that, for open polydisc $\Omega \subset \Delta$, $t^0 \in \Delta$ and $d \in \mathbb{N}^n_*$ with $d_n \neq \infty$,

$$(**) \begin{cases} \operatorname{Ker}\left(H^{\nu}_{\alpha}(\Omega, t^0, d) \longrightarrow \Gamma(\Omega, \mathcal{H}^{\nu}[t^0, d])\right) \\ \qquad \subset \operatorname{Ker}\left(H^{\nu}_{\alpha}(\Omega, t^0, d) \longrightarrow H^{\nu}_{\alpha+\nu-p+3}(\Omega, t^0, d)\right) . \end{cases}$$

When $n = 1$, (**) is clearly satisfied for $\nu = p$. So we can assume that $n \geqq 2$. Since $s \geqq p+n$, $\mathcal{H}^{p+1}$ is coherent on Δ by (II). Now we modify the argument of (II) to avoid the use of (**) when $\nu = p$. We pick up the argument of (II) at the point where

$$\sum_{i=1}^{k} b_i \bar{\xi}_i \in \Gamma(\Delta(\rho''), \mathfrak{m}_{t^0}^{\lambda} \mathcal{H}^p) .$$

By (10.5), for ρ'' sufficiently strictly small, there exists λ' depending on λ such that

i) $\lambda' \longrightarrow \infty$ as $\lambda \longrightarrow \infty$

ii) the image of $\sum_{i=1}^{k} b_i \xi_i$ in $H^p_{\alpha+3n-2}(\Delta(\rho''))$ belongs to $\Gamma(\Delta(\rho''), \mathfrak{m}_{t^0}^{\lambda'}) H^p_{\alpha+3n-2}(\Delta(\rho''))$.

By $(B)^p_n$, for ρ' sufficiently strictly small, the image of $\sum_{i=1}^{k} b_i \xi_i$ in $H^{\nu}_{\alpha+3n-2}(\Delta(\rho'))$ equals the image of $\sum_{i=1}^{k} c_i \xi_i$ in $H^{\nu}_{\alpha+3n-2}(\Delta(\rho'))$ for some

$$c_i \in \Gamma(\Delta(\rho'), \mathfrak{m}_{t^0}^{\lambda'}) \qquad (1 \leqq i \leqq k) .$$

As in (II), we conclude that

$$\mathcal{R}_{t^0} \subset \mathfrak{m}_{t^0}^{\lambda'} \, {}_n\mathcal{O}^k_{t^0} + {}_n\mathcal{O}^k_{t^0} \Gamma(\Delta(\rho'), \mathcal{R}) .$$

Since $\lambda' \longrightarrow \infty$ as $\lambda \longrightarrow \infty$, it follows from the arbitrariness of λ that

$$\mathcal{R}_{t^0} = {}_n\mathcal{O}^k_{t^0} \Gamma(\Delta(\rho'), \mathcal{R}) .$$

Hence $\mathcal{R}$ is coherent on $\Delta(\rho)$. Q.E.D.

(10.7) Proof of Main Theorem.

For every $s_0 \in S$ one can find a proper holomorphic map σ with finite fibers from an open neighborhood U of s_0 into an open subset G of $\mathbb{C}^n$. We have

$$R^0\sigma_*\left(R^k(\pi_a^b)_*(\mathcal{F}|X_a^b \cap \pi^{-1}(U))\right) = R^k(\sigma\circ\pi_a^b)_*\left(\mathcal{F}|X_a^b \cap \pi^{-1}(U)\right) .$$

It is easy to see that an analytic sheaf $\mathcal{G}$ on U is coherent if and only if $R^0\sigma_*\mathcal{G}$ is coherent on G . Moreover, if M is an $\mathcal{O}_{S,s}$-module for some $s \in U$, then

$$\operatorname{codh}_{\mathcal{O}_{S,s}} M = \operatorname{codh}_{{}_n\mathcal{O}_{\sigma(s)}} M$$

where M is regarded naturally as an ${}_n\mathcal{O}_{\sigma(s)}$-module. In particular, if $\operatorname{codh}_{\mathcal{O}_{S,s}} M \geqq n$, then M is a flat ${}_n\mathcal{O}_{\sigma(s)}$-module (see (A.8)-(A.12) of the Appendix). Hence for the proof of the Main Theorem we can assume without loss of generality that $S = \Delta$ and $\mathcal{F}$ is π-flat. Moreover, in the course of the proof, any replacement of Δ by $\Delta(\rho)$ (with $\rho \in \mathbb{R}_+^n$) does not result in any loss of generality.

In (6.5) we have constructed a sequence of complexes $E_\alpha^\bullet$ of trivial Banach bundles on Δ . By (5.2) and the results of Andreotti-Grauert [1] these complexes $E_\alpha^\bullet$ satisfy $(E)^\nu$, $(M)^{\nu+1}$, $(F)^\nu$, and (10.1) a), b) for $p \leqq \nu < r - q - n$. By (10.6), $\mathcal{H}^\nu$ is coherent on Δ for $p \leqq \nu < r - q - n - 1$.

By (7.3) and the bumping techniques of Andreotti-Grauert 1 , it is easy to see that

$$\mathcal{H}^{\nu} \approx R^{\nu}(\pi_a^b)_* \mathcal{F}$$

for $p \leqq \nu < r-q-n$, $a_* \leqq a < a_\#$, and $b_\# < b < b_*$. Moreover for $p < \nu < r-q-n$,

$$\mathcal{H}^{\nu} \approx R^{\nu}(\pi_a^b)_* \mathcal{F}$$

for $a_* \leqq a < a_\#$ and $b_\# < b \leqq b_*$.

PART III APPLICATIONS

§11. Coherent Sheaf Extension

For the definition and properties of gap-sheaves $\mathcal{F}^{[n]}$ used here, refer to the Appendix.

(11.1) Theorem (Coherent Sheaf Extension on Ring Domains). Suppose $0 \leqq a < b$ in $\mathbb{R}^N$, D is an open subset of $\mathbb{C}^n$, $\mathcal{F}$ is a coherent analytic sheaf on $D \times G^N(a,b)$ such that $\mathcal{F}^{[n+1]} = \mathcal{F}$. Then $\mathcal{F}$ extends uniquely to a coherent analytic sheaf $\tilde{\mathcal{F}}$ on $D \times \Delta^N(b)$ such that $\tilde{\mathcal{F}}^{[n+1]} = \tilde{\mathcal{F}}$.

Proof. The uniqueness of $\tilde{\mathcal{F}}$ follows from the extension theory of sections of gap-sheaves (see (A.18) of the Appendix).

For the existence of $\tilde{\mathcal{F}}$, we consider first the special case where

i) D is bounded and Stein

ii) $\operatorname{codh} \mathcal{F} \geqq n + 3$

iii) $\mathcal{F}$ is flat with respect to the natural projection $\tilde{\pi}: D \times G^N(a,b) \longrightarrow D$.

For $m \in \mathbb{N}$ sufficiently large, there exist

$$0 < \alpha < \beta \quad \text{in } \mathbb{R}$$

$$a < a' < b' < b \quad \text{in } \mathbb{R}^N$$

such that

$$G^N(a',b') \subset\subset \{z \in \mathbb{C}^N \,|\, \alpha < \sum_{i=1}^N |z_i|^{2m} < \beta\} \subset\subset G^N(a,b) .$$

For $\varepsilon > 0$ sufficiently small, $D \times G^N(a',b')$ is contained in

$$\{(t,z) \in D \times \mathbb{C}^N \,|\, \alpha + \varepsilon < \varepsilon|t|^2 + \sum_{i=1}^N |z_i|^{2m} < \beta - \varepsilon\} .$$

Let

$$\alpha' = \alpha + \frac{\varepsilon}{2}$$

$$\alpha'' = \alpha + \varepsilon$$

$$\beta' = \beta - \frac{\varepsilon}{2}$$

$$\beta'' = \beta - \varepsilon$$

$$\varphi(t,z) = \varepsilon|t|^2 + \sum_{i=1}^N |z_i|^{2m} .$$

Take $\alpha < \beta_1 < \alpha'$ and $\beta' < \beta_2 < \beta$. Let

$$X = \{(t,z) \in D \times \mathbb{C}^N \,|\, \alpha < \varphi(t,z) < \beta\}$$

$$X_i = X \cap \{\varphi < \beta_i\} \quad (i = 1,2)$$

$$\pi = \tilde{\pi}|X$$

$$\pi^i = \pi|X_i \quad (i = 1,2) .$$

Fix $t^0 \in D$. For any $(t^0,z^0) \in X_2 - X_1$, there exists a linear function $f(t,z)$ such that $f(t^0,z^0) = 0$ and $f(t,z)$

is nowhere zero on X_1. Let $\theta : \mathcal{F} \longrightarrow \mathcal{F}$ on X_2 be defined by multiplication by f. Since f is nowhere zero on X_1, Supp Ker θ and Supp Coker θ are subvarieties of the Stein space

$$(D \times \mathbb{C}^N) \cap \{\varphi < \beta_2\} \cap \{f = 0\}$$

and hence are Stein. From the cohomology sequence of the short exact sequence

$$0 \longrightarrow \operatorname{Ker}\theta \longrightarrow \mathcal{F} \xrightarrow{\theta} \operatorname{Im}\theta \longrightarrow 0 ,$$

we conclude that θ induces an isomorphism

$$\zeta : (R^1\pi^2_*\mathcal{F})_{t^0} \longrightarrow \left(R^1\pi^2_*(\operatorname{Im}\theta)\right)_{t^0} .$$

Consider the following exact sequence

$$(R^0\pi^2_*\mathcal{F})_{t^0} \xrightarrow{\xi} \left(R^0\pi^2_*(\operatorname{Coker}\theta)\right)_{t^0}$$

$$\longrightarrow \left(R^1\pi^2_*(\operatorname{Im}\theta)\right)_{t^0} \xrightarrow{\eta} (R^1\pi^2_*\mathcal{F})_{t^0}$$

coming from the short exact sequence

$$0 \longrightarrow \operatorname{Im}\theta \hookrightarrow \mathcal{F} \longrightarrow \operatorname{Coker}\theta \longrightarrow 0 .$$

We are going to prove that ξ is surjective. It suffices to show that $\operatorname{Ker}\eta = 0$. Consider the following commutative diagram

$$\begin{array}{ccc} (R^1\pi^2_*\mathcal{F})_{t^0} & \xrightarrow{\theta_2} & (R^1\pi^2_*\mathcal{F})_{t^0} \\ g\downarrow & & \downarrow g \\ (R^1\pi^1_*\mathcal{F})_{t^0} & \xrightarrow{\theta_1} & (R^1\pi^1_*\mathcal{F})_{t^0} \end{array}$$

where θ_1, θ_2 are induced by θ and g is the restriction map. Since $\theta_2 = \eta\zeta$ and ζ is an isomorphism it suffices to show that $\mathrm{Ker}\,\theta_2 = 0$. Since θ is an isomorphism on X_1, θ_1 is an isomorphism. By applying the Main Theorem to the map $\pi: X \longrightarrow D$ together with the function φ and the sheaf $\mathcal{F}$ on X, we conclude that g is an isomorphism. Hence $\mathrm{Ker}\,\theta_2 = 0$ and ξ is surjective. Since Supp Cokerξ is Stein, the image of the natural map

$$\left(R^0\pi^2_*(\mathrm{Coker}\,\theta)\right)_{t^0} \longrightarrow (\mathrm{Coker}\,\theta)_{(t^0,z^0)}$$

generates $(\mathrm{Coker}\,\theta)_{(t^0,z^0)}$ over ${}_{n+N}\mathcal{O}_{(t^0,z^0)}$. Since ξ is surjective, by Nakayama's Lemma, the image of the natural map

$$(R^0\pi^2_*\mathcal{F})_{t^0} \longrightarrow \mathcal{F}_{(t^0,z^0)}$$

generates $\mathcal{F}_{(t^0,z^0)}$. By letting z^0 vary, for some open neighborhood U' of t^0 in D we can find a sheaf-epimorphism $\sigma: {}_{n+N}\mathcal{O}^p \longrightarrow \mathcal{F}$ on

$$\{(t,z) \in U' \times \mathbb{C}^N \,|\, \alpha' < \varphi(t,z) < \beta'\}.$$

By applying the same argument to $\operatorname{Ker}\sigma$ instead of $\mathcal{F}$, for some open neighborhood U'' of t^0 in U' we can find a sheaf-epimorphism

$$\tau : {}_{n+N}\mathcal{O}^q \longrightarrow \operatorname{Ker}\sigma$$

on

$$\{(t,z) \in U'' \times \mathbb{C}^N \mid \alpha'' < \varphi(t,z) < \beta''\} .$$

By Hartogs' Theorem, τ extends to a sheaf-homomorphism

$$\tilde{\tau} : {}_{n+N}\mathcal{O}^q \longrightarrow {}_{n+N}\mathcal{O}^p$$

on $\{(t,z) \in U'' \times \mathbb{C}^N \mid \varphi(t,z) < \beta''\}$. Then

$$\tilde{\mathcal{F}} := \left(\operatorname{Coker}\tilde{\tau} \Big/ 0_{[n+2]}\operatorname{Coker}\tilde{\tau} \right)^{[n+1]}$$

extends $\mathcal{F} \mid U'' \times G^N(a',b')$ and, hence, extends $\mathcal{F} \mid U'' \times G^N(a,b')$ (see (A.18) of the Appendix). By the arbitrariness of t^0 and the uniqueness of extension, the special case follows.

For the general case we use induction on n. Let S be the set of points of $D \times G^N(a,b)$ where $\operatorname{codh}\mathcal{F} \leq n+2$. Let

$$T_1 = \{(t,z) \in S \mid \dim_{(t,z)} S \cap (\{t\} \times G^N(a,b)) \geq 1\} .$$

Let T_2 be the set of points of $D \times G^N(a,b)$ where $\mathcal{F}$ is not $\tilde{\pi}$-flat. Take $a < a' < b' < b$ in $\mathbb{R}^N$. Let

$$A = \tilde{\pi}\left((T_1 \cup T_2) \cap \overline{(D \times G^N(a',b'))}\right) .$$

By applying the special case to $\mathcal{F} \mid U \times G^N(a',b')$ for bounded

Stein open subsets U of $D-A$, we conclude that $\mathcal{F}|(D-A)\times G^N(a,b)$ can be extended to a coherent analytic sheaf $\hat{\mathcal{F}}$ on $(D-A)\times\Delta^N(b)$ satisfying $\hat{\mathcal{F}}^{[n+1]} = \hat{\mathcal{F}}$. Since $\dim S \leq n$, rank $\tilde{\pi}|T_1 \cup T_2 < n$ (cf (A.13) of the Appendix). Since $A = \emptyset$ when $n = 0$, the case $n = 0$ is proved. Take arbitrarily $t^0 \in A$. After a coordinates transformation, we can assume that $t^0 = 0$ and there exist $0 < \alpha < \beta$ in $\mathbb{R}$ and $\gamma > 0$ in $\mathbb{R}^{n-1}$ such that $\Delta^{n-1}(\gamma)\times G^1(\alpha,\beta)$ is disjoint from A. By induction hypothesis, the sheaf on

$$\left(\Delta^{n-1}(\gamma)\times G^1(\alpha,\beta)\times\Delta^N(b)\right) \cup \left(\Delta^{n-1}(\gamma)\times\Delta^1(\beta)\times G^N(a,b)\right)$$

which agrees with $\hat{\mathcal{F}}$ on

$$\Delta^{n-1}(\gamma)\times G^1(\alpha,\beta)\times\Delta^N(b)$$

and agrees with $\mathcal{F}$ on

$$\Delta^{n-1}(\gamma)\times\Delta^1(\beta)\times G^N(a,b)$$

can be extended to a coherent analytic sheaf $\mathcal{F}^*$ on $\Delta^{n-1}(\gamma)\times\Delta^1(\beta)\times\Delta^N(b)$ satisfying $(\mathcal{F}^*)^{[n+1]} = \mathcal{F}^*$. The general case now follows from the arbitrariness of t^0 and the uniqueness of extension. Q.E.D.

§12. Blow-downs

(12.1) A holomorphic map $\pi: X \longrightarrow S$ is said to be strongly 1-pseudoconvex if there exist a C^2 function $\varphi: X \longrightarrow (-\infty, c_*) \subset (-\infty,\infty)$ and a real number $c < c_*$ such that

i) $\pi | \{\varphi \leq c\}$ is proper for $c < c_*$

ii) φ is strongly 1-pseudoconvex on $\{\varphi > c_\#\}$.

(When the additional condition $\{\varphi \leq c\} = \{\varphi < c\}^-$ for $c < c_*$ is added, this definition agrees with a special case of strongly (p,q)-pseudoconvex-pseudoconcave maps.)

For $f \in \Gamma(X, \mathcal{O}_X)$ and $x \in X$ let $f(x)$ denote the image of the germ of f at x under the natural map

$$\mathcal{O}_{X,x} \longrightarrow \mathcal{O}_{X,x}/\mathfrak{m}_{X,x} = \mathbb{C}.$$

We are going to prove the following result concerning blowing down. If $\pi: X \longrightarrow S$ is strongly 1-pseudoconvex and S is Stein, then X is holomorphically convex (that is, for every discrete sequence $\{x_\nu\}$ in X there exists $f \in \Gamma(X, \mathcal{O}_X)$ such that $f(x_\nu) \longrightarrow \infty$ as $\nu \longrightarrow \infty$). Once we have the holomorphic convexity of X, we can blow down X by the Reduction Theorem of Remmert (whose generalization to the unreduced case can be proved in a way analogous to the reduced case [30]).

For $c_\# < c < c_*$ let $X^c = \{\varphi < c\}$ and $\pi^c = \pi | X^c$.

The result on blowing down will be proved by using the finite generation of $(R^1\pi_*^c\mathcal{O}_X)_s$ for $s \in S$ and $c_\# < c < c_*$. For such a finite generation, it suffices to consider the case where $S = \Delta$ and $s = 0$. Strictly speaking this finite generation does not follow from the Main Theorem, because $\mathcal{O}_X$ in general is not π-flat. However, this can be obtained from the argument used in proving the Main Theorem. In the proof of the Main Theorem, the flatness is needed to get a sequence of complexes $E_\alpha^\bullet$ satisfying $(E)^\nu$, $(M)^{\nu+1}$, $(F)^\nu$ for $p \leq \nu < r-q-n$. Such a sequence of complexes is needed for getting right inverses of coboundary maps (§8) and global isomorphisms (§9) which, in turn, are essential for proving the coherence of the direct image sheaves under consideration. However, when only the finite generation of the stalks of the direct image sheaves under consideration is needed, it can be proved directly by the arguments of §7 without using the sequence of complexes, provided that $\mathcal{F}$ is π-flat on $\{\varphi < a_\#\}$. Of course, some modifications are needed and, when $\nu = p$, one can only obtain the finite generation of $(R^\nu(\pi_a^b)_*\mathcal{F})_0$ for $a_* < a < a_\#$ and $b_\# < b < b_*$. As before, we replace X by the graph of π. We prove the finite generation by induction on n. By replacing X by X_a^b with $a_* < a < a_\#$ and $b_\# < b < b_*$, we can assume without loss of generality that there exists $m \in \mathbb{N}_0$ such that t_n is not a zero-divisor for any stalk of $t_n^m\mathcal{F}$. By applying the induction hypothesis to $\mathcal{F}/t_n^m\mathcal{F}$ and considering the exact sequence

$$\left(R^{\nu}(\pi_a^b)_*(t_n^m\mathcal{F})\right)_s \longrightarrow \left(R^{\nu}(\pi_a^b)_*\mathcal{F}\right)_s \longrightarrow \left(R^{\nu}(\pi_a^b)_*(\mathcal{F}/t_n^m\mathcal{F})\right)_s ,$$

we can without loss of generality replace $\mathcal{F}$ by $t_n^m\mathcal{F}$ and assume that $m = 0$. Use the notations of (6.5). Now, to get the finite generation, one need only replace $E_\alpha^\nu(\rho)$ by

$$\{\xi \in C^\nu(\Delta(\rho)\times \mathfrak{U}_\alpha, \mathcal{F}) \,|\, \|\xi\|_{\mathfrak{U}_\alpha,\rho} < \infty\}$$

and use the parenthetical statement at the end of (5.2). We note that the case of the strongly 1-pseudoconvex map corresponds to the case where $p = 1$ and $\{\varphi < a_\#\} = \emptyset$. Hence no flatness of $\mathcal{O}_X$ is needed for the finite generation of $(R^1\pi_*^c\mathcal{O}_X)_s$.

(12.2) <u>Lemma</u>. <u>If</u> $\pi: X \longrightarrow S$ <u>is a strongly</u> 1-<u>pseudoconvex holomorphic map and</u> S <u>is a single point, then</u> X <u>is holomorphically convex</u>.

<u>Proof</u>. Take $c_\# < c < c_*$ and take a discrete sequence $\{x_\nu\}$ in $\{\varphi > c\}$. Let $\mathcal{J}$ be the ideal-sheaf of the subvariety $\{x_\nu\}$. The exact sequence

$$0 \longrightarrow \mathcal{J} \longrightarrow \mathcal{O}_X \longrightarrow \mathcal{O}_X/\mathcal{J} \longrightarrow 0$$

yields the commutative diagram with exact rows

$$\begin{array}{ccccccc} \Gamma(X,\mathcal{O}_X) & \xrightarrow{\eta} & \Gamma(X,\mathcal{O}_X/\mathcal{J}) & \longrightarrow & H^1(X,\mathcal{J}) & \longrightarrow & H^1(X,\mathcal{O}_X) \\ & & & & \sigma\downarrow & & \downarrow \\ & & & & H^1(X^c,\mathcal{J}) & \xrightarrow{\tau} & H^1(X^c,\mathcal{O}_X) \end{array}$$

Since $\mathcal{I} = \mathcal{O}_X$ on X^c, τ is an isomorphism. By the results of Andreotti-Grauert [1], σ is an isomorphism. It follows that η is surjective. There exists $f \in \Gamma(X,\mathcal{O}_X)$ such that $f(x_\nu) \longrightarrow \infty$. Q.E.D.

(12.3) Lemma. Suppose $\pi: X \longrightarrow S$ is a strongly 1-pseudoconvex map. Then for $s \in S$ and $c_\# < c < c_*$ there exists $k \in \mathbb{N}$ such that

$$(R^0\pi_*^c\mathcal{O}_X)_s \longrightarrow \left(R^0\pi_*^c(\mathcal{O}_X/\mathfrak{m}_{S,s}\mathcal{O}_X)\right)_s$$

has the same image as

$$\left(R^0\pi_*^c(\mathcal{O}_X/\mathfrak{m}_{S,s}^k\mathcal{O}_X)\right)_s \longrightarrow \left(R^0\pi_*^c(\mathcal{O}_X/\mathfrak{m}_{S,s}\mathcal{O}_X)\right)_s .$$

Proof. Use induction on $\dim_s S$. For some open neighborhood U of s in S, there exists $f \in \Gamma(U,\mathcal{O}_S)$ such that $f(s) = 0$ and f_s is not a zero-divisor of $\mathcal{O}_{S,s}$. Since $(R^1\pi_*^c\mathcal{F})_s$ is finitely generated over $\mathcal{O}_{S,s}$, there exists $m \in \mathbb{N}_0$ such that

i) f_s is not a zero-divisor of $f_s^m(R^1\pi_*^c\mathcal{O}_X)_s$,

ii) f_x (when f is naturally regarded as an element of $\Gamma(\pi^{-1}(U),\mathcal{O}_X)$) is not a zero-divisor of $f_x^m\mathcal{O}_X$ for $x \in X^c \cap \pi^{-1}(U)$.

The commutative diagram

$$\begin{array}{ccccccccc}
0 & \longrightarrow & f^{2m+1}\mathcal{O}_X & \longrightarrow & \mathcal{O}_X & \longrightarrow & \mathcal{O}_X/f^{2m+1}\mathcal{O}_X & \longrightarrow & 0 \\
 & & \big\downarrow & & \| & & \big\downarrow & & \\
0 & \longrightarrow & f\mathcal{O}_X & \longrightarrow & \mathcal{O}_X & \longrightarrow & \mathcal{O}_X/f\mathcal{O}_X & \longrightarrow & 0
\end{array}$$

yields the following commutative diagram

$$\begin{array}{ccccccc}
(R^0\pi_*^c\mathcal{O}_X)_s & \longrightarrow & (R^0\pi_*^c(\mathcal{O}_X/f^{2m+1}\mathcal{O}_X))_s & \xrightarrow{a} & (R^1\pi_*^c(f^{2m+1}\mathcal{O}_X))_s & \xrightarrow{c} & (R^1\pi_*^c\mathcal{O}_X)_s \\
\| & & \eta\big\downarrow & & \big\downarrow b & & \\
(R^0\pi_*^c\mathcal{O}_X)_s & \xrightarrow{\xi} & (R^0\pi_*^c(\mathcal{O}_X/f\mathcal{O}_X))_s & \longrightarrow & (R^1\pi_*^c(f\mathcal{O}_X))_s & &
\end{array}$$

We are going to show that $ba = 0$. Consider the following commutative diagram

$$\begin{array}{ccccc}
 & & (R^1\pi_*^c(f^{2m+1}\mathcal{O}_X))_s & \xrightarrow{b} & (R^1\pi_*^c(f\mathcal{O}_X))_s \\
 & \swarrow c & \big\uparrow \beta & & \\
(R^1\pi_*^c\mathcal{O}_X)_s & & (R^1\pi_*^c(f^m\mathcal{O}_X))_s & & \big\uparrow \tau \\
 & \nwarrow \alpha & \big\downarrow \theta & & \\
 & & (R^1\pi_*^c\mathcal{O}_X)_s & \xrightarrow{\sigma} & (R^1\pi_*^c\mathcal{O}_X)_s
\end{array}$$

where

i) α and β are defined by multiplication by f^{m+1}

ii) θ is induced by the inclusion map $f^m\mathcal{O}_X \hookrightarrow \mathcal{O}_X$

iii) σ is defined by multiplication by f^m

iv) τ is defined by multiplication by f .

By the choice of m , β is an isomorphism and $\operatorname{Ker}\alpha \subset \operatorname{Ker}\sigma$. It follows that

$$b(\operatorname{Ker} c) = b(\ker \alpha\theta\beta^{-1}) \subset b(\operatorname{Ker}\sigma\theta\beta^{-1})$$

$$\subset b(\operatorname{Ker}\tau\sigma\theta\beta^{-1}) = b(\operatorname{Ker} b) = 0 .$$

Hence $ba = 0$. It follows that $\operatorname{Im}\xi = \operatorname{Im}\eta$.

By induction hypothesis, there exists $k \in \mathbb{N}$ such that

$$(R^0\pi_*^c\mathcal{O}_X/f^{2m+1}\mathcal{O}_X)_s \longrightarrow \left(R^0\pi_*^c(\mathcal{O}_X/\mathfrak{m}_{S,s}\mathcal{O}_X)\right)_s$$

and

$$\left(R^0\pi_*^c\left(\mathcal{O}_X/(\mathfrak{m}_{S,s}^k\mathcal{O}_X + f^{2m+1}\mathcal{O}_X)\right)\right)_s \longrightarrow \left(R^0\pi_*^c(\mathcal{O}_X/\mathfrak{m}_{S,s}\mathcal{O}_X)\right)_s$$

have the same image. Then k satisfies the requirement.

Q.E.D.

(12.4) Lemma. *Suppose* $\pi: X \longrightarrow S$ *is a strongly* 1-*pseudoconvex holomorphic map*, $s \in S$, *and* K *is a compact subset of* $\pi^{-1}(s)$. *Then there exists an open neighborhood* G *of* K *in* X *such that* $(G, \mathcal{O}_X|G)$ *is holomorphically convex*.

Proof. Choose $c_\# < c < c_*$ such that $K \subset X^c$. There exists $k \in \mathbb{N}$ satisfying the condition of (12.3). By (12.2), $(\pi^{-1}(s), \mathcal{O}_X/\mathfrak{m}_{S,s}^k\mathcal{O}_X)$ is holomorphically convex. There exist

$$f_1, \dots, f_\ell \in \Gamma(\pi^{-1}(s), \mathcal{O}_X/\mathfrak{m}^k_{S,s}\mathcal{O}_X)$$

and an open neighborhood U of K in $\pi^{-1}(s) \cap X^c$ such that K is contained in

$$\tilde{K} := \{x \in U \,|\, |f_i(x)| \leq 1 \text{ for } 1 \leq i \leq \ell\}.$$

By the choice of k, for some open neighborhood D of s in S, there exist

$$g_1, \dots, g_\ell \in \Gamma(\pi^{-1}(D) \cap X^c, \mathcal{O}_X)$$

such that f_i and g_i have the same image in $\Gamma(\pi^{-1}(s), \mathcal{O}_X/\mathfrak{m}_{S,s}\mathcal{O}_X)$. Let W be an open neighborhood of $\tilde{K}$ in X^c such that $W^- \cap \pi^{-1}(s) \subset U$. There exists a Stein open neighborhood Q of s in D such that the restriction of π to

$$\{x \in W \cap \pi^{-1}(Q) \,|\, |g_i(x)| \leq 1, 1 \leq i \leq \ell\}$$

is a proper map into Q. It follows that

$$\{x \in W \cap \pi^{-1}(Q) \,|\, |g_i(x)| < 1 \text{ for } 1 \leq i \leq \ell\}$$

is a holomorphically convex neighborhood of K in X. Q.E.D.

(12.5) Lemma. Suppose $\pi: X \longrightarrow S$ is a strongly 1-pseudoconvex holomorphic map, S is Stein, and $c_\# < c < c_*$. Then X is Stein if and only if for every compact subset K of X^c there exists a strongly plurisubharmonic function on an open neighborhood of K.

<u>Proof</u>. Only the "if" part requires a proof. S is the countable union of relatively compact Stein open subsets S_k such that $S_k \subset\subset S_{k+1}$ and the restriction map

$$\Gamma(S_{k+1}, \mathcal{O}_S) \longrightarrow \Gamma(S_k, \mathcal{O}_S)$$

has dense image. It suffices to show that $\pi^{-1}(S_k)$ is Stein for each k. Let ψ_k be a strongly plurisubharmonic exhaustion function on S_k. Take $c_{\#} < e < b < a < c_*$. By assumption there exists a strongly plurisubharmonic function θ on $\pi^{-1}(S_{k+1}) \cap X^a$. Choose a nonnegative C^2 function τ on X whose support is contained in X^a and which is identically 1 on X^b. There exists a C^2 function σ on $(-\infty, c_*)$ such that

i) $\operatorname{Supp} \sigma \subset (c_{\#}, c_*)$

ii) the first and second derivatives of σ are $\geqq 0$ and are > 0 on (e, c_*)

iii) $\sigma(\lambda) \longrightarrow \infty$ as $\lambda \longrightarrow c_*$ from the left.

For some positive number A, the function

$$\psi_k \circ \pi + \tau\theta + A\,\sigma\circ\varphi$$

is a strongly plurisubharmonic exhaustion function on $\pi^{-1}(S_k)$. Q.E.D.

(12.6) <u>Lemma</u>. <u>Suppose</u> D <u>is an open subset of</u> $\mathbb{C}^n$, $\{U_i\}_{i \in I}$ <u>is a locally finite open covering of</u> D <u>and</u> τ_i

$(i \in I)$ _is a_ C^2 _nonnegative function on_ D _with support in_ U_i _such that all first-order derivatives of_ τ_i _vanish on the zero-set of_ τ_i _and_ $\sum_{i \in I} \tau_i > 0$ _on_ D. _Let_ ψ _be a strongly plurisubharmonic function on_ D. _Suppose_ G _is an open subset of_ $\mathbb{C}^N$ _and_ σ_i $(i \in I)$ _is a strongly plurisubharmonic function on_ $U_i \times G$. _Let_ K _be a compact subset of_ $D \times G$ _and let_ $f: D \times G \longrightarrow D$ _be the natural projection. Then there exists_ $A_1 \in \mathbb{R}$ _and a function_ $B_1: \mathbb{R} \longrightarrow \mathbb{R}$ _such that_

$$\sum_{i \in I} (\tau_i \circ f) e^{A\sigma_i} + B(\psi \circ f)$$

is strongly plurisubharmonic on some open neighborhood of K _when_ $A \geq A_1$ _and_ $B \geq B_1(A)$.

Proof. For a C^2 function h on an open subset Ω of $\mathbb{C}^m$ and for $x \in \Omega$ and $a \in \mathbb{C}^m$, let $L(h;x,a)$ denote

$$\sum_{i,j=1}^{m} \left(\frac{\partial^2 h}{\partial z_i \partial \bar{z}_j}\right)(x) a_i \bar{a}_j$$

and let $\partial(h;x,a)$ denote

$$\sum_{i=1}^{m} \frac{\partial h}{\partial z_i}(x) a_i .$$

Let $\tilde{\tau}_i = \tau_i \circ f$ and $\tilde{\psi} = \psi \circ f$ and let S be the unit sphere in $\mathbb{C}^{n+N}$. It suffices to show that for fixed $x \in K$ and $a \in S$ there exist $A'(x,a) \in \mathbb{R}$ and $B'(\cdot,x,a) \in \mathbb{R}$ such that if $A \geqq A'(x,a)$ and $B \geqq B'(A,x,a)$, then

$$L\Big(\sum_{i \in I} \tilde{\tau}_i e^{A\sigma_i} + B\tilde{\psi};x,a\Big) > 0 .$$

Direct computation shows that

$$L\Big(\sum_{i \in I} \tilde{\tau}_i e^{A\sigma_i} + B\tilde{\psi};x,a\Big) = \Big[\sum_{i \in I} \tilde{\tau}_i A e^{A\sigma_i} L(\sigma_i;x,a)\Big]$$

$$+ \Big[\sum_{i \in I} \tilde{\tau}_i A^2 e^{A\sigma_i} L(\tilde{\tau}_i;x,a) + BL(\tilde{\psi};x,a)\Big]$$

$$+ \Big[\sum_{i \in I} \tilde{\tau}_i A^2 e^{A\sigma_i} |\partial(\sigma_i;x,a)|^2$$

$$+ \sum_{i \in I} 2Ae^{A\sigma_i} \operatorname{Re}\partial(\sigma_i;x,a)\overline{\partial(\tilde{\tau}_i;x,a)}\Big] .$$

The first bracketed term is > 0 when $A > 0$. When the first n components of a are all zero, the second bracketed term is 0. When the first n components of a are not all zero, there exists $B^*(\cdot) \in \mathbb{R}$ such that the second bracketed term is > 0 when $B \geqq B^*(A)$. The third bracketed term is at least as great as

$$J := \sum_{i \in I} Ae^{A\sigma_i} |\partial(\sigma_i;x,a)| \Big(\tilde{\tau}_i A |\partial(\sigma_i;x,a)| - 2|\partial(\tilde{\tau}_i;x,a)|\Big)$$

where the only nonzero terms are those with $\tilde{\tau}_i(x) \neq 0$ and $\partial(\sigma_i;x,a) \neq 0$. Therefore there exists $A^* \in \mathbb{R}$ such that $J > 0$ when $A \geqq A^*$. Q.E.D.

(12.7) Lemma. Suppose $\pi: X \longrightarrow S$ is a strongly 1-pseudoconvex holomorphic map and S is Stein. If every point of S has an open neighborhood such that $\pi^{-1}(U)$ is Stein, then X is Stein.

Proof. Let K be an arbitrary compact subset of X. By (12.5), it suffices to prove the existence of a strongly plurisubharmonic function on some open neighborhood of K. By embedding an open neighborhood of $\pi(K)$ as a subspace of $\mathbb{C}^n$, we can assume that $S = \mathbb{C}^n$. Cover $\pi(K)$ by a finite number of open balls B_i such that $\pi^{-1}(B_i)$ is Stein. For each i, choose a nonnegative C^∞ function τ_i on B_i with compact support such that the first-order derivatives of τ_i vanish whenever τ_i vanishes and $\sum_i \tau_i > 0$ on $\pi(K)$. Since every point of K has an open neighborhood D which can be embedded into $\mathbb{C}^{n+N}$ by a (nonproper) holomorphic map whose composite with the natural projection $\mathbb{C}^{n+N} \longrightarrow \mathbb{C}^n$ equals $\pi|D$, by using the fact that every strongly plurisubharmonic function on a subspace can be extended locally to a strongly plurisubharmonic function on the ambient space, we conclude from (12.6) that, if ψ (respectively σ_i) is a strongly plurisubharmonic function on S (respectively $\pi^{-1}(B_i)$), then

$$\sum_i (\tau_i \circ \pi) e^{A\sigma_i} + B(\psi \circ \pi)$$

is strongly plurisubharmonic on an open neighborhood of K for some $A, B > 0$. Q.E.D.

(12.8) Theorem. If $\pi: X \longrightarrow S$ is a strongly 1-pseudoconvex holomorphic map and S is Stein, then X is holomorphically convex.

Proof. By (12.4), every point s of S admits an open neighborhood U such that

$$G: = \pi^{-1}(U) \cap X^c$$

is holomorphically convex for some $c_\# < c < c_*$. Let $\eta: G \longrightarrow R$ be the Remmert quotient of G (that is,

i) R is Stein

ii) η is a proper surjective holomorphic map

iii) $R^0 \eta_* \mathcal{O}_G = \mathcal{O}_R$

iv) $\eta^{-1}(x)$ is connected for every $x \in R$).

Since no compact irreducible positive-dimensional subvariety of G can intersect $\{\varphi > c_\#\}$ by virtue of the **maximum** principle for strongly plurisubharmonic functions, η maps $G \cap \{\varphi > c_\#\}$ biholomorphically onto its image. Hence we can piece $\pi^{-1}(U) \cap \{\varphi > c_\#\}$ and R through η and obtain a complex space $\tilde{R}$. π induces a strongly 1-pseudoconvex holomorphic map $\tilde{R} \longrightarrow U$. Since there exists a strongly plurisubharmonic function on R, by (12.4), $\tilde{R}$ is Stein. By the uniqueness of Remmert quotients, as s varies, we can piece together all the Remmert quotients $\tilde{R}$ together and form a complex space X'. π induces a strongly 1-pseudoconvex map $\pi': X' \longrightarrow S$. Since, for every $s \in S$,

$(\pi')^{-1}(U) = \tilde{R}$ is Stein, by (12.7) X' is Stein. Since there is a proper holomorphic map σ from X to X', it follows that X is holomorphically convex. Q.E.D.

(12.9) Theorem. Suppose $\pi: X \longrightarrow S$ is a strongly 1-pseudoconvex holomorphic map and $\mathcal{F}$ is a coherent analytic sheaf on X. Then for $\nu \geqq 1$

i) $R^\nu \pi_* \mathcal{F}$ is coherent on S

ii) $R^\nu \pi_* \mathcal{F} \longrightarrow R^\nu \pi_*^c \mathcal{F}$ is an isomorphicm for $c_\# < c < c_*$

iii) $H^\nu(\pi^{-1}(U), \mathcal{F}) \longrightarrow \Gamma(U, R^\nu \pi_* \mathcal{F})$ is an isomorphism for any Stein open subset U of S.

Proof. Let π', X', σ be as in the proof of (12.8). Since σ is proper, by Grauert's direct image theorem (for the proper case), for $\mu \geqq 0$, $R^\mu \sigma_* \mathcal{F}$ is coherent on X' and

$$H^\mu(\sigma^{-1}(W), \mathcal{F}) \longrightarrow \Gamma(W, R^\mu \sigma_* \mathcal{F})$$

is an isomorphism for any Stein open subset W of X'. Since σ maps $\{\varphi > c_\#\}$ biholomorphically onto its image, for $\nu \geqq 1$

$$\operatorname{Supp} R^\nu \sigma_* \mathcal{F} \subset \sigma\{\varphi \leq c_\#\} .$$

By applying Grauert's direct image theorem (for the proper case) to $\pi' \,|\, \operatorname{Supp} R^\nu \sigma_* \mathcal{F}$, it follows that, for $\nu \geqq 1$, $R^0(\pi')_*(R^\nu \sigma_* \mathcal{F})$ is coherent and

$$\Gamma(U, R^0(\pi')_*(R^\nu \sigma_* \mathcal{F})) \approx \Gamma((\pi')^{-1}(U), R^\nu \sigma_* \mathcal{F}) \approx H^\nu(\pi^{-1}(U), \mathcal{F})$$

for every Stein open subset U of S. It follows that

$$R^\nu \pi_* \mathcal{F} \approx R^0(\pi')_*(R^\nu \sigma_* \mathcal{F}) .$$

Q.E.D.

Y-T. Siu

§13. Relative Exceptional Sets

(13.1) Suppose $\pi: X \longrightarrow S$ is a holomorphic map. A subvariety A of X is said to be *proper nowhere discrete above* S if $\pi|A$ is proper and every fiber of $\pi|A$ is positive-dimensional at any of its points. A subvariety A of X which is proper nowhere discrete above S is said to be *exceptional relative to* S if there exists a commutative diagram of holomorphic maps

$$\begin{array}{ccc} X & \xrightarrow{\Phi} & Y \\ & \pi\searrow \quad \swarrow\sigma & \\ & S & \end{array}$$

such that

i) Φ is proper

ii) every fiber of $\sigma|\Phi(A)$ has dimension ≤ 0

iii) Φ maps $X-A$ biholomorphically onto $Y-(A)$

iv) $R^0\Phi_*\mathcal{O}_X = \mathcal{O}_Y$.

The following result on relative exceptional sets is a consequence of (12.8).

(13.2) Theorem. *Suppose* $\pi: X \longrightarrow S$ *is a holomorphic map and* A *is a subvariety of* X *which is proper nowhere discrete above* S. *Then* A *is exceptional relative to* S *if and only if for every* $s \in S$ *there exist an open neighborhood*

U <u>of</u> s <u>and an open neighborhood</u> W <u>of</u> $A \cap \pi^{-1}(U)$ <u>in</u> $\pi^{-1}(U)$ <u>such that</u> $\pi | W: W \longrightarrow U$ <u>is strongly</u> 1-<u>pseudoconvex</u> <u>and</u> $A \cap \pi^{-1}(U)$ <u>is maximum among all subvarieties of</u> W <u>which are proper nowhere discrete above</u> U .

§14. Projectivity Criterion

(14.1) Suppose $\pi: X \longrightarrow S$ is a proper holomorphic map and $p: V \longrightarrow X$ is a holomorphic vector bundle. V is said to be weakly negative relative to S if for every $s \in S$ there exist an open neighborhood U of s and an open neighborhood D of the zero-section of $V|\pi^{-1}(U)$ such that $\pi \circ p|D: D \longrightarrow U$ is strongly 1-pseudoconvex.

Let N be the zero-section of V and $\mathcal{J}$ be the ideal-sheaf of N. We identify N with X and consider $\mathcal{J}^k/\mathcal{J}^{k+1}$ as an $\mathcal{O}_X$-sheaf. It is easy to see that, where V is a line bundle, $\mathcal{J}^k/\mathcal{J}^{k+1}$ equals $\mathcal{O}((L^*)^k)$ where L^* is the dual of L.

(14.2) Theorem. Suppose $p: V \longrightarrow X$ is a weakly negative holomorphic vector bundle relative to S. If $\mathcal{F}$ is a coherent analytic sheaf on X and K is a compact subset of S, then there exists $k_0 \in \mathbb{N}_0$ such that $R^\nu\pi_*(\mathcal{F} \otimes_{\mathcal{O}_X} \mathcal{J}^k/\mathcal{J}^{k+1})$ is zero on K for $\nu \geq 1$ and $k \geq k_0$.

Proof. The case where S is a single point was proved by Grauert [7]. The proof of the general case is completely analogous to that of the special case. It follows from the coherence of $R^\nu(\pi \circ p|D)_*(p^*\mathcal{F})$ and the fact that

$$\bigoplus_{m=0}^{k} R^\nu\pi_*(\mathcal{F} \otimes_{\mathcal{O}_X} (\mathcal{J}^m/\mathcal{J}^{m+1}))\Big|U$$

is a subsheaf of $R^{\nu}(\pi \circ p|D)_*(p^*\mathcal{F})$, where $\nu \geq 1$, $k \in \mathbb{N}_0$, and U, D are as in (14.1). Q.E.D.

(14.3) Theorem. If there exists a weakly negative holomorphic line bundle $p: L \longrightarrow X$ relative to S and S is Stein, then for every relatively compact open subset T of S there exists a holomorphic embedding σ of $\pi^{-1}(T)$ into $\mathbb{P}_N \times T$ such that the composite of σ and the natural projection $\mathbb{P}_N \times T \longrightarrow T$ equals π.

Proof. Let $\mathcal{L} = \mathcal{O}(L^*)$. Let Ω be a relatively compact Stein open neighborhood of T^- in S. Take two distinct points x, y of X. Consider the two exact sequences

$$0 \longrightarrow \mathfrak{m}_{X,x,y} \otimes \mathcal{L}^k \longrightarrow \mathcal{L}^k \longrightarrow (\mathcal{O}_X/\mathfrak{m}_{X,x,y}) \otimes \mathcal{L}^k \longrightarrow 0$$

$$0 \longrightarrow \mathfrak{m}^2_{X,x} \otimes \mathcal{L}^k \longrightarrow \mathfrak{m}_{X,x} \otimes \mathcal{L}^k \longrightarrow (\mathfrak{m}_{X,x}/\mathfrak{m}^2_{X,x}) \otimes \mathcal{L}^k \longrightarrow 0,$$

where $\mathfrak{m}_{X,x,y} = \mathfrak{m}_{X,x} \cap \mathfrak{m}_{X,y}$. By (14.2), for k sufficiently large, both $R^1\pi_*(\mathfrak{m}_{X,x,y} \otimes \mathcal{L}^k)$ and $R^1\pi_*(\mathfrak{m}^2_{X,x} \otimes \mathcal{L}^k)$ vanish on Ω. It follows that the two maps

$$R^0\pi_*\mathcal{L}^k \longrightarrow R^0\pi_*\left((\mathcal{O}_X/\mathfrak{m}_{X,x,y}) \otimes \mathcal{L}^k\right)$$

$$R^0\pi_*(\mathfrak{m}_{X,x} \otimes \mathcal{L}^k) \longrightarrow R^0\pi_*\left((\mathfrak{m}_{X,x}/\mathfrak{m}^2_{X,x}) \otimes \mathcal{L}^k\right)$$

are both surjective on Ω. Since Ω is Stein, it follows that there exist enough holomorphic sections of L^* over $\pi^{-1}(\Omega)$ to construct an embedding σ. Q.E.D.

§15. Extension of Complex Spaces

(15.1) Theorem. Suppose X is a complex space with $\mathcal{O}_X^{[n+1]} = \mathcal{O}_X$ and S is a Stein space of dimension $\leq n$. Suppose $\pi: X \longrightarrow S$ is a holomorphic map and $\varphi: X \longrightarrow (a_*, b_*) \subset (-\infty, \infty)$ is a strongly plurisubharmonic function such that $\pi | \{a \leq \varphi \leq b\}$ is proper for $a_* < a < b < b_*$. Then there exist uniquely a Stein space $\tilde{X}$ and a holomorphic map $\tilde{\pi}: \tilde{X} \longrightarrow S$ such that

i) $\mathcal{O}_{\tilde{X}}^{[n+1]} = \mathcal{O}_{\tilde{X}}$

ii) X is an open subset of $\tilde{X}$ which intersects every branch of $\tilde{X}$

iii) $\pi = \tilde{\pi} | X$

iv) the restriction of $\tilde{\pi}$ to $\tilde{X} - \{\varphi > a\}$ is proper for every $a_* < a < b_*$.

Proof (sketch). First we prove the case $n = 0$. It follows from $\mathcal{O}_X^{[1]} = \mathcal{O}_X$ that, outside a subvariety of dimension ≤ 0 in X, codh $\mathcal{O}_X \geq 3$. We can choose $a_* < a < b' < b < b_*$ such that codh $\mathcal{O}_X \geq 3$ on $\{\varphi = a\}$. For $a_* \leq c < d \leq b_*$ let $X_c^d = \{c < \varphi < d\}$. Let $\mathcal{J}$ be an arbitrary coherent ideal-sheaf on X such that $\mathcal{J} = \mathcal{O}_X$ on $X_a^{b'}$. Consider the following commutative diagram

$$\begin{array}{ccccccc}
\Gamma(X_a^b,\mathcal{O}_X) & \xrightarrow{\theta} & \Gamma(X_a^b,\mathcal{O}_X/\mathcal{J}) & \longrightarrow & H^1(X_a^b,\mathcal{J}) & \longrightarrow & H^1(X_a^b,\mathcal{O}_X) \\
 & & & & \sigma\downarrow & & \downarrow \\
 & & & & H^1(X_a^{b'},\mathcal{J}) & \xrightarrow{\tau} & H^1(X_a^{b'},\mathcal{O}_X)
\end{array}$$

coming from

$$0 \longrightarrow \mathcal{J} \longrightarrow \mathcal{O}_X \longrightarrow \mathcal{O}_X/\mathcal{J} \longrightarrow 0 .$$

By the results of Andreotti-Grauert [1], σ is an isomorphism. Since $\mathcal{J} = \mathcal{O}_X$ on $X_a^{b'}$, τ is an isomorphism. It follows that θ is an epimorphism. From the arbitrariness of $\mathcal{J}$, we conclude that there exist a holomorphic map

$$f: X_a^b \longrightarrow \mathbb{C}^N$$

such that, for some $0 < \alpha < \beta < \gamma$ in $\mathbb{R}^N$ and some $a < a' < b'$,

i) f maps $f^{-1}(G^N(\alpha,\beta))$ biholomorphically onto a subspace V of $G^N(\alpha,\beta)$.

ii) $X_a^{a'} \subset f^{-1}(\Delta^N(\alpha))$

iii) $X_{b'}^b \subset f^{-1}(G^N(\beta,\gamma))$.

By (11.1), V can be extended to a complex subspace $\tilde{V}$ of $\Delta^N(\beta)$ satisfying $\mathcal{O}_{\tilde{V}}^{[1]} = \mathcal{O}_{\tilde{V}}$. By gluing $f^{-1}(G^N(\alpha,\beta))$ and V by means of f, we can piece together $\tilde{V}$ and $f^{-1}(G^N(\alpha,\gamma)) \cup X_{b'}^{b_*}$ to form a complex space $\tilde{X}$. By (12.5), $\tilde{X}$ is Stein. $\tilde{X}$ satisfies $\mathcal{O}_{\tilde{X}}^{[1]} = \mathcal{O}_{\tilde{X}}$ and extends $X_{b'}^{b_*}$. By the property of gap-sheaves (cf. [A.16] of the Appendix), any complex space satisfying these three conditions is isomorphic to $\tilde{X}$ and the isomorphism is unique. Choose arbitrarily $a < b'' < b'$. By repeating the preceding argument with b'' (instead of b'), we obtain a complex space $\hat{X}$ (instead of $\tilde{X}$). Since $\hat{X}$ extends $X^{b_*}_{b''}$, in particular $\hat{X}$ extends $X^{b_*}_{b'}$ and is therefore uniquely isomorphic to $\tilde{X}$. From the arbitrariness of b'', we conclude that X is an open subset of $\tilde{X}$.

Now, consider the case $n > 0$. The uniqueness of $\tilde{X}$ is a consequence of the properties of gap-sheaves (cf. [A.16] of the Appendix). Because of (12.7), it suffices to prove the existence of $\tilde{X}$ locally (with respect to S). By locally representing S as an analytic cover, we can assume without loss of generality that $S = \Delta$. First consider the special case where $\mathcal{O}_X$ is π-flat and codh $\mathcal{O}_X \geq n+3$. By the finite generation of $(R^1(\pi^b_a)_*\mathcal{O}_X)_s$ for $a_* < a < b < b_*$ and $s \in S$, we use the methods of §12 and of the case $n = 0$ to get the local existence of $\tilde{X}$.

For the general case, as in the case of sheaf extension, we use a ring domain to avoid the bad set and appeal to induction on n. More precisely, we choose $0 < \alpha < \beta$ in $\mathbb{R}$ and $\gamma \in \mathbb{R}^{n+1}_+$ and $a_* < a < b < b_*$ such that

i) on $\pi^{-1}(\Delta^{n-1}(\gamma) \times G^1(\alpha,\beta)) \cap X^b_a$, $\mathcal{O}_X$ is π-flat and codh $\mathcal{O}_X \geq n+3$

ii) the extension $\hat{\pi}: \hat{X} \longrightarrow \Delta^{n-1}(\gamma) \times G^1(\alpha,\beta)$ of $\pi^{-1}(\Delta^{n-1}(\gamma) \times G^1(\alpha,\beta)) \cap X_a^b$ satisfies $\operatorname{codh} \mathcal{O}_{\hat{X}} \geq n+2$.

Now we apply the induction hypothesis to the holomorphic map

$$\pi^*: \hat{X} \cup \left(\pi^{-1}(\Delta^{n-1}(\gamma) \times \Delta^1(\beta)) \cap X_a^b\right) \longrightarrow \Delta^{n-1}(\gamma)$$

which is induced by the composites of the natural projection $\Delta^{n-1}(\gamma) \times \Delta^1(\beta) \longrightarrow \Delta^{n-1}(\gamma)$ with π and with $\hat{\pi}$.

Q.E.D.

APPENDIX

This appendix contains materials on homological codimension, flatness, and gap-sheaves which are used in these lecture notes. For more details, see [28].

(A.1) Suppose M is a finitely generated module over a local ring $(R, \mathfrak{m})$. An M-sequence is a sequence $f_1, \ldots, f_k \in \mathfrak{m}$ such that f_j is not a zero-divisor of $M / \sum_{i=1}^{j-1} f_i M$ for $1 \leq j \leq k$. All maximal M-sequences have the same length. Define the homological codimension of M over R (denoted by $\mathrm{codh}_R M$ or simply by $\mathrm{codh}\, M$) as the common length of all maximal M-sequences. If $S \longrightarrow R$ is an epimorphism of local rings, then $\mathrm{codh}_S M = \mathrm{codh}_R M$ when M is regarded as over S. When R is regular of dimension n, $\mathrm{codh}\, M$ agrees with the maximum of $n-\ell$ such that there exists an exact sequence

$$0 \longrightarrow R^{p_\ell} \longrightarrow \ldots \longrightarrow R^{p_1} \longrightarrow R^{p_0} \longrightarrow M \longrightarrow 0 .$$

• Suppose $\mathcal{F}$ is a coherent analytic sheaf on a complex space X. We define $\mathrm{codh}\, \mathcal{F}$ as the function

$$x \longrightarrow \mathrm{codh}_{\mathcal{O}_{X,x}} \mathcal{F}_x .$$

Let $S_k(\mathcal{F})$ denote the set of points of X where $\mathrm{codh}\, \mathcal{F} \leq k$.

(A.2) Lemma (Frenkel). Suppose D is a Stein domain in $\mathbb{C}^n$ and D' is a nonempty Stein subdomain of D. If $0 \leqq a < b$ in $\mathbb{R}^N$, then, for $1 \leqq \nu < N-1$,

$$H^\nu\Big((D' \times \Delta^N(b)) \cup (D \times G^N(a,b)),\ {}_{n+N}\mathcal{O}\Big) = 0 .$$

This lemma is proved by Laurent series expansion. For details, see [1, pp.217-219]. As a corollary, we have the following.

(A.3) Proposition (Scheja). Suppose A is a subvariety of dimension $\leqq d$ in a complex space X and $\mathcal{F}$ is a coherent analytic sheaf on X with $\operatorname{codh} \mathcal{F} \geqq r$. Then $\mathcal{H}^\nu_A \mathcal{F} = 0$ for $0 \leqq \nu < r-d$. Hence $H^\nu(X,\mathcal{F}) \longrightarrow H^\nu(X-A,\mathcal{F})$ is bijective for $0 \leqq \nu < r-d-1$ and injective for $\nu = r-d-1$.

We can assume without loss of generality that X is an open subset of $\mathbb{C}^n$. When $\mathcal{F} = {}_n\mathcal{O}$ and A is regular, it is a direct consequence of Frenkel's lemma, and, when the singular set A' of A is nonempty, it follows from the long exact sequence

$$0 \longrightarrow \mathcal{H}^0_{A'}({}_n\mathcal{O}) \longrightarrow \mathcal{H}^0_{A}({}_n\mathcal{O}) \longrightarrow \mathcal{H}^0_{A-A'}({}_n\mathcal{O}) \longrightarrow \mathcal{H}^1_{A'}({}_n\mathcal{O}) \longrightarrow \cdots$$

$$\longrightarrow \mathcal{H}^\nu_{A'}({}_n\mathcal{O}) \longrightarrow \mathcal{H}^\nu_{A}({}_n\mathcal{O}) \longrightarrow \mathcal{H}^\nu_{A-A'}({}_n\mathcal{O}) \longrightarrow \mathcal{H}^{\nu+1}_{A'}({}_n\mathcal{O}) \longrightarrow \cdots$$

When $\mathcal{F} \neq {}_n\mathcal{O}$, we use a local finite free resolution of $\mathcal{F}$.

Now we define relative gap-sheaves with respect to a subvariety. Suppose A is a subvariety of a complex space

X and $\mathcal{F} \subset \mathcal{G}$ are coherent analytic sheaves on X. Define the sheaf $\mathcal{F}[A]_{\mathcal{G}}$ by the presheaf

$$U \longmapsto \{s \in \Gamma(U,\mathcal{G}) \,\big|\, (s|U-A) \in \Gamma(U-A,\mathcal{F})\} .$$

The following is a consequence of the Nullstellensatz.

(A.4) <u>Proposition</u>. <u>If $\mathcal{J}$ is the ideal sheaf of A, then $\mathcal{F}[A]_{\mathcal{G}} = \bigcup_{k=1}^{\infty} (\mathcal{F}:\mathcal{J}^k)_{\mathcal{G}}$ and is therefore coherent, where $(\mathcal{F}:\mathcal{J}^k)_{\mathcal{G}}$ is the subsheaf of $\mathcal{G}$ whose stalk at x is the set of all $s \in \mathcal{G}_x$ such that $\mathcal{J}_x^k s \subset \mathcal{F}_x$.</u>

(A.5) <u>Proposition</u>. <u>$S_m(\mathcal{F})$ is a subvariety of dimension $\leq m$.</u>

<u>Proof</u>. We can assume without loss of generaltiy that $\mathcal{F}$ is defined on an open subset of $\mathbb{C}^n$. By taking a local finite free resolution of $\mathcal{F}$ and considering the rank of the matrix defining the extreme left sheaf-homomorphism of the resolution, we see easily that $S_m(\mathcal{F})$ is a subvariety. For the dimension estimate, the case $m = 0$ is obtained by considering $\mathcal{F}/0[\{x\}]_{\mathcal{F}}$ for $x \in S_0(\mathcal{F})$. The general case follows from induction on m and considering the quotient of $\mathcal{F}$ by the subsheaf generated by a holomorphic function whose germ at some $x \in S_m(\mathcal{F})$ is not a zero-divisor of $\mathcal{F}_x$. Q.E.D.

Now we define the dth <u>relative gap-sheaf</u>. Suppose $\mathcal{F} \subset \mathcal{G}$ are coherent analytic sheaves on a complex space X. Define the subsheaf $\mathcal{F}_{[d]\mathcal{G}}$ of $\mathcal{G}$ by the presheaf

$$U \longmapsto \{s \in \Gamma(U,\mathcal{G}) \mid (s|U-A) \in \Gamma(U-A,\mathcal{F}) \text{ for some subvariety } A \text{ of } U \text{ of dimension } \leq d\} .$$

(A.6) Proposition. $\mathcal{F}_{[d]\mathcal{G}} = \mathcal{F}[S_d(\mathcal{G}/\mathcal{F})]_{\mathcal{G}}$. Hence $\mathcal{F}_{[d]\mathcal{G}}$ is coherent and $\dim \operatorname{Supp}(\mathcal{F}_{[d]\mathcal{G}}/\mathcal{F}) \leq d$.

Proof. Follows from (A.3) and (A.5). Q.E.D.

(A.7) Lemma. Suppose $\mathcal{F} \subset \mathcal{G}$ are coherent analytic sheaves on a complex space X and $x \in X$ such that $\mathcal{F}_x$ is a primary submodule of $\mathcal{G}_x$ whose radical P is of dimension d . Then $\dim_x \operatorname{Supp} \mathcal{G}/\mathcal{F} \leq d$ and $(\mathcal{F}_{[d-1]\mathcal{G}})_x = \mathcal{F}_x$.

Proof. For some open neighborhood U of x in X there exists a coherent ideal sheaf $\mathcal{I}$ on U whose stalk at x is P . There exists $k \in \mathbb{N}$ such that $P^k\mathcal{G}_x \subset \mathcal{F}_x$. Hence $\mathcal{I}^k\mathcal{G} \subset \mathcal{F}$ on some open neighborhood and

$$\dim_x \operatorname{Supp} \mathcal{G}/\mathcal{F} \leq \dim_x \operatorname{Supp} \mathcal{O}_X/\mathcal{I} = d .$$

Let $Y = \operatorname{Supp}(\mathcal{F}_{[d-1]\mathcal{G}}/\mathcal{F})$ and let $\mathcal{J}$ be the ideal sheaf of Y . Since $\dim Y < d$, there exists $f \in \mathcal{J}_x - P$. By the Nullstellensatz,

$$f^\ell(\mathcal{F}_{[d-1]\mathcal{G}})_x \subset \mathcal{F}_x$$

for some $\ell \in \mathbb{N}$. Since $f \notin P$, it follows that

$$\mathcal{F}_x = \{s \in \mathcal{G}_x \mid fs \in \mathcal{F}_x\} .$$

Hence $(\mathcal{F}_{[d-1]\mathcal{G}})_x = \mathcal{F}_x$. Q.E.D.

(A.8) Proposition. Suppose $\mathcal{F}$ is a coherent analytic sheaf on a complex space X, $x \in X$, and $f \in \Gamma(X, \mathcal{O}_X)$. Then f_x is not a zero-divisor of $\mathcal{F}_x$ if and only if $\dim_x V(f) \cap \operatorname{Supp} O_{[k]}\mathcal{F} < k$ for all $k \in \mathbb{N}_*$, where $V(f) = \operatorname{Supp} \mathcal{O}_X / f\mathcal{O}_X$.

Proof. Suppose f_x is a zero-divisor. Then $f_x s_x = 0$ for some $s \in \Gamma(U, \mathcal{F})$ with $s_x \neq 0$, where U is an open neighborhood of x. Let $k = \dim_x \operatorname{Supp} s$. Then $\operatorname{Supp} s \subset V(f) \cap \operatorname{Supp} O_{[k]}\mathcal{F}$.

Suppose f_x is not a zero-divisor. By considering the kernel of the sheaf-homomorphism $\mathcal{F} \longrightarrow \mathcal{F}$ defined by multiplication by f, we conclude that, for some open neighborhood U of x, f_y is not a zero-divisor of $\mathcal{F}_y$ for $y \in U$. If $\dim_x V(f) \cap \operatorname{Supp} O_{[k]}\mathcal{F} = k$ for some k, then, for some open subset W of U,

$$\emptyset \neq W \cap \operatorname{Supp} O_{[k]}\mathcal{F} \subset V(f),$$

which, because of the Nullstellensatz, contradicts f_y being a non-zero-divisor of $\mathcal{F}_y$ for $y \in U$. Q.E.D.

(A.9) Proposition. Suppose $\mathcal{F}$ is a coherent analytic sheaf on a complex space X. Then the d-dimensional component of $\operatorname{Supp} O_{[d]}\mathcal{F}$ equals the d-dimensional component of $S_d(\mathcal{F})$. Equivalently, for $x \in X$, $\dim_x \operatorname{Supp} O_{[d]}\mathcal{F} < d$ if and only if $\dim_x S_d(\mathcal{F}) < d$.

Proof. We prove the equivalent statement. The "if" part follows from

$$O_{[d]}\mathcal{F} = O[S_d(\mathcal{F})]_{\mathcal{F}} .$$

For the "only if" part, we assume that $\dim_x \operatorname{Supp} O_{[d]}\mathcal{F} < d$ and $\dim_x S_d(\mathcal{F}) = d$. There exists an open subset U of $X - \operatorname{Supp} O_{[d]}\mathcal{F}$ such that $\dim U \cap S_d(\mathcal{F}) = d$. After replacing U by a smaller open subset, we can assume that there exists $f \in \Gamma(U, \mathcal{O}_X)$ such that $V(f) := \operatorname{Supp} \mathcal{O}_X/f\mathcal{O}_X$ contains $U \cap S_d(\mathcal{F})$ and $V(f)$ does not contain any k-dimensional branch of $U \cap \operatorname{Supp} O_{[k]}\mathcal{F}$ for any $k > d$. By (A.8), f_y is not a zero-divisor of $\mathcal{F}_y$ for $y \in U$. Hence $S_d(\mathcal{F}) \cap U = S_{d-1}(\mathcal{F}/f\mathcal{F}) \cap U$, contradicting $\dim U \cap S_d(\mathcal{F}) = d$.

Q.E.D.

(A.10) Corollary. $O_{[d]}\mathcal{F} = 0$ <u>if and only if</u> $\dim S_{k+1}(\mathcal{F}) \leq k$ <u>for</u> $k < d$.

(A.11) Corollary. <u>For</u> $f \in \Gamma(X, \mathcal{O}_X)$, f_x <u>is not a zero-divisor of</u> $\mathcal{F}_x$ <u>if and only if</u> $\dim_x (\operatorname{Supp} \mathcal{O}_X/f\mathcal{O}_X) \cap S_k(\mathcal{F}) < k$ <u>for</u> $k \geq 0$.

(A.12) Suppose $\pi: X \longrightarrow Y$ is a holomorphic map of complex spaces and $\mathcal{F}$ is a coherent analytic sheaf on X. $\mathcal{F}$ is said to be π-<u>flat</u> at $x \in X$ if $\mathcal{F}_x$ is a flat $\mathcal{O}_{Y,\pi(x)}$-module. $\mathcal{F}$ is said to be π-<u>flat</u> on (or at) a subset G of X

if $\mathcal{F}$ is π-flat at every point of G.

When $Y = \mathbb{C}^n$, $\mathcal{F}$ is π-flat at x if and only if $t_j - t_j^0$ is not a zero-divisor for $\mathcal{F}_x / \sum_{i=1}^{j-1} (t_i - t_i^0)\mathcal{F}_x$, where $(t_1^0, \ldots, t_n^0) = \pi(x)$.

(A.13) Proposition. *Suppose $\mathcal{F}$ is a coherent analytic sheaf on a complex space X and $\pi: X \longrightarrow \mathbb{C}^n$ is a holomorphic map. Let Z be the set of all points of $x \in X$ such that $\mathcal{F}_x$ is not π-flat. Then Z is a subvariety of X and the rank of $\pi|Z$ is $< n$.*

Proof. By (A.11),

$$Z = \bigcup_{k=0}^{\infty} \{x \in X \mid \pi^{-1}\pi(x) \cap S_k(\mathcal{F}) > k-n\} .$$

Hence Z is a subvariety of X. Let T_k be the set of all $x \in S_k(\mathcal{F})$ such that $\operatorname{rank}_x \pi | S_k(\mathcal{F}) < n$. Since $Z \subset \bigcup_{k=0}^{\infty} T_k$, it follows that $\operatorname{rank} \pi|Z < n$. Q.E.D.

Now we define absolute gap-sheaves. Suppose $\mathcal{F}$ is a coherent analytic sheaf on a complex space X. The dth *absolute gap-sheaf* $\mathcal{F}^{[d]}$ of $\mathcal{F}$ is defined by the presheaf

$$U \longmapsto \operatorname*{ind\ lim}_{A \in \mathcal{O}_d(U)} \Gamma(U-A, \mathcal{F})$$

where $\mathcal{O}_d(U)$ is the directed set of all subvarieties of U of dimension $\leq d$.

(A.14) Proposition. Suppose $\mathcal{F}$ is a coherent analytic sheaf on a complex space X. Then the following three conditions are equivalent.

i) $\mathcal{F}^{[d]}$ is a coherent on X.

ii) $\dim \operatorname{Supp} \mathcal{O}_{[d+1]\mathcal{F}} \leq d$.

iii) $\dim S_{d+1}(\mathcal{F}) \leq d$.

Proof. The equivalence of ii) and iii) is (A.9).

To show iii) $\Longrightarrow$ i), we can assume that X is an open subset of $\mathbb{C}^n$ and there exist exact sequences

$$(*) \quad 0 \longrightarrow \mathcal{K} \longrightarrow {}_n\mathcal{O}^{p_{n-d-2}} \longrightarrow \cdots \longrightarrow {}_n\mathcal{O}^{p_1} \longrightarrow {}_n\mathcal{O}^{p_0} \longrightarrow \mathcal{F} \longrightarrow 0$$

$$(**) \quad {}_n\mathcal{O}^{q_{n-d-2}} \longrightarrow \cdots \longrightarrow {}_n\mathcal{O}^{q_1} \longrightarrow {}_n\mathcal{O}^{q_0} \longrightarrow \mathcal{H}om_{{}_n\mathcal{O}}(\mathcal{K}, {}_n\mathcal{O}) \longrightarrow 0$$

on X. Let $A = S_{d+1}(\mathcal{F})$. For a coherent analytic sheaf $\mathcal{G}$ on X, let $R^\nu_A\mathcal{G}$ be the νth direct image of $\mathcal{G}|X-A$ under the inclusion map $X-A \hookrightarrow X$. By (A.3) and (*), $\mathcal{F}^{[d]} \approx R^0_A\mathcal{F} \approx R^{n-d-2}_A\mathcal{K}$. By applying $\mathcal{H}om_{{}_n\mathcal{O}}(\cdot, {}_n\mathcal{O})$ to (**), we obtain an exact sequence

$$0 \longrightarrow \mathcal{K} \longrightarrow {}_n\mathcal{O}^{q_0} \longrightarrow {}_n\mathcal{O}^{q_1} \longrightarrow \cdots \longrightarrow {}_n\mathcal{O}^{q_{n-d-3}} \xrightarrow{\alpha} {}_n\mathcal{O}^{q_{n-d-2}}$$

on X-A, because $\mathcal{K}$ is locally free on X-A. It follows from (A.3) that

$$R_A^{n-d-2}\mathcal{K} \approx R_A^0(\operatorname{Im}\alpha) \approx (\operatorname{Im}\alpha)[A]_{{}_n\mathcal{O}^{q_{n-d-2}}}$$

is coherent.

To show i) $\Longrightarrow$ ii), we suppose $\dim \operatorname{Supp} 0_{[d+1]}\mathcal{F} = d+1$ and are going to derive a contradiction. Let D be the unit open 1-disc. Without loss of generality we can assume the following.

$$X = D^n$$

$$0_{[d]}\mathcal{F} = 0$$

$$\operatorname{Supp} 0_{[d+1]}\mathcal{F} = D^{d+1} \times 0 .$$

Let $A = D^d \times 0$ and let $\mathcal{G}$ be the zeroth direct image of $0_{[d+1]}\mathcal{F}|D^{d+1} \times 0 - A$ under the inclusion map $D^{d+1} \times 0 - A \hookrightarrow D^{d+1} \times 0$. Since

$$\mathcal{G} = (0_{[d+1]}\mathcal{F})^{[A]}{}_{(\mathcal{F}^{[d]})},$$

$\mathcal{G}$ is coherent. Since the sheaf-homomorphism $\mathcal{G} \longrightarrow \mathcal{G}$ defined by multiplication by t_{d+1} is 0, it follows from Nakayama's Lemma that $\mathcal{G}_0 = 0$. Let $\{x_\nu\}_{\nu=1}^{\infty}$ be a sequence in $D^{d+1} \times 0 - A$ approaching 0. Since $D^{d+1} \times 0 - A$ is Stein, there exists

$$s \in \Gamma(D^{d+1} \times 0 - A, 0_{[d+1]}\mathcal{F})$$

such that $s_x \neq 0$ for all ν. It follows that s defines a non-zero element of $\mathcal{G}_0$, contradicting $\mathcal{G}_0 = 0$. Q.E.D.

(A.15) Proposition. Suppose $\mathcal{F}$ is a coherent analytic sheaf on a complex space X . Then the natural sheaf-homomorphism $\mathcal{F} \longrightarrow \mathcal{F}^{[d]}$ is an isomorphism if and only if $\dim S_{k+2}(\mathcal{F}) \leqq k$ for all $k < d$.

Proof. i) The "if" part. Use induction on d . The case $d = -1$ is trivial. Suppose A is a subvariety of dimension $\leqq d$ in an open subset U of X . Since $\operatorname{codh} \mathcal{F} \geqq d+2$ on $U - (A \cap S_{d+1}(\mathcal{F}))$, it follows from (A.3) that

$$\Gamma\big(U-(A \cap S_{d+1}(\mathcal{F})), \mathcal{F}\big) \approx \Gamma(U-A, \mathcal{F}) .$$

Since $\dim S_{d+1}(\mathcal{F}) \leqq d-1$, by induction hypothesis

$$\Gamma(U,\mathcal{F}) \approx \Gamma\big(U-(A \cap S_{d+1}(\mathcal{F})), \mathcal{F}\big).$$

Hence $\mathcal{F} \approx \mathcal{F}^{[d]}$.

ii) The "only if" part. From the definition of $\mathcal{F}^{[d]}$, we conclude that $0_{[d]}\mathcal{F} = 0$. By (A.14), $\dim \operatorname{Supp} 0_{[d+1]}\mathcal{F} \leqq d$. Hence $0_{[d+1]}\mathcal{F} = 0$. For $x \in X$ there exists $f \in \Gamma(U, \mathcal{O}_X)$ for some open neighborhood U of x such that $\operatorname{Supp} \mathcal{O}_X / f\mathcal{O}_X$ contains $S_{d+1}(\mathcal{F})$ but contains no k-dimensional branch of $U \cap \operatorname{Supp} 0_{[k]}\mathcal{F}$ for any $k > d+1$. Let $\mathcal{G} = \mathcal{F}/f\mathcal{F}$. Then $0_{[d]}\mathcal{G} = 0$ and

$$\dim U \cap S_{k+2}(\mathcal{F}) \leqq \dim S_{k+1}(\mathcal{G}) \leqq k$$

for $k < d$. Q.E.D.

(A.16) Proposition. Suppose $0 \leqq a < b$ in $\mathbb{R}^N$, D is a domain in $\mathbb{C}^n$, and D' is a nonempty open subset of D. If $\mathcal{F}$ is a coherent analytic sheaf on $D \times \Delta^N(b)$ such that $\mathcal{F}^{[n-1]} = \mathcal{F}$, then the restriction map

$$\theta : \Gamma(D \times \Delta^N(b), \mathcal{F}) \longrightarrow \Gamma\big((D \times G^N(a,b)) \cup (D' \times \Delta^N(b)), \mathcal{F}\big)$$

is an isomorphism.

Proof. For any open subset U of D, the restriction map

$$\alpha : \Gamma(U \times \Delta^N(b), \mathcal{F}) \longrightarrow \Gamma(U \times G^N(a,b), \mathcal{F})$$

is injective, because the support V of any nonzero element s of $\operatorname{Ker} \alpha$ would be a subvariety of $U \times \Delta^N(b)$ disjoint from $U \times G^N(a,b)$ and, by considering $(\{x\} \times \Delta^N(b)) \cap V$ for $x \in U$, we conclude that $\dim V \leqq n$, contradicting $0_{[n]}\mathcal{F} = 0$. In particular, θ is injective.

We are going to prove the surjectivity of θ by induction on n. Take $s \in \Gamma\big((D \times G^N(a,b)) \cup (D' \times \Delta^N(b)), \mathcal{F}\big)$. We have to show that $s \in \operatorname{Im} \theta$.

Consider first the special case where $\operatorname{codh} \mathcal{F} \geqq n+1$ on $D \times \Delta^N(b)$. Let Ω be the largest open subset of D such that $s \mid \Omega \times G^N(a,b)$ extends to an element of $\Gamma(\Omega \times \Delta^N(b), \mathcal{F})$. To show that Ω is closed in D, take an element x of the boundary of Ω in D. Let $P' \subset P$ be nonempty open polydiscs in D such that $x \in P$ and $P' \subset \Omega$. There exists an exact sequence

$$0 \longrightarrow {}_{n+N}\mathcal{O}^{p_{N-1}} \longrightarrow \cdots \longrightarrow {}_{n+N}\mathcal{O}^{p_1} \longrightarrow {}_{n+N}\mathcal{O}^{p_0} \longrightarrow \mathcal{F} \longrightarrow 0$$

on $P \times \Delta^N(b)$. It follows from

$$H^\nu\Big((P \times G^N(a,b)) \cup (P' \times \Delta^N(b)), {}_{n+N}\mathcal{O}\Big) = 0 \quad (1 \leqq \nu < N-1)$$

that $P \subset \Omega$. Hence Ω is closed in D and $\Omega = D$.

Now consider the general case. Take $a < b' < b$ in $\mathbb{R}^N$. Let $\pi: D \times \Delta^N(b) \longrightarrow D$ be the natural projection. Let $A = \pi\big(S_n(\mathcal{F}) \cap (D \times \overline{\Delta^N(b')})\big)$. By the special case, $s \mid (D-A) \times G^N(a,b')$ can be extended to $s' \in \Gamma\big((D-A) \times \Delta^N(b'), \mathcal{F}\big)$. To finish the proof, it suffices to show that, for any given $x \in A$, there exists an open neighborhood U of x in D such that $s \mid U \times G^N(a,b)$ extends to an element of $\Gamma(U \times \Delta^N(b), \mathcal{F})$. Since $\dim S_n(\mathcal{F}) \leqq n-2$, we can assume without loss of generality that $x = 0$ and there exist $0 < \gamma$ in $\mathbb{R}^{n-2}$ and $0 < \alpha < \beta$ in $\mathbb{R}^2$ such that $\Delta^{n-2}(\gamma) \times \Delta^2(\beta) \subset D$ and $\Delta^{n-2}(\gamma) \times G^2(\alpha,\beta)$ is disjoint from A. By induction hypothesis, the restriction map

$$\Gamma(\Delta^{n-2}(\gamma) \times \Delta^2(\beta) \times \Delta^N(b), \mathcal{F}) \longrightarrow$$
$$\Gamma\Big((\Delta^{n-2}(\gamma) \times \Delta^2(\beta) \times G^N(a,b)) \cup (\Delta^{n-2}(\gamma) \times G^2(\alpha,\beta) \times \Delta^N(b)), \mathcal{F}\Big)$$

is surjective. Hence the element of

$$\Gamma\Big((\Delta^{n-2}(\gamma) \times \Delta^2(\beta) \times G^N(a,b)) \cup (\Delta^{n-2}(\gamma) \times G^2(\alpha,\beta) \times \Delta^N(b)), \mathcal{F}\Big)$$

which agrees with s on $\Delta^{n-2}(\gamma) \times \Delta^2(\beta) \times G^N(a,b)$ and agrees with s' on $\Delta^{n-2}(\gamma) \times G^2(\alpha,\beta) \times \Delta^N(b')$ can be extended to an element of $\Gamma(\Delta^{n-2}(\gamma) \times \Delta^2(\beta) \times \Delta^N(b), \mathcal{F})$. Q.E.D.

(A.17) Lemma. Suppose $\mathcal{F}$, $\mathcal{G}$ are coherent analytic sheaves on a complex space X. If $\mathcal{F} = \mathcal{F}^{[n]}$, then $\left(\mathcal{H}om_{\mathcal{O}_X}(\mathcal{G},\mathcal{F})\right)^{[n]} = \mathcal{H}om_{\mathcal{O}_X}(\mathcal{G},\mathcal{F})$.

Proof. Because of the local nature, we can assume that there exists an exact sequence

$$\mathcal{O}_X^q \longrightarrow \mathcal{O}_X^p \longrightarrow \mathcal{G} \longrightarrow 0$$

on X. Hence

$$0 \longrightarrow \mathcal{H}om_{\mathcal{O}_X}(\mathcal{G},\mathcal{F}) \longrightarrow \mathcal{H}om_{\mathcal{O}_X}(\mathcal{O}_X^p,\mathcal{F}) \longrightarrow \mathcal{H}om_{\mathcal{O}_X}(\mathcal{O}_X^q,\mathcal{F})$$

is exact on X. The result follows from the fact that $\mathcal{H}om_{\mathcal{O}_X}(\mathcal{O}_X^\nu,\mathcal{F})$ is isomorphic to the direct sum of ν copies of $\mathcal{F}$. Q.E.D.

(A.18) Proposition. Suppose $0 \leq a \leq a' < b$ in $\mathbb{R}^N$, D is an open subset of $\mathbb{C}^n$, and $\mathcal{F}_i$ is a coherent analytic sheaf on $D \times G^N(a,b)$ such that $\mathcal{F}_i = \mathcal{F}_i^{[n]}$ $(i = 1,2)$. Then every sheaf-isomorphism $\mathcal{F}_1 \longrightarrow \mathcal{F}_2$ on $D \times G^N(a',b)$ extends uniquely to a sheaf-isomorphism $\mathcal{F}_1 \longrightarrow \mathcal{F}_2$ on $D \times G^N(a,b)$.

Proof. We can assume without loss of generality that $a_j = a_j'$ for $2 \leq j \leq n$. By (A.17) and (A.16), both restriction maps

$$\Gamma\big(D\times G^N(a,b), \mathcal{H}om_{n+N}\mathcal{O}(\mathcal{F}_1,\mathcal{F}_2)\big) \longrightarrow \Gamma\big(D\times G^N(a',b), \mathcal{H}om_{n+N}\mathcal{O}(\mathcal{F}_1,\mathcal{F}_2)\big)$$

$$\Gamma\big(D\times G^N(a,b), \mathcal{H}om_{n+N}\mathcal{O}(\mathcal{F}_2,\mathcal{F}_1)\big) \longrightarrow \Gamma\big(D\times G^N(a',b), \mathcal{H}om_{n+N}\mathcal{O}(\mathcal{F}_2,\mathcal{F}_1)\big)$$

are bijective. Q.E.D.

REFERENCES

[1] A. Andreotti and H. Grauert, "Théorèmes de finitude pour la cohomologie des espaces complexes", Bull. Soc. Math. France 90 (1962), 193-259.

[2] A. Douady, "Le problème des modules pour les sous-espaces analytiques compacts d'un espace analytique donné", Ann. Inst. Fourier (Grenoble) 16 (1966), 1-95.

[3] O. Forster and K. Knorr, "Ein Beweis des Grauertschen Bildgarbensatz nach Ideen von B. Malgrange", Manuscripta Math. 5 (1971), 19-44.

[4] __________, "Relativ-analytische Räume und die Kohärenz von Bildgarben", Invent. Math. 16 (1972), 113-160.

[5] J. Frisch and G. Guenot, "Prolongement de faisceaux analytiques cohérents", Invent. Math. 7 (1969), 321-343.

[6] H. Grauert, Ein Theorem der analytischen Garbentheorie und die Modulräume komplexer Strukturen, I.H.E.S. No. 5 (1960). Berichtigung. I.H.E.S. No. 16 (1963), 35-36.

[7] __________, "Über Modifikationen und exzeptionelle analytische Mengen", Math. Ann. 146 (1962), 331-368.

[8] R. C. Gunning and H. Rossi, Analytic Functions of Several Complex Variables, Englewood Cliffs, N.J.: Prentice-Hall, 1965.

[9] C. Houzel, "Espaces analytiques relatifs et théorème de finitude", Preprint, Nice, 1972.

[10] R. Kiehl, "Variationen über den Kohärenzsatz von Grauert", Preprint, Frankfurt, 1970.

[11] ________, "Relativ analytische Räume", Invent. Math. 16 (1972), 40-112.

[12] ________ and J.-L. Verdier, "Ein einfacher Beweis des Kohärenzsatzes von Grauert", Math. Ann. 195 (1971), 24-50.

[13] K. Knorr, "Der Grauertsche Projektionssatz", Invent. Math. 12 (1971), 118-172.

[14] ________, "Noch ein Theorem der analytischen Garbentheorie", Preprint, Regensburg, 1970.

[15] ________ and M. Schneider, "Relativexzeptionelle analytische Mengen", Math. Ann. 193 (1971), 238-254.

[16] H.-S. Ling, "Extending families of pseudoconcave complex spaces", Math. Ann. 204 (1973), 13-48.

[17] A. Markoe and H. Rossi, "Families of strongly pseudoconvex manifolds", Proc. Conf. Several Complex Variables (Park City, Utah, 1970) pp. 182-207, Lecture Notes in Math. 184, Berlin-Heidelberg-New York: Springer 1971.

[18] R. Narasimhan, Grauert's Theorem on Direct Images of Coherent Sheaves, Séminaire de Mathématiques Supérieures, 1969, Montréal: Université de Montréal, 1971.

[19] J.-P. Ramis and G. Ruget, "Résidue et dualité", Preprint, Strasbourg, 1973.

[20] H. Rossi, "Attaching analytic spaces to an analytic space along a pseudoconcave boundary", Proc. Conf. Complex Analysis (Minneapolis 1964), pp. 242-256, Berlin-Heidelberg-New York: Springer, 1965.

[21] P. Siegfried, "Un théorème de finitude pour les morphismes q-convexes", Preprint, Geneva, 1972.

[22] Y.-T. Siu, "Extending coherent analytic sheaves", Ann. of Math. **90** (1969), 108-143.

[23] __________, "A Hartogs type extension theorem for coherent analytic sheaves", Ann. of Math. **93** (1971), 166-188.

[24] __________, "A pseudoconcave generalization of Grauert's direct image theorem I,II", Ann. Scuola Norm. Sup. Pisa **24** (1970), 278-330; 439-489.

[25] __________, "The 1-convex generalization of Grauert's direct image theorem", Math. Ann. **190** (1971), 203-214.

[26] __________, "A pseudoconvex-pseudoconcave generalization of Grauert's direct image theorem", Ann. Scuola Norm. Sup. Pisa **25** (1972), 649-664.

[27] Y.-T. Siu, Techniques of Extension of Analytic Objects, University of Paris VII Lecture Notes, 1972 (to appear in the Marcel Dekker Lecture Notes Series).

[28] ________ and G. Trautmann, Gap-sheaves and Extension of Coherent Analytic Subsheaves, Lecture Notes in Math. 172, Berlin-Heidelberg-New York: Springer 1971.

[29] G. Trautmann, "Ein Kontinuitätssatz für die Fortsetzung kohärenter analytischer Garben", Arch. Math. 18 (1967), 188-196.

[30] K.-W. Wiegmann, "Über Quotienten holomorph-knonvexer komplexer Räume", Math. Z. 97 (1967), 251-258.

Department of Mathematics
Yale University
New Haven, Connecticut 06520
U.S.A.

EDITORIALE GRAFICA—ROMA

CENTRO INTERNAZIONALE MATEMATICO ESTIVO (C.I.M.E.)

INTERNATIONAL MATHEMATICAL SUMMER CENTER

VOLUMES ON C.I.M.E. SESSIONS

1960

1. **Sistemi dinamici e teoremi ergodici.** (Lessons of P. R. Halmos, E. Hopf and J. L. Massera) . . . . $ 4.—

1961

1. **Geometria del calcolo delle variazioni.** — Second volume. Directed by Prof. Enrico Bompiani and V. V. Wagner . . . » 5.—
3. **Onde superficiali.** (Lessons of H. Bremmer, L. Felsen, F. J. Zucker, etc.) . . . » 10.—

1962

3. **Magnetofluidodinamica.** (Lessons of C. Agostinelli, V. Ferrario, R. Nardini, A. G. Pacholczy and U Schmidt) . . . . » 7.—

1963

3. **Proprietà di media e teoremi di confronto in fisica matematica.** (Lessons of B. Coleman, J. Serrin, H. Ziegler, C. Agostinelli, D. Graffi and G. Grioli) . . . » 6.—

1964

1. **Relatività generale.** Directed by Prof. Carlo Cattaneo. (Lessons of L. Bel, J. Ehlers, G. Caricato, G. Ferrarese and L Mariot) » 6.—
2. **Dinamica dei gas rarefatti.** Directed by Prof. Carlo Ferrari . . . . . » 10.—
3. **Equazioni differenziali non-lineari.** Directed by Prof. Guido Stampacchia . . . » 8.—
4. **Questioni di analisi numerica.** Directed by Prof. Aldo Ghizzetti . . . . . » 10.—

1965

1. **Non Linear Continuum Theories.** Directed by Prof. C. Truesdell . . . . » 8.—
3. **Mathematical Optimization in Economics.** Directed by Prof. Bruno de Finetti . » 7.—

1966

1. **Calculus of Variations, Classical and Modern.** Directed by Prof. R. Conti . . . » 8.—
2. **Economia matematica.** Directed by Prof. Bruno de Finetti . . . . . . » 4.—
3. **Classi caratteristiche e questioni connesse.** Directed by Prof. E. C. Martinelli . » 6.—
4. **Some aspects of diffusion theory.** Directed by Prof. Pignedoli . . . . . » 13.—

1967

1. **Modern Questions of Celestial Mechanics.** Directed by Prof. G. Colombo . . . $ 7.—
2. **Numerical Analysis of Partial Differential Equations.** Directed by Prof. J. L. Lions » 12.—
3. **Geometry of homogeneous bounded domains.** Directed by Prof. Vesentini . . » 8.—

1968

1. **Controllability and Observability.** Directed by Prof. G. Evangelisti . . . » 8.—
2. **Pseudo-differential Operators.** Directed by Prof. L. Nirenberg . . . . . » 10.—
3. **Aspects of Mathematical Logic.** Directed by Prof. E. Casari . . . . . . » 8.—

1969

1. **Potential Theory.** Directed by Prof. M. Brelot . . . . » 8.—
2. **Non-linear Continuum Theories in Mechanics and Physics and their applications.** Directed by Prof. R. S. Rivlin . . . » 10.—
3. **Questions on Algebraic Varieties.** Directed by Prof. E. Marchionna . . . . » 10.—

1970

1. **Relativistic Fluid Dynamics.** Directed by Prof. C. Cattaneo . . . . . . » 13.—
2. **Theory of Group Representations and Fourier Analysis.** Directed by Prof. F. Gherardelli . . . . » 10.—
3. **Functional Equations and Inequalities.** Directed by Prof. B. Forte. . . . . . » 13.—
4. **Problems in non-linear Analysis.** Directed by Prof. G. Prodi . . . . » 16.—

1971

1. **Stereodynamics.** Directed by Prof. G. Grioli . . . . » 10.—
2. **Constructive Aspects of Functional Analysis.** Directed by Prof. G. Geymonat. (Being printed) . . . . . . » 25.—
3. **Categories and Commutative Algebra.** Directed by Prof. P. Salmon . . . » 10.—

1972

1. **Non-linear Mechanics.** Directed by Prof. D. Graffi . . . . » 14.—
2. **Finite geometric structures and their applications.** Directed by Prof. A. Barlotti » 9.—
3. **Geometric measure theory and minimal surfaces.** Directed by Prof. G. Bombieri. » 7.—

1973

1. **Complex Analysis.** Directed by Prof. F. Gherardelli . . . . » 14.—
2. **New variational techniques in mathematical physics.** Directed by Prof. G. Capriz and Prof. G. Stampacchia . . . » 10.—
3. **Spectral Analysis.** Directed by Prof. J. Cecconi . . . . » 9.—

EDIZIONI CREMONESE
Via della Croce, 77
00187 ROMA (Italia)